Actual Problems of Radiophysics International Conference (APR 2021)

AA007373

Journal of Physics: Conference Series Volume 2140

Tomsk, Russia
20 – 22 October 2021

ISBN: 978-1-7138-5111-0
ISSN: 1742-6588

Printed from e-media with permission by:

Curran Associates, Inc.
57 Morehouse Lane
Red Hook, NY 12571

Some format issues inherent in the e-media version may also appear in this print version.

This work is licensed under a Creative Commons Attribution 3.0 International Licence.
Licence details: http://creativecommons.org/licenses/by/3.0/.

No changes have been made to the content of these proceedings. There may be changes to pagination and minor adjustments for aesthetics.

Printed with permission by Curran Associates, Inc. (2023)

For permission requests, please contact the Institute of Physics at the address below.

Institute of Physics
Dirac House, Temple Back
Bristol BS1 6BE UK

Phone: 44 1 17 929 7481
Fax: 44 1 17 920 0979

techtracking@iop.org

Additional copies of this publication are available from:

Curran Associates, Inc.
57 Morehouse Lane
Red Hook, NY 12571 USA
Phone: 845-758-0400
Fax: 845-758-2633
Email: curran@proceedings.com
Web: www.proceedings.com

TABLE OF CONTENTS

9 International Conference «Actual Trends in Radiophysics» in 2021 was held in Tomsk, Russia from 20th to 22th october 2021. International Conference became worthwhile platform for researchers to present their finding in the areas «Physics of radio waves: radiation, reception and use», «radio electronics and electrodynamics of microwave, extremely high frequencies and hyper high frequencies», «Solid-state electronics, micro- and nanoelectronic», «Laser and optoelectronic systems: development, creation, application», «Quantum electronics and photonics», «Modern measuring instruments and technologies», «Modern problems and technologies for training specialists in the field of radiophysics, radio engineering and optics». It aims to provide an international cooperation and exchange platform for experts, scholars and enterprise managers in the fields of the application of radiophysics to share research results, discuss existing problems and challenges and explore cutting-edge technologies.

COVID-19 Update: For reasons of Covid 19, the conference changed its time from September 20-22 to October 20-22 and was held in full-time format.

This scientific event brings together more than 10 national and international researchers in radiophysics. On top of the local participants coming from different national universities, international participants are also registered from different countries. During the conference, the conference model was divided into seven sessions, including oral presentations, keynote speeches, and Q&A discussion. In the first part, keynote speakers were each allocated 20-30 minutes to hold their speeches. Then in the second part, some scholars, whose submissions were selected as the excellent papers, were given about 10 minutes to perform their oral presentations one by one.

Program Committee

Co-Chairmen:

Buznik V.M. – Academician of RAS (Moscow, Russia).

Dyomin V.V. – Vice-Rector for Academic Affairs of TSU, PhD in Physics and Mathematics (Tomsk, Russia).

Committee members:

Kabanov M.V. – Corresponding Member of RAS (Tomsk, Russia).

Bellucci S. – Prof., Prof., Doctor of Physics and Mathematics (Frascati, Italy).

Nathan Blaunstein – Prof., Doctor of Physics and Mathematics (Beer-Sheva, Israel).

Villa T. – Prof., Doctor of Physics and Mathematic (Verona, Italy).

Izhnin I.I. – Prof., Doctor of Physics and Mathematic (Lviv, Ukraine).

Lambin F. – Prof., Doctor of Physics and Mathematic (Namur, Belgium).

Maksimenko S. A. – Prof., Doctor of Physics and Mathematic (Minsk, Belarus).

Mitev V. – доктор наук (Neuenburg, Switzerland)

Sato M. – Prof., Doctor of Physics and Mathematic (Sendai, Japan).

Svirko U. – Prof. (Joensuu, Finland).

Swiatek Z. – Prof. (Institute of Metallurgy and Material Science PAN, Krakow, Poland)

Smith P. – Prof., Doctor of Physics and Mathematic (Loughborough, United Kingdom).

Cezard A. – Prof. (Nancy, France).

Chang R. – Prof., Doctor of Physics and Mathematic (Taiwan).

Siemens E. – Prof., Doctor of Physics and Mathematic (Anhalt, Germany).

Voitsekhovskii A.V. – Prof., Doctor of Physics and Mathematic (Tomsk, Russia).

Prudaev I.A. – PhD in Physics and Mathematics (Tomsk, Russia).

Content from this work may be used under the terms of the Creative Commons Attribution 3.0 licence. Any further distribution of this work must maintain attribution to the author(s) and the title of the work, journal citation and DOI.
Published under licence by IOP Publishing Ltd

Dunaevskii G.E. – Prof., Doctor of Technical Sciences (Tomsk, Russia).

Yevtushenko N.V. – Prof., Doctor of Physics and Mathematic (Tomsk, Russia).

Samokhvalov I.V. – Prof., Doctor of Physics and Mathematic (Tomsk, Russia).

Yakubov V.P. – Prof., Doctor of Physics and Mathematic (Tomsk, Russia).

Belyaev B. A. – Prof., Doctor of Technical Sciences (Krasnoyarsk, Russia).

Ivonin I.V. – Prof., Doctor of Physics and Mathematic (Tomsk, Russia).

Kistenev Y.V. – Prof., Doctor of Physics and Mathematic (Tomsk, Russia).

Koshelev V.I. – Prof., Doctor of Physics and Mathematic (Tomsk, Russia).

Krutikov V.A. – Prof., Doctor of Physics and Mathematic (Tomsk, Russia).

Lukin V.P. – Prof., Doctor of Physics and Mathematic (Tomsk, Russia).

Matvienko G.G. – Doctor of Physics and Mathematic (Tomsk, Russia).

Pchelyakov O.P. – Prof., Doctor of Physics and Mathematic (Novosibirsk, Russia).

Sredin V.G. – Prof., Doctor of Physics and Mathematic (Москва, Россия).

Tarasenko V.F. – Prof., Doctor of Physics and Mathematic (Tomsk, Russia).

Tolbanov O.P. – Prof., Doctor of Physics and Mathematic (Tomsk, Russia).

Shchukin G.G. – Prof., Doctor of Physics and Mathematic (Saint-Petersburg, Russia).

Kolesnik S.A. – Assistant professor, PhD in Physics and Mathematics (Tomsk, Russia).

Minin I.V. – Prof., Doctor of Physics and Mathematic (Novosibirsk, Russia).

Minin O.V. – Prof., Doctor of Physics and Mathematic (Novosibirsk, Russia).

Filimonov S.N. – Assistant professor, PhD in Physics and Mathematics (Tomsk, Russia).

Organizing Committee

Chair:
Korotaev A.G. – Assistant professor, PhD in Physics and Mathematics (Tomsk, Russia).
Co-chair:
Yurchenko A.V. – Professor, Doctor of Technical Sciences (Tomsk, Russia)
Vice-chairmans:
Shipilov S.E. – Professor, PhD in Physics and Mathematics (Tomsk, Russia)
Belichenko V.P. – Prof., Doctor of Physics and Mathematic (Tomsk, Russia)
Kohanenko A.P. – Prof., Doctor of Physics and Mathematic (Tomsk, Russia)
Olshukov A.S. – Associate Professor, PhD in Physics and Mathematics (Tomsk, Russia)
Kochetkova T.D. – Assistant professor, PhD in Physics and Mathematics (Tomsk, Russia)
Satarov R.N. – Research associate, PhD in Physics and Mathematics (Tomsk, Russia)
Klokov A.V. – Assistant professor, PhD in Physics and Mathematics (Tomsk, Russia)
Badin A.V. – Assistant professor, PhD in Physics and Mathematics (Tomsk, Russia)
Bryukhanova V.V. – Assistant professor, PhD in Physics and Mathematics (Tomsk, Russia)
Khmelev V.L. – Engineer, graduate student of the RFF TSU (Tomsk, Russia)
Roslyakov S.N. – graduate student of the RFF TSU (Tomsk, Russia)
Krasilova E.A. – Senior Lecturer (Tomsk, Russia)
Kharapudchenko O.V. – Senior Lecturer (Tomsk, Russia)

Editorial Committe

Nathan Blaunshtein, Ben-Gurion University of the Negev, (Israel)
Mekhtiyev Ali, Karaganda State Technical University (Kazahstan)
Siemens Eduard, University for Applied Sciences (Germany)
Izhnin I.I. - Professor, Scientific Research Company "Electron-Carat" (Lviv,Ukraine)

Invited Speakers
Blaunstein Nathan Sh. - Professor, Ben-Gurion University of the Negev (Beer Sheva, Israel) , Izhnin I.I. - Professor, Scientific Research Company "Electron-Carat" (Lviv, Ukraine).
Siemens Eduard, Professor, University for Applied Sciences (Germany)

Journal of Physics: Conference Series **2140** (2021) 011001 doi:10.1088/1742-6596/2140/1/011001

Mekhtiyev Ali, Saken Seifullin Kazakh Agrotechnical University (Kazahstan)
Tarasenko Victor, Professor, Institute of High Current Electronics of Siberian Branch of the Russian Academy of Sciences, (Tomsk, Russian Federation)
Dunaevsky Grigory, Professor, Tomsk State University, (Tomsk, Russian Federation)

Organizing Institutions

National Research Tomsk State University

Center for collective use "Centre of radiophysical measurements, diagnosis and research of parameters of natural and artificial materials" of TSU

Radiovision Ltd.

Sponsors

Rohde & Schwarz
Keysight Technologies
LCC "Scientific devices and sysytems"

Peer review declaration

All papers published in this volume of Journal of Physics: Conference Series have been peer reviewed through processes administered by the Editors. Reviews were conducted by expert referees to the professional and scientific standards expected of a proceedings journal published by IOP Publishing.

Type of peer review:
Double-blind

- **Conference submission management system: Information System "Conference"**
- **Number of submissions received: 84**
- **Number of submissions sent for review: 65**
- **Number of submissions accepted: 39**
- **Acceptance Rate (Number of Submissions Accepted / Number of Submissions Received X 100): 46,4%**
- **Average number of reviews per paper: 2**
- **Total number of reviewers involved: 4**
- **Any additional info on review process: uses double-blind review, which means that both the reviewer and author identities are concealed from the reviewers, and vice versa, throughout the review process.**
- **Contact person for queries:**
Name : Yurchenko Alexey
Affiliation: Professor, National Research Tomsk State University, 36, Lenin Avenue Tomsk RUSSIA
Email : niipp@inbox.ru (apr@mail.tsu.ru)

Content from this work may be used under the terms of the Creative Commons Attribution 3.0 licence. Any further distribution of this work must maintain attribution to the author(s) and the title of the work, journal citation and DOI.
Published under licence by IOP Publishing Ltd

Journal of Physics: Conference Series 2140 (2021) 012001 doi:10.1088/1742-6596/2140/1/012001

Modelling of epitaxial growth of two-dimensional film

V P Vinarskiy, A P Kokhanenko and K A Lozovoy

Tomsk State University, Tomsk, Russia

E-mail: vinarskiy2017@gmail.com

Abstract. The study is devoted to the theory of epitaxial growth of two-dimensional materials (in particular, the group IV elements) and its applications in creating a virtual program for modeling this growth by the molecular beam epitaxy (MBE) method.

1. Introduction

Graphene, silicene and germanene are three representers of group IV two-dimensional materials with different structures, properties and physical characteristics. For instance, the pure graphene cannot be a semiconductor in real life because of loss of energy bandgap, while germanene and silicene can [1]. Therefore, the last two materials are more promising materials for various instrument applications [1, 2].

In this work the growth of two-dimensional and zero-dimensional nanostructures are studied. The applicability of the work is due to the fact that modeling the epitaxial synthesis of two-dimensional materials of the IV group allows us to create diodes and transistors with characteristics useful in nanoelectronics. Special attention is paid to four main processes, listed in the order of growth stages: island nucleation, independent island growth, coalescence and three-dimensional film growth. Determination of time and conditions of growth of materials depending on desired parameters of nanostructures are of particular interest. This work is dedicated to the modeling of these dependences.

Among the various methods for the growth of two-dimensional materials, special attention is paid to the method of molecular beam epitaxy (MBE). This is explained by the fact that the MBE method has proven itself well in the growth of two-dimensional materials of group IV due to the capabilities of analysis and control of synthesis processes, as well as the ability to interrupt film growth. To analyze the morphology of growing structures, for example, the chambers of the Katun-100 setup can be equipped with a fast electron diffractometer and an automatic high-speed ellipsometer. Precise control of film thickness is ensured by quick-acting shutters. The presence of modules in the installation also makes it possible to fully automate the technological process of growth.

2. Methods and materials

In order to construct a physical and mathematical model of the growth of nanostructures the thermodynamic model [3] has been taken. While simulating the processes of epitaxial growth of two-dimensional materials of group IV elements, the presence of three zones in the installation chamber was taken into account: a zone of generation of molecular beams, a zone of mixing of evaporated elements, and a zone of crystallization.

As films grow with a thickness of less than 100 Å, it is necessary to take into account adsorption, desorption, dissociation of molecules, and migration of atoms.

There are three growth modes of materials:

Content from this work may be used under the terms of the Creative Commons Attribution 3.0 licence. Any further distribution of this work must maintain attribution to the author(s) and the title of the work, journal citation and DOI.
Published under licence by IOP Publishing Ltd

1. 'layer-by-layer growth' (Frank–Van der Merwe growth mode);
2. 'island growth' (Volmer–Weber growth mode);
3. 'layer-plus-island growth' (Stranski–Krastanov growth mode).

The mathematical formulation of each of the growth modes is based on solving the general equation:

$$\sigma_s = \sigma_{s/f} + \sigma_f \cos\varphi, \tag{1}$$

where φ is wetting angle (Figure 1), σ_s, σ_f and $\sigma_{s/f}$ are surface tension coefficients between substrate and vacuum, between film and vacuum and between substrate and a film correspondingly.

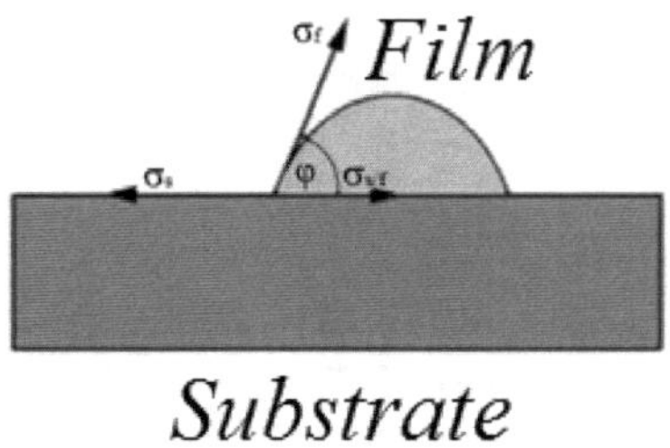

Figure 1. Three-dimensional island on a substrate [3].

Since there are certain stages in each growth process, the growth process of two-dimensional materials has been divided into the following points:

1. Nucleation;
2. Independent island growth;
3. Island coalescence;
4. Three-dimensional growth of the film.

As a result of research, the boundaries of changes in the structural and thermodynamic parameters of deposited material and substrate, at which the growth processes are realized according to the modes of Frank-van der Merwe, Volmer-Weber, and Stranski-Krastanov, have been determined. Besides, it was found that two-dimensional materials have synthesis conditions that are somewhat different from three-dimensional materials. The influence of curvatures in the lattice structure can lead to significant changes in the properties of the material, which was shown for the example of silicene and graphene.

3. Model

Let us consider a parameter *E*, which defines atom condensation mode:

$$E = 4\alpha^3 e^{-a/\ln(\Phi+1)}, \tag{2}$$

where α is a kinetic parameter, *a* is an energetic parameter, Φ is a vapour-phase supersaturation. According to the equation (1), if the condition $E >> 1$ is true, the complete condensation mode in the MBE method is realized. If this is not the case – it is incomplete condensation mode.

Parameter Φ_{max} is a maximal vapour-phase supersaturation which is found for optimal parameters of the system and which is performed at (3):

$$\Phi_{\max}(t) = \frac{V\tau_a}{\theta_{eq}}. \tag{3}$$

Here *V* stands for material settling rate, τ_a for lifetime of an adatom on the surface and θ_{eq} for equilibrium filling of a surface with adatoms.

Vapour-phase supersaturation $\Phi(t_r)$ is defined with transcendental equation:

$$f(\Phi_r)=\frac{a\sqrt{\pi}\left(\frac{\sqrt{\sigma}}{l_d}\right)^2}{\Phi_r^{\,2}(\Phi_r+1)^3\cdot\ln(\Phi_r+1)^{9/2}}\cdot\frac{(\Phi_{\max}-\Phi_r)^3}{(\Phi_{\max}-1)^3}\cdot\frac{(V\cdot t_d)^3}{\theta_{eq}^5}\cdot e^{a/\ln(\Phi_r+1)}-1\,. \quad (4)$$

Here σ is an adsorption cell area, l is a diffusion jump length, V is a material settling rate and t_d is a surface diffusion time.

From equation (3) we can find a maximal supersaturation $\Phi(t)$ and nucleation rate $I(t)$:

$$\Phi(t)=(\Phi_{\max}+1)(1-e^{-t/\tau_a})-1, \quad (5)$$

$$I(t)=I(\Phi_r)e^{(t-t_r/\Delta t)-e^{(t-t_r/\Delta t)}}, \quad (6)$$

Here $I(\Phi_r)$ is the nucleation rate at maximal supersaturation of vapor-phase, t_r is the time of maximal supersaturation, also it is called the time of nucleation delay, Δt is the nucleation stage endurance.

Adatom supersaturation in general form can be performed using formula:

$$\zeta(\mathrm{t})=\Phi_r+\frac{\Phi_r*\left(\left(\frac{t-t_r}{\Delta t}\right)-e^{\frac{t-t_r}{\Delta t}}\right)}{\Gamma}. \quad (7)$$

where big theory parameter Γ has a form:

$$\Gamma=\frac{\Phi_r\cdot a}{(\Phi_r+1)\cdot\ln(\Phi_r+1)^2}. \quad (8)$$

4. Results

For the calculations the following parameters and their values have been used: $T_c = 2\,200$ K, $T = 853$ K, $\nu_a = 10^{12}$ Hz, $\nu_d = 3\cdot10^9$ Hz, $k_e = 8.617\cdot10^{-5}$ eV/K, $k = 1.38\cdot10^{-23}$ J/K, $E_d = 0.8$ eV, $E_a = 2.1$ eV, $\varepsilon = 5.7\cdot10^{-6}$ J/m, $V = 0.1$ ML/s, $\sigma = 16\cdot10^{-20}$ m^2, $\Gamma = 9.442$. Here the next parameter values have been obtained: $t_r = 0.162$ s and $\Delta t = 0.011$ s correspondingly. Using these values we can find that Φ_r and Φ_{max} are equal to 1.727 units and 43,475 units correspondingly, therefore these values mathematically describe complete condensation mode in the MBE method. At the Figure 2 the dependence of supersaturation on time is performed.

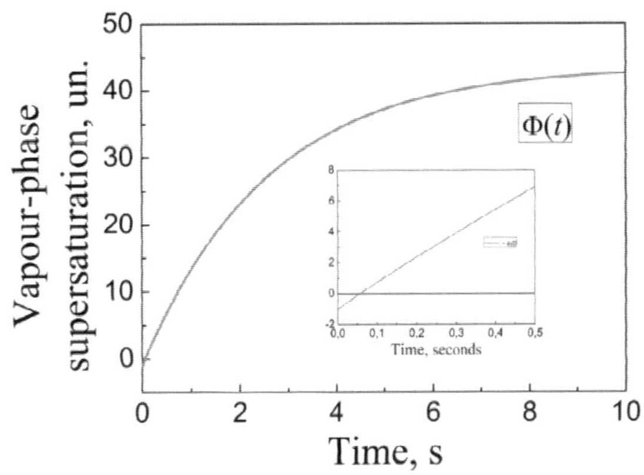

Figure 2. Dependence of supersaturation on time.

It is evident from formula (4) that near the time of maximum supersaturation t_r the nucleation rate has approximately Gaussian form with a half-width $\sqrt{2}/\Delta t$, performed in the Figure 3.

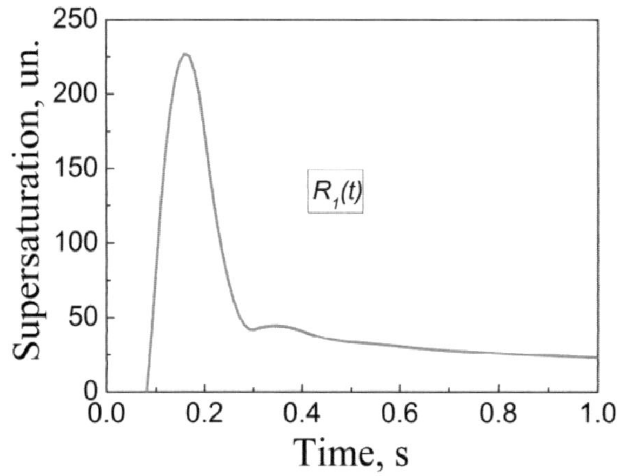

Figure 3. Dependence of reduced supersaturation of adatoms $\zeta(t)/\zeta(t_r)$ on time.

It is evident from formula (4) that the flex point of a function graph corresponds to the moment t_r, which is proved with Figures 3 and 4.

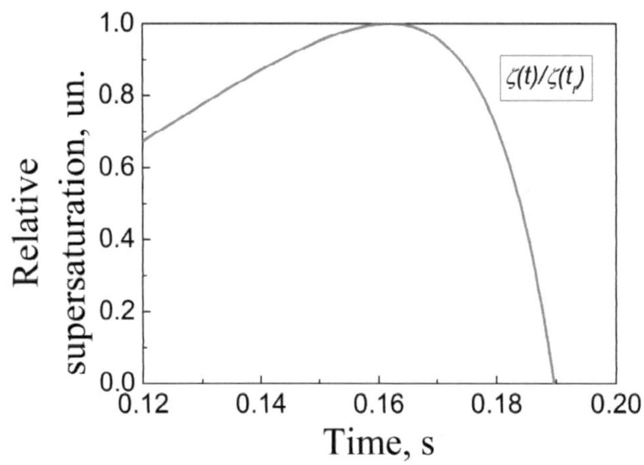

Figure 4. Dependence of reduced supersaturation of adatoms $\zeta(t)/\zeta(t_r)$ on time at flex point.

5. Conclusion

As a result of the research the different process stages of growth of two-dimensional films and nano islands have been considered. The dependences of supersaturation, nucleation rate and island density on time have been defined using epitaxial growth at complete condensation mode.

Acknowledgments

The reported study was supported by grant from the Russian Science Foundation No. 21-72-10031, https://rscf.ru/project/21-72-10031/.

References

[1] An Y, Tian Y, Wei C *et al.* 2020 *Energy Storage Materials* **32** 115–150
[2] Izhnin I, Kurbanov K, Lozovoy K *et al. Applied Nanoscience (Switzerland)* **10** 4375–4383
[3] Dubrovskii V G 2014 *Nucleation theory and growth of nanostructures* (Berlin: Springer) p. 601.

Journal of Physics: Conference Series **2140** (2021) 012002 doi:10.1088/1742-6596/2140/1/012002

The use of optical fiber for the creation of security systems for electrical cables of distribution networks of 0.4 kV

V V Yugay [1], A D Mekhtiyev [2,3], P Sh Madi [1,3], A D Alkina [1] and E G Neshina [1]

[1]Karaganda Technical University, Karaganda, Republic of Kazakhstan
[2]Kazakh Agrotechnical University name of Saken Seifullin, Nur-Sultan, Republic of Kazakhstan
[3]Tomsk Polytechnic University, Tomsk, Russia

E-mail: perizat1@tpu.ru

Abstract. The article presents the results of research aimed at solving important production issues related to the theft of power cables and reducing the reliability of electrical networks with a voltage of 0.4 kV. To implement the security system, a single-mode optical fiber and a hardware and software complex for data processing were used. The results of the work can be used in security systems for various purposes, including telecommunications and construction systems. During the experiments, it was found that the photoelasticity effect occurs during mechanical action on the optical fiber, which causes changes in the refractive indices of the shell and core, while the properties of light change. In contrast to the known security systems and methods of fixing violations, a high-resolution television matrix is used, and not a photodetector device that records only changes in intensity. The television matrix is installed at the output of the optical fiber and records all changes in the light spot, and the hardware and software complex converts these changes into numerical values.

1. Introduction

One of the problems of voltage distribution networks was the theft of power cables. Attackers are attracted by the presence of non-ferrous metal and to implement their plans, they are organized into small mobile groups that can use a significant arsenal of tools and special equipment. Especially often, wires are stolen from overhead lines located in rural areas, but there are cases when power lines with a voltage of 35–110 kV become the victim. The theft of cables and wires made of non-ferrous metals allows some residents of remote rural areas to get a substantial income, which encourages this type of activity as the most profitable. They steal not only wires, but also disassemble the metal structures of the supports, which leads to their fall. Trucks, cranes, tractors and excavators are used for this purpose. People who steal electrical conductors are quite well aware of what an electric current is and how it can be turned off. As a result, tens of kilometers of electric cable are stolen, causing significant damage to energy supply companies. This problem is noted not only in the Republic of Kazakhstan, but also in the independent countries of the former USSR. Every year, thousands of kilometers of electric cables and wires of power transmission lines are stolen in Russia. Not only material losses are caused from the very fact of the theft of an electric cable, but also technological equipment of enterprises is put out of order, emergency situations are created, a power outage is created, and much more. There are cases of people dying when hospitals are turned off, which is very tragic. Thousands of people are left without electricity, and the kidnappers themselves put their lives in danger, and

Content from this work may be used under the terms of the Creative Commons Attribution 3.0 licence. Any further distribution of this work must maintain attribution to the author(s) and the title of the work, journal citation and DOI.
Published under licence by IOP Publishing Ltd

sometimes endanger the lives of other people who may accidentally be hit by an electric current, due to various situations. Unfortunately, there are children and teenagers among the victims. Especially many of them die during the theft of non-ferrous metal from transformer substations. There are egregious cases when throws attracted by deception were used as" voltage indicators". At the same time, the financial resources of energy supply companies are spent not on the development and modernization of the production base, as well as carrying out routine repairs of electrical equipment, but on the installation of new cable and overhead lines to replace the stolen ones. Solving this problem requires a serious state approach, since energy companies are not able to solve this problem on their own, with the help of their own security service. Power engineers create mobile special teams that protect the most vulnerable sections of the electric network. Current conductors are exposed to theft not only of existing power transmission lines and substations, but also backup lines that have to be put under voltage. Law enforcement agencies, primarily the police, are involved in solving this problem, but it is not yet possible to reverse the situation in favor of society [1]. It will not be possible to completely solve this problem only at the expense of the police forces or improve the laws regulating the turnover of non-ferrous metals and toughen the responsibility for their theft. This is confirmed by the publication of reports about thefts of electrical conductors in the media [2]. Employees of the police and the administration of populated areas close the reception points for the collection of non-ferrous metal, but they open again. We need a systematic approach to solving this problem by developing technical means and methods to combat this phenomenon.

2. The relevance of conducting scientific research and setting the task

The relevance of conducting scientific research aimed at solving an important industrial problem is indisputable. One of the solutions is a timely warning of the on-duty personnel of energy companies and law enforcement agencies about the attempted theft, which will allow the security service to go to the place where the attackers have deployed their activities. It is necessary to have a distributed type of electrical cable security system that allows you to record not only the fact of intrusion into the protected area, but also to accurately determine the place where the thieves are located. To do this, it is necessary to develop a security system capable of ensuring the safety of the electric cable in the channel, since practice has shown that intruders bypass the existing security alarm system that responds to penetration into the premises of a transformer substation (TS) or a well [3]. For example, a tunnel is organized that allows you to get into the cable channel, through which intruders enter the TS room where the current-carrying parts are stolen. The excavation is carried out manually and has sufficient secrecy, it is difficult to detect it even for police and security service detachments circling the patrol area, so cables from power supply systems located even inside the city are stolen. Backup cable lines on which there is no voltage are particularly affected. The factor of collusion between employees of energy companies and the kidnappers is not excluded, since the latter are sufficiently informed about the location of cable channels and the presence of existing or backup cable lines. There is also a leak of information about the methods of protecting energy facilities, since the attackers are well informed about the ways to relieve tension and neutralize the alarm system. Things are even worse with cable lines located far from the city and providing power to remote areas, heavy equipment is already being used there that allows opening the cable channel and stealing the electric cable. As the analysis of cases related to the theft of electrical cables has shown, this is too late a reaction on the part of the security services and the police, as well as the lack of technical means to counter this phenomenon. This allows criminal communities to flourish and continue their activities with almost impunity. There are many security systems of different technical levels and costs, which are based on different principles: infrared, vibroacoustic, magnetometric, capacitive, seismic and other types of systems. They are already partially used to protect transformer substations and cable wells, but as noted above, penetration and theft of electric cable is carried out through the cable channel. This circumstance makes the existing security systems ineffective. Also, departmental or citywide video surveillance systems are not fused with this problem. For the most part, violators are detected and detained by the security service or the police [3].

One of the solutions to this problem is the development of methods and means of protection of electrical distribution networks with a voltage of 0.4 kV. One of the solutions is to use an optical fiber (OF) as a sensitive sensor. Fiber-optic sensors have long been used in various fields of industry, and security systems using optical fiber as a sensitive element are also being produced [4]. The relevance of the use of these security systems is extremely high, interest in the use of optical fibers as sensors with high metrological characteristics is growing every year. Based on the optical fiber, it is possible to create distributed security systems that can not only establish the fact of an invasion, but also determine the exact location of the invasion [5]. The task is also to create a fundamentally new fiber-optic security system that has a significantly lower cost compared to existing analogues, but is fully adapted to the climatic conditions of the city of Nur-Sultan and the operating conditions of the Astana REC. Another wish of the production workers is a certain share of the Kazakh content to receive benefits and support from the state in the implementation of plans for the introduction of fiber-optic security systems.

3. Analysis of foreign experience and the choice of the direction of solving the task

The analysis of the literature has shown a significant interest in the use of optical fiber in security systems, for example, security systems of various companies are already being produced and sold on the world market. The first discovered patent for security systems of this type was registered in the United States in 1977, which became the basis for subsequent developments. The basis is a radiation source, a fiber-optic sensor, a photodetector and a data processing device [4]. There are a number of successful developments and leaders in this field that have been present on the market for several decades, for example, Future Fiber Technologies FFT (Australia); Remsdaq (England); TRANS Security Systems and Technology (TSS) (Israel); Fiber SenSys (USA); Magal (Israel); Senstar-Stellar (Canada). There are from Russia, for example, the NGO Applied Radiophysics "Voron" (Russia); "Dunai" (Russia); "Gyurza" (Russia), you can give a dozen more examples to what extent successful projects in the field of fiber-optic security systems. Separately, we can note the success of the developers of Yangtze Optical Fiber and Cable Company Ltd from China, who produce fairly inexpensive systems based on the developments of Fiber SenSys (USA), but having a lower cost [5, 6]. Each manufacturer uses its own scientific developments, methods and means. When analyzing the available sources, these developments are not sufficiently covered in the open press, and some are classified, it can be concluded that there are a number of certain similarities between different security systems that unite them into several groups. There are also more general characteristic similarities, this is the use of the principle of an optical interferometer. Manufacturers strive to improve the technical level of their products and achieve a certain advantage over competitors, but there are circumstances that cause a number of common problems that limit their application possibilities and lead to a serious increase in price [7–9]. For example, the use of the principles of an optical interferometer causes a number of problems when interference occurs due to temperature influences on the optical fiber. Therefore, the known problems of distributed fiber-optic security systems that have not been completely solved and the lack of available detailed information about their device serve as a basis for conducting additional research and familiarizing the general public with the principles of their functioning. It is also necessary to conduct research to improve them and reduce the cost, which will make them more accessible. All the advantages and disadvantages of fiber-optic systems can be found in the sources [10,11]. The main advantage of such systems is their secrecy. The optical fiber is not affected by electromagnetic interference [9]. Therefore, the use of optical fibers for the construction of passive systems for the protection of perimeters and boundaries of various objects is an extremely promising direction.

Optical fiber and components are getting cheaper on the market every year, for example, the cost of optical fiber is less than 10 US dollars. An optical fiber is intrinsically safe if the light energy transmitted through the fiber is at or below a certain power level. They are not affected by noise, lightning, interference from electromotive force, electromagnetic interference common in the mining industry. Fiber-optic communication is ideal for use in underground work, but the use of fiber can be expanded and applied to measure various parameters. Fiber-optic cables must be appropriately armored so that they remain operational under the following conditions: the movement of underground vehicles, the collapse of an underground roof, flooding of underground water, the impact of a pressure wave resulting from underground explosions. As a result, we can say that an optical fiber can serve simultaneously as a guiding system for transmitting information, and a measuring sensor, of almost any parameters [12–14]. Signal transmission over a distance of 100 km via fiber is not a problem, while energy costs are not comparably low compared to control cables [15,16]. Ensuring the protection of objects of limited access and dangerous objects of state significance or simply private territories from unauthorized access, which occupy significant areas, unlike local objects, requires high costs and complex communications for building a security system and monitoring the perimeter, this circumstance significantly increases the cost of security systems. For example, the security system "VORON-3M-K "(Russia) is 10.65 million rubles. - the length of the protected perimeter: 2400 m, the length of the zone: 15 m. Fiber SenSys products (USA) cost more than twice as much [17]. Accordingly, an important point is the search for new solutions to reduce the cost of fiber-optic signaling [18,19]. To solve the tasks set, a proprietary method has been developed that allows solving this problem. Having analyzed the cost of the distributed VOSS "Voron" [20], we can say that the most profitable option for the customer is the length of the protected perimeter from 15 to 30 km, since the cost of one kilometer is significantly reduced. Accordingly, it is desirable to have extended sections in order to reduce the cost of investments [21].

4. Creating an experimental sample

The appearance of the FOSS prototype is shown in Figure 1, where 1 is a sensor, 2 is a drum with a cable, 3 is a FOSS unit for data processing and pulse generation, 4 is a monitor with a program window, 5 is a personal computer with a hardware and software complex.

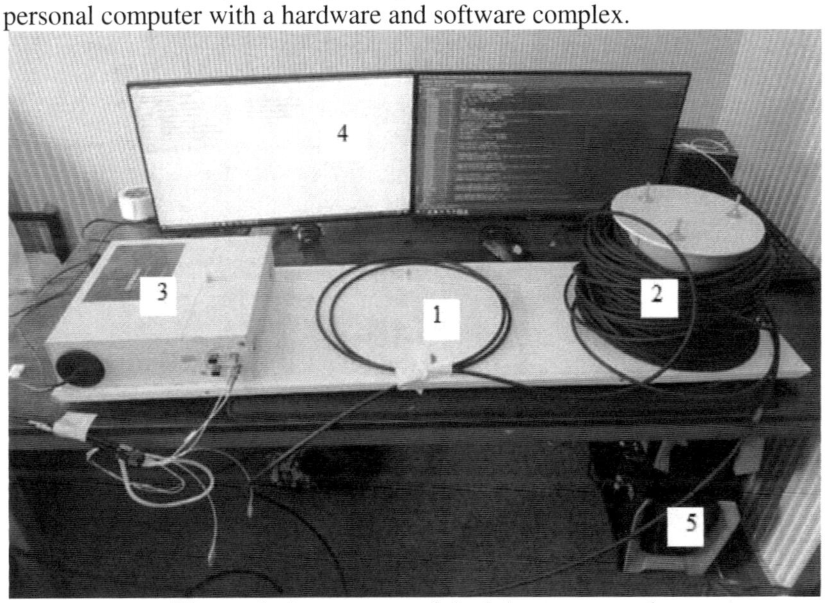

Figure 1. Appearance of the laboratory stand.

The experimental sample of the fiber-optic security system of the FOSS structurally consists of a measuring part, which is represented by a piece of fiber-optic cable 1, which perceives the impact, it can be placed on a silt fence in the ground. To communicate from the sensor to the data processing unit, the same fiber-optic cable is used, wound on a drum 2, which is a guide communication system, this allows transmitting information from the sensor for 30 km or more. The cable is connected by means of optical convectors and adapters to the data preprocessing and pulse generation unit of the FOSS, where all the main data processing units are located, determining the distance to the intrusion point, an optical switch with a switch, a radiation source. On monitor 4, the program window is located (on the left), and on the right, the source code of the program, the monitor provides information about the status of the FOSS. The hardware and software complex and the server for storing information about the operation of the FOSS are organized on the basis of a personal computer.

To test the design of the FOSS and the method of processing the data received from optical fiber sensors, a laboratory stand was developed, its appearance is shown in Figure 1, as well as the main blocks are indicated.

When creating an experimental sample of a fiber-optic security system of the FOSS, all the features associated with the operation of power cables with a voltage of 0.4 kV are taken into account. The experimental sample structurally consists of a measuring part, which is represented by a piece of fiber-optic cable that perceives the impact, it can be placed on a fence or in the ground, as well as the data processing unit and the hardware and software complex. The data processing unit contains a high-resolution television matrix and a coherent laser radiation source with a wavelength of 650 nm. The power of the semiconductor laser is 30 mW. To communicate from the sensor to the data processing unit, the same fiber-optic cable is used, wound on a drum, which is a guide communication system, this allows transmitting information from the sensor for 30 km or more. The cable is connected by means of optical convectors and adapters to the data preprocessing and pulse generation unit of the FOSS, where all the main data processing units are located, determining the distance to the intrusion point, an optical switch with a switch, a radiation source.

5. Experimental results

The program works in several modes, in nominal mode, when there is no intrusion into the protected perimeter, there are only noise and temperature disturbances that the program has learned to recognize during the initial setup. The second state is the detection of a weak vibroacoustic effect on the fiber-optic sensor, which is located in the cable channel. When motion is detected, the yellow display lights up and a warning signal is heard. If the impact on fiber-optic sensors continues, then the hardware and software complex and its decision-making body, after performing certain logical operations, decides on the activation of the FOSS. The alarm is canceled by the actions of the operator, who takes measures to protect the territory from penetration according to the standard instructions. Figure 2 shows two FOSS states (a) normal state; b) alarm state), in the normal state the screen remains black and the number of white pixels is minimal. White pixels are present on a black background in the normal state of the FOSS, but this is just a fixation of noise. The more noise the radiation source makes, the more random disturbances occur.

Figure 2, in the upper part, shows a picture of changes in the pixels of the light spot, and in the lower part, a graph of changes in the intensity parameters is shown. When exposed to a fiber-optic sensor, the pixel pattern changes and the program analyzes the received data and makes a decision about triggering. The analysis is performed using artificial intelligence algorithms with pre-prescribed weight coefficients. As can be seen in Figure 2*a*, the noise level is not significant and in the absence of an impact on the fiber-optic sensor, the FOSS does not work. When exposed, as shown in Figure 2*b*, the number of white pixels increases sharply and the picture of the light spot changes, in this case, the FOSS is triggered. During the experiments, it was noted that when the optical fiber is mechanically affected, a photoelastic effect occurs, which in turn causes changes in the refractive index and changes the parameters of the light wave. The television matrix installed at the output of the optical fiber records all changes, and the hardware and software complex converts these changes into numerical values. There is also a change in the picture in pixels from black to white. The stronger the impact on the optical fiber, the more white pixels will be on the screen.

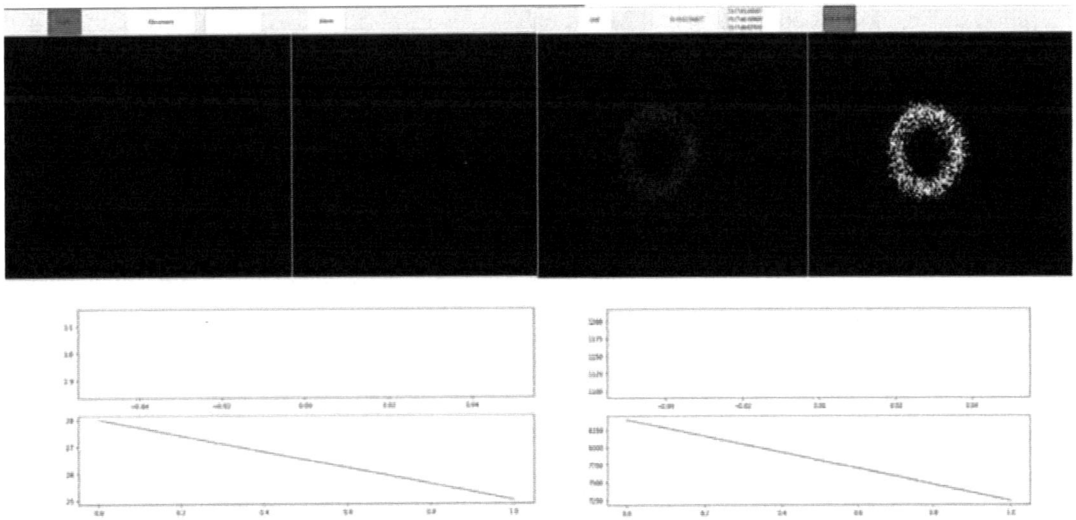

a) b)

Figure 2. The status of the FOSS displayed in the program window

Figure 3 shows a short-term effect on a fiber-optic sensor, this effect is reflected in the diagram by some dip in the lower graph and several peaks in the upper graph.

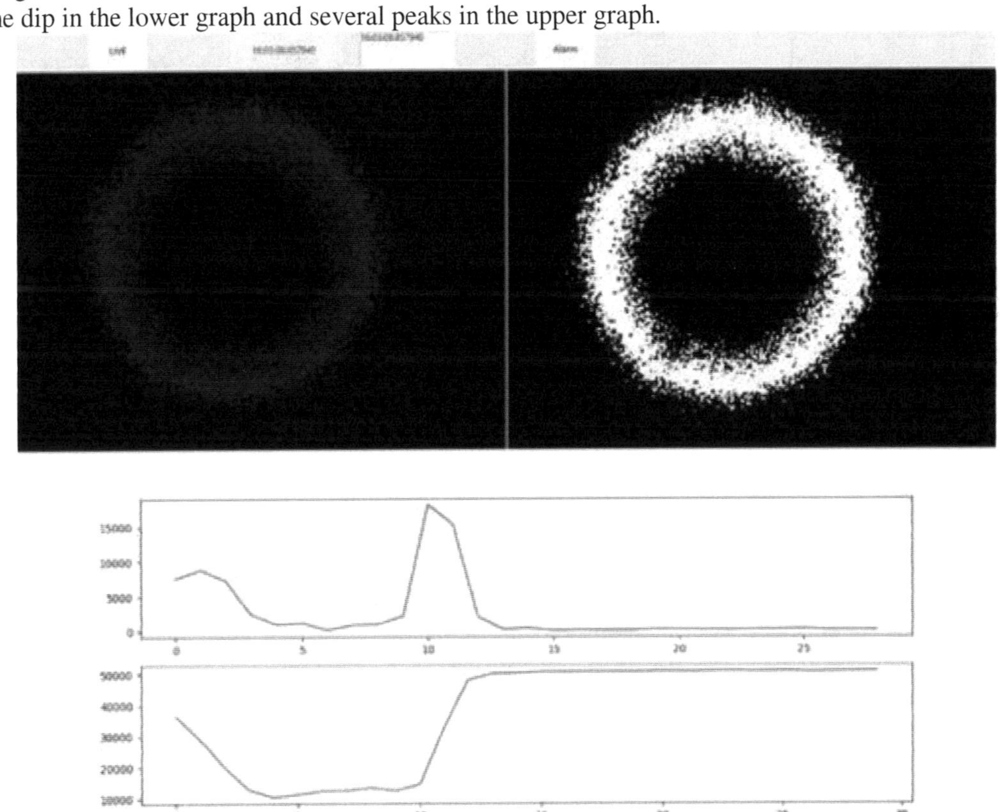

Figure 3. Short-term impact on the fiber-optic sensor that does not lead to triggering.

Since the yellow scoreboard has lit up, the system only warns without triggering the alarm. This signal should be paid attention to by the staff on duty and, if possible, double-check the information. If the impact is accidental and is of a short-term nature, the FOSS returns to normal operation after a few seconds, but if the effects do not stop, the FOSS will work and give an alarm signal.

The results of laboratory experiments have been tested in practice and at the moment the prototype of the VOSS is being tested. As a sensor, a fiber-optic conductor of the brand is used: КС - ОККО - А - 2 -G.652.D - UCF (4039) или КС - FTTH - А - 2 -G.652.D - UCF (7203).. For the installation of the sensor, the cable channel was partially opened. The sensors were stacked on top of the electrical cables. FOSS is able to record an attempt to undermine the channel and break through its walls. Also, in the case of an intrusion into the cable channel, attackers touch fiber-optic sensors (FOS). The algorithm of the FOSS operation provides for the supply of sound signals on the operator's console at various impacts on the FOS. Information about the intrusion is transmitted automatically in the form of a message via cellular communication channels. As mentioned earlier, FOSS itself sends a message to the operator's console of the Astana REC security service, and all data is stored on the hard disk of a personal computer that is a server.

6. Discussion of the results

The proposed FOSS showed a sufficiently high sensitivity and high noise immunity of the channels in the event of various kinds of noise associated with the operation of the radiation source. An important point of stable operation of the FOSS is the coherence of the radiation source, fluctuations in the radiation frequency leads to a change in the wavelength, which causes noise to appear in the measuring channel, which is displayed on the computer screen by increasing the number of white pixels. The hardware and software complex is able to assess the growth of the number of white pixels with noise pollution and with mechanical impact. The hardware and software complex performs averaging of the received data in a certain period of time, the number of impacts, the frequency of impacts and the rate of change of the impact. This allows you to significantly reduce the number of false positives. The use of a high-resolution television matrix, in contrast to existing FOSS analogues that use a photodetector, allows you to increase sensitivity and reduce the influence of noise on the operation of the measuring channel.

7. Conclusion and conclusions

The work is aimed at solving the most preferable from a functional and financial point of view production problems associated with the theft of power cables and reducing the reliability of electrical networks with a voltage of 0.4 kV. The results of the work can be used not only for the construction of a security system for power cables, but also for fiber-optic guide communication systems, as well as for fiber-optic means for the protection of extended objects. During the experiments, it was found that the photoelasticity effect occurs during mechanical action on the optical fiber, which causes changes in the refractive indices of the shell and core, while the properties of light change. In contrast to the known security systems and methods of fixing violations, a high-resolution television matrix is used, and not a photodetector device that records only changes in intensity. The television matrix is installed at the output of the OF and records all changes in the light spot, and the hardware and software complex converts these changes into numerical values. This method is based on controlling the change of the picture in pixels from black to white. The stronger the impact on the optical fiber, the more white pixels will be on the screen.

Acknowledgments

The publications are based on the results of scientific research carried out at the expense of the IPN project No. AP09562222 "Information and measurement system of a new generation based on fiber-optic sensors" (contract 254/12-2 of 15.07.21). Scientific research was carried out using the laboratories of the Kazakh Agrotechnical University name of S. Seifullin.

References

[1] Cable for 2 million tenge stolen in Astana. Available at: https://365info.kz/2016/04/kabel-na-2-mln-tenge-ukrali-v-astane
[2] Police detained suspected cable stealing in the amount of 1.5 million tenge. Available at: https://astana.gov.kz/ru/news/news/3724
[3] All rights reserved. Use an active reference to inform.kz https://www.inform.kz/en/podozrevaemye-v-krazhe-kabelya-na-devyat-mln-tenge-zaderzhany-v-nur-sultane_a3782959
[4] Perimeter detection tools based on the "Voron" technology. Available at: https://www.grand-prix.ru/upload/iblock/73e/voron.pdf
[5] Yurchenko A, Mekhtiyev A, Bulatbayev F, Neshina Y, Alkina A 2018 *Russian Journal of Nondestructive Testing* **54**(7) 528–533 doi:10.1134/S1061830918070094
[6] Juarez J and Taylor H 2007 *Applied Optics* **46**(11) 1968–1971 doi:10.1364/AO.46.001968
[7] Shih-Chu Huang and Hermann Lin 2006 *Applied Optics* **45**(35) 8832–8838 doi:10.1364/AO.45.008832
[8] Polyakov A, Ksenofontov M 2007 *Proceeding of the International Conference on Laser, Applications and Technologies* 93
[9] Iqbal M, Fathallah H, Belhadj N 2011 *Proceeding of the 8th International Conference on High-capacity Optical Networks and Emerging Technologies*n 164–168 doi: 10.1109/HONET.2011.6149809
[10] Peng Y, Sun Z, Du S *et al.* 2011 *Opt. Eng.* **50**(8) 085002 doi: 10.1117/1.3607412
[11] Madi P, Kalytka V, Alkina A and Nurmaganbetova M 2019 *Journal of Physics: Conf. Series* **1327** 012036 doi:10.1088/1742-6596/1327/1/012036
[12] Madi P, Gorokhov D, Mekhtiyev R and Nurmaganbetova M 2021 *Journal of Physics: Conf. Cep.* **1843** 012016 doi:10.1088/1742-6596/1843/1/012016
[13] Mekhtiyev A *et al* 2020 *Russian Physics Journal* **63**(2) 323–331 doi: 10.1007/s11182-020-02038-y
[14] Mekhtiyev A, Neshina Y, Madi P, Gorokhov D 2021 *Occupational safety in industry* **4** 19–26 doi:10.24000/0409-2961-2021-4-19-26
[15] Winzer P, Gnauck A, Konczykowska A *et al.* 2011 *Proc. European Conf. on Optical Communication (ECOC)* Tu.5.B. 7 doi:10.1364/ECOC.2011.Tu.5.B.7
[16] Bruno G, Camera M M 2014 *Proc. of the Fotonica AEIT Italian Conference Of PhotonicsTechnologies* 1–3 doi: 10.1109/Fotonica.2014.6843840
[17] Ivanov I V 2000 *Perimeter Security-2* (Moscow: Graf Parity)
[18] Yurchenko A V, Mekhtiyev A D, Bulatbaev F N, Alkina A D, Kokkoz M M 2017 *Journal of Physics: Conference Series* **881**(1) 012034
[19] Lavrinenko A V 2001 *Special equipment* (5)
[20] Vvedensky B 2006 *Mir and Security* **4-5**
[21] Svirsky Yu 2001 *Security systems* (38)

Numerical simulation of the frequency response and phase response of combined UWB antennas in the receiving-transmitting mode

Yu A Andreev[1], E A Kuznetzova[2] and V V Plisko[1]

[1] Institute of high current electronics SB RAS, Tomsk, Russia
[2] National Research Tomsk State university, Tomsk, Russia

E-mail: andreev@lhfe.hcei.tsc.ru

Abstract. Results of a simulation and experimental study of a combined ultra-wideband antenna, which is a combination of electric and magnetic dipoles, are presented. The input signal recovery of the transceiver path of combined antennas made by the MathCad code and the experimental one were compared. The reconstruction took place according to the frequency and phase responses of the transceiver path of these antennas.

1. Introduction

Combined ultra-wideband (UWB) antennas are widely used as radiators in high-power UWB radiation sources [1, 2]. Such sources can be used in underground radar and in the search for mines, in the radar of forests and aerospace objects, for electromagnetic compatibility tests of different objects and environments, including biological ones. In relation to combined antennas, in a number of publications, the authors apply frequency response and phase response as parameters of a single antenna [3, 4]. Using these parameters, the authors define the antenna bandwidth as the frequency range in which the conditions for low-distortion signal transmission are met, i.e., the change in the frequency response of the antenna relative to the average value within ± 1.5 dB and the deviation of the phase response from the linear one within $\pm \pi/16$ [3]. However, in the traditional theory there are no such parameters for single antennas [5]. At the same time, these parameters are traditionally used in radio engineering to describe multipolars. It is known from radio engineering that, knowing the frequency response and phase response of the circuit and the pulse at its output (input), it is possible to restore the incoming pulse (output pulse) [6]. In this paper, we will consider the application of the terms frequency response and phase response to a single combined antenna and the possibility of restoring the signal from these terms of a single antenna.

2. Geometry of the UWB combined antenna

The geometry of UWB combined antenna is shown in Figure 1. The antenna can be considered as a combination of the electric-type radiator made as a TEM horn with one active and two passive magnetic dipoles. Two identical antennas were used in these studies. The combined antennas chosen as objects of the study have the following dimensions: length L = 16 cm, height H = 15 cm, and width W = 15 cm. The thickness of aluminium plates of which the antenna was made (except the rear wall) is 1.7 mm. These antennas are optimized for excitation by bipolar voltage pulses with a duration of 1 ns. These

Content from this work may be used under the terms of the Creative Commons Attribution 3.0 licence. Any further distribution of this work must maintain attribution to the author(s) and the title of the work, journal citation and DOI.
Published under licence by IOP Publishing Ltd

antennas (or antennas of similar geometry) were used in high-power UWB radiation sources [1, 2]. The characteristics of these antennas were studied in [3, 4, 7, 8].

Figure 1. Photograph of the UWB combined antenna.

In addition, combined antenna models were created for simulation in the CST Studio Suite environment. The dimensions and design of the model corresponded to the real antenna (Figure 2).

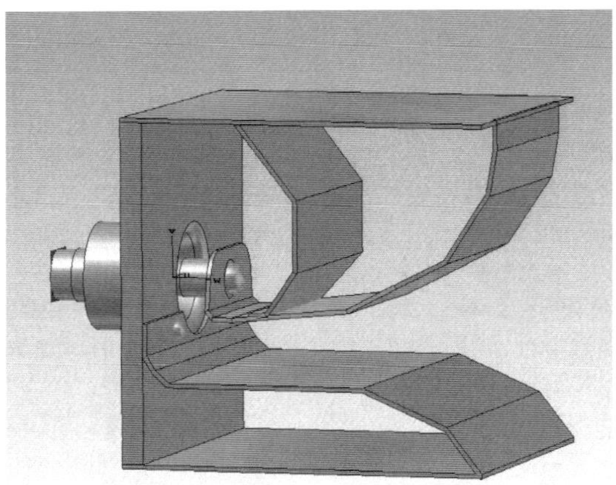

Figure 2. Model of the UWB combined antenna.

3. Comparison of the characteristics of the combined antenna and its model

As a research method for frequency response measurement, a scheme of two identical antennas was chosen, where one is transmitting and other is receiving. To verify the identity of the receiving and transmitting antennas, their VSWR were measured. The measurements were carried out on the Agilent PNA N5227A network analyzer (provided by shared research facilities of Tomsk Scientific Center SB RAS). In addition, numerical modeling was carried out and the dependencies of the VSWR of the combined antenna model on the frequency were obtained in the CST Studio Suite environment. The results of measurements and simulations are presented in Figure 3. Based on the results obtained, we can say that the antennas are identical and the numerical simulation corresponds to the real experiment (Figure 3).

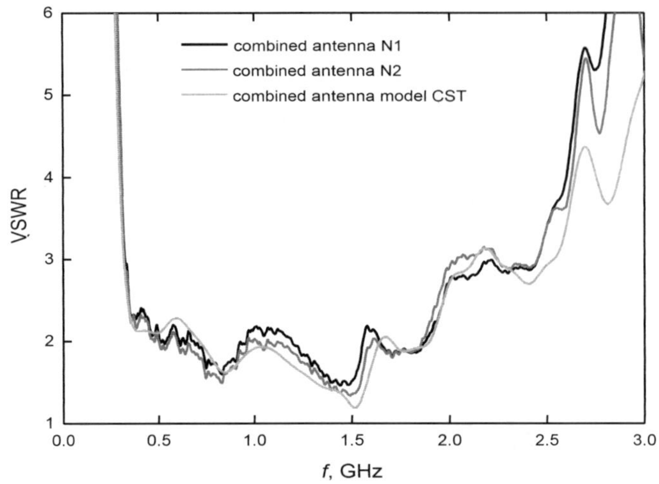

Figure 3. VSWR of combined antennas and the model.

The normalized radiation power patterns of the combined antenna and antenna model were measured as well. Measurements were made in the *E*- and *H*-planes at the fixed frequencies: 0.3, 0.5, 1, 1.5 and 1.8 GHz (Figure 4). The experimental patterns is shown in Figure 4 in black and simulated model patterns is shown in gray. The measurement results showed a good agreement of the model patterns with the experimental ones.

4. Results

The frequency response and phase response measurements were carried out using the method of two identical antennas at different distances between the antenna apertures. The calculated and measured frequency response for the distance between the antenna apertures of 80 cm is shown in Figure 5. The calculated and measured deviation of the phase response from the linear for the distance between the antenna apertures of 80 cm is shown in Figure 6. One can see that the simulated data are in good agreement with measurements in a wide frequency band. Significant differences are observed only in the high frequency region ($f > 2$ GHz), which is probably due to the using absorber plates but not installing the experimental setup inside an anechoic chamber.

To recover the signal from known frequency characteristics (frequency response and phase response), a MathCad code was written. As an example, Figure 7 shows the incident signal $U_g(t)$ at the input of the transmitting antenna, measured by a Le Croy 830Zi oscilloscope with a 30 GHz band (provided by shared research facilities of Tomsk Scientific Center SB RAS), and reconstructed one in the MathCad code. As can be seen from Figure 7, the accuracy of pulse recovery is very high. The same graph shows the pulse waveform at the output of the receiving antenna $U_a(t)$. The amplitude of the pulse $U_a(t)$ is normalized to the amplitude $U_g(t)$ for clarity.

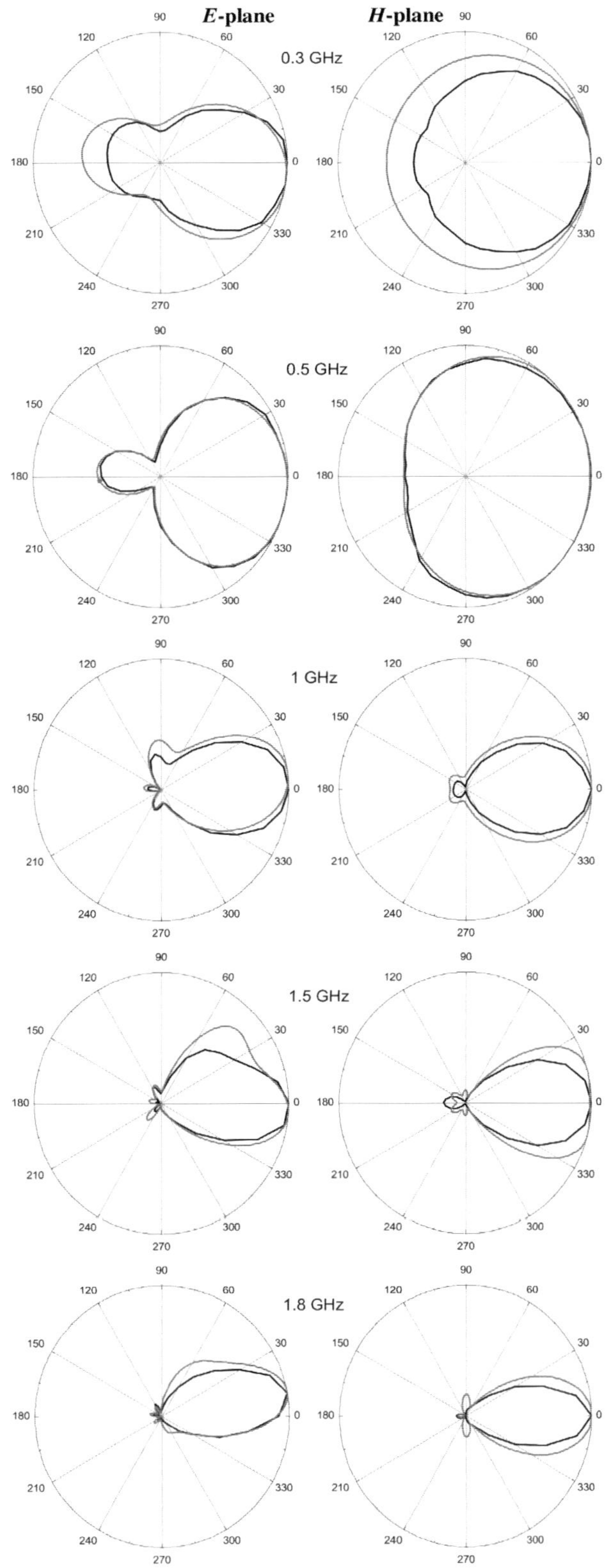

Figure 4. Combined antenna power patterns.

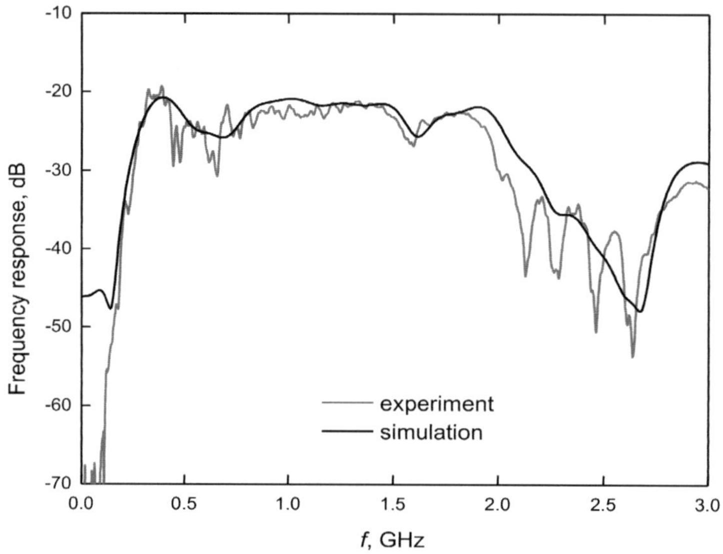

Figure 5. Calculated and measured dependences of the frequency response of the two identical antennas at a distance of 80 cm between their apertures.

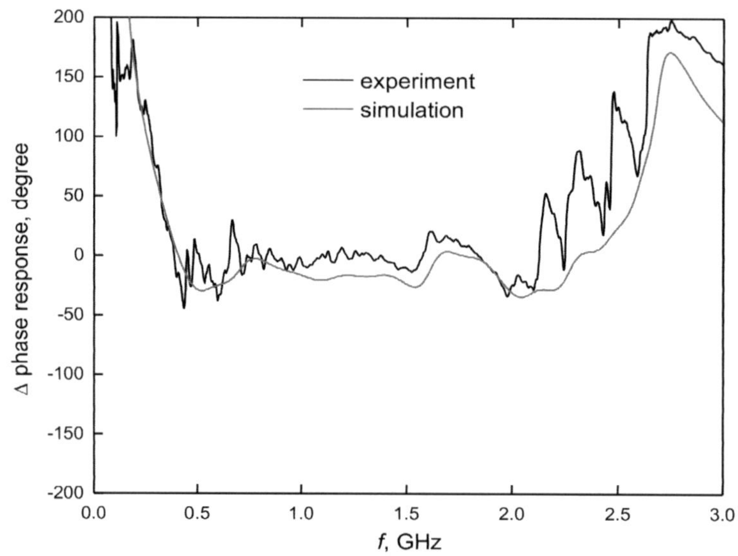

Figure 6. Calculated and measured dependences of the deviation of phase response from the linear of the two identical antennas at a distance of 80 cm between their apertures.

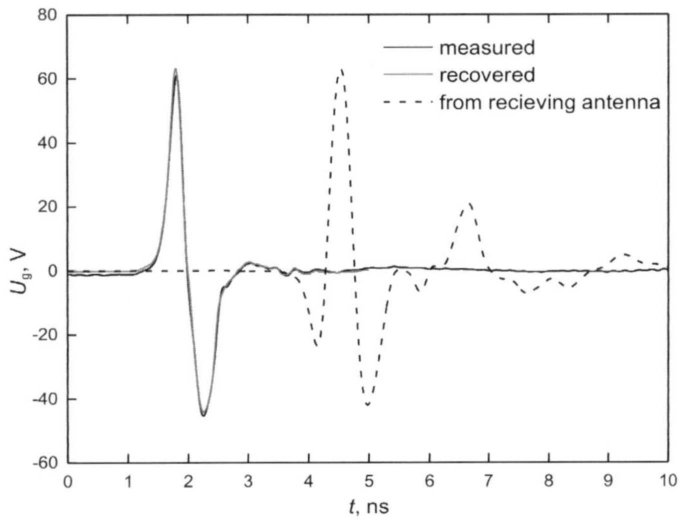

Figure 7. Reconstructed and measured input pulses of the transmitting-receiving line and receiving pulse at the receiving antenna output.

5. Conclusion

Comparison of the characteristics of UWB combined antennas with the simulated ones in CST Studio was carried out. Good agreement of the data was obtained. Based on the knowledge of the experimental frequency and phase responses of the transmitting-receiving path of the combined antennas and the output pulse of this path, the reconstruction of the input signal was made. To check a possibility of reconstruction of the radiated pulse with frequency and phase responses of a single antenna and the output pulse will be the next part of the study.

Acknowledgments

The work was performed under State Assignment of the Ministry of Science and Higher Education of the Russian Federation (project No. FWRM-2021-0002). The authors also would like to thank Prof. S. Shipilov and Dr. E. Balzovsky for the fruitful discussions.

References

[1] Andreev Yu A, Gubanov V P, Efremov A M *et al.* 2003 *Laser and Particle Beams* **21**(02) 211–217 doi: 10.1017/S0263034603212088
[2] Efremov A M, Koshelev V I, Kovalchuk B M *et al.* 2014 *Laser and Particle Beams* **32**(03) 413–418 doi: 10.1017/S0263034614000299
[3] Andreev Yu A, Buyanov Yu I, Koshelev V I 2005 *Journal of Communications Technology and Electronics* **50**(05) 535–543
[4] Andreev Yu A, Kornienko V N, Liu Sh 2017 *Journal of Communications Technology and Electronics* **62**(09) 976–983 doi: 10.1134/S1064226917080010
[5] Balanis A C 2016 *Antenna Theory Analysis and Design 4th ed* (New Jersey: John Wiley & Sons, Inc.) p 1065
[6] Denisenko A N 2005 *Signals. Theoretical radio engineering* (Moscow: Hotline-Telecom) p 704
[7] Koshelev V I, Andreev Y A, Buyanov Y I *et al.* 2003 Ultrawideband transmitting antennas, arrays, and high-power radiation sources *Ultra-Wideband Short-Pulse Electromagnetics 6* ed Mokole E L (New York: Kluwer Academic) pp 357–367
[8] Andreev Yu A, Kornienko V N, Liu Sh 2018 *IEEE Transactions on Antennas and Propagation* **66**(08) 4269–4276 doi: 10.1109/TAP.2018.2839888

The effect of shape and sizes of particles of wet quartz powders on complex dielectric permittivity in the frequency range of 10 kHz –10 GHz

P P Bobrov, E S Kroshka and O V Rodionova

Omsk State Pedagogical University, Omsk, Russia

E-mail: pavel.bobrov.2012@gmail.com

Abstract. The results of experimental studies of complex dielectric permeability of river sand and powders of granules of fused quartz with narrow distributions of particles in size at the frequencies from 10 kHz to 10 GHz are presented. The granule particles are spheres and the sand particles are irregularly shaped. The samples were moistened with distilled water and NaCl salt solution with conductivity of 0.1 and 0.77 S/m. It has been shown that the shape of the particles affects the complex dielectric permittivity (CDP) in the low frequency part of the range only when proportion of the solution is small and its concentration is weak. At full saturation of the samples with the solution and its high concentration, as well as in all cases at frequencies above 100 MHz, the influence of the particle shape is small. In the mid-frequency part of the range (from units to tens of megahertz) in a sample of quartz granules with small particles, there is a strong relaxation process, leading to a significant increase in the real part of the CDP. A similar, but slight increase is observed in samples of sand with larger particles. In the high-frequency range, the effect of the shape and particle sizes is very weak.
Keywords: dielectric permittivity, specific electrical conductivity, saline sands, dielectric relaxation processes, broadband dielectric measurements.

1. Introduction

Dielectric spectroscopy is an effective tool for studying the structure of soils and rocks. The analysis of spectra of the complex dielectric permittivity (CDP) $\varepsilon^*=\varepsilon'-i\varepsilon''$ in a wide frequency range can give information not only on water content, but also on other properties of rocks: clay content, wettability, porosity, morphology of pores, a saturation level and etc. At present, relatively narrow frequency ranges are used to study natural objects. At low frequencies from millihertz to kilohertz, the electrical properties of rocks are well studied in view of the application of the phenomenon of induced polarization (IP) to study geological environments. The types of physical mechanisms that generate IP in rocks are studied, the obtained expressions were validated using data from the literature [1]. Measuring complex electrical conductivity allows you to evaluate the hydraulic permeability of rocks [2].

Expansion of the frequency range of measurements, allows you to obtain additional information about the rock properties. Garrouch and Sharma [3] found that the dielectric permittivity in the frequency range from 10 Hz to 10 MHz, correlates with the clay volume fraction, the cation exchange capacity, and the electrochemical potential of the rock and strongly depends on the saturating brine salinity. Breede et al. [4] investigated the IP properties of sand and sand-clay mixtures with various saturation in the frequency range from 1 mHz to 45 kHz. In order to describe rock behavior in a wide frequency range, some authors supplement the models by the Maxwell-Wagner relaxation phenomena,

Content from this work may be used under the terms of the Creative Commons Attribution 3.0 licence. Any further distribution of this work must maintain attribution to the author(s) and the title of the work, journal citation and DOI.
Published under licence by IOP Publishing Ltd

for example, [5], [6]. Loewer et al. [7] also considered possible relaxation processes that occur in wet rocks in the frequency range from a few millihertz to tens of gigahertz. However, the frequency range from 10-100 kHz to 10-100 MHz is poorly investigated. Due to the wide range of pore and particle sizes of the multiphase medium, relaxation phenomena caused by various reasons overlap each other. This complicates determination of the relaxation processes parameters of and identification of its relations with the physical rock properties. In the work [8] the effects of the sizes of almost spherical sand particles (0.6 to 6.3 mm) on the electrical characteristics in the frequency range 42 Hz-1 MHz are experimentally investigated. It is shown that values of soil electrical properties increase with increasing water content and decreasing grain size.

We have developed a method for the dielectric permittivity measurement in a wide frequency range [9] that has made it possible to study relaxation processes at the frequencies from 10 kHz to 8–10 GHz. It has been shown that the existence of a relaxation process in the frequency range from 1 to 30–100 MHz leads to an increase in the real part of the CDP when the moisture is decreased and the relaxation time depends on the size of the sand particles [10]. In this paper, we present the results of the CDP measurement of sand samples and quartz granule powders with sizes from 2–9 to 220–300 μm of different shapes saturated with distilled water and brine. The purpose of the work is to identify the conditions in which the shape of particles has a significant impact on the CDP.

2. Materials and method

To investigate the effect of particle shape over a wide frequency range, we used cross-sieved fractions of quartz granule powders of almost spherical shape (sample # 1) and river sand (sample # 2) with particle sizes of 63-74, 90-100, 150-180 and 220-300 μm. Sample # 3 contains quartz spheres ranging in size from 20 to 74 μm and fragments of quartz of approximately the same size (Figure 1). Dry density of samples during measurements was changed in samples # 1 and # 3 from 1.42 to 1.47 g/cm^3, in sample # 2 from 1.29 to 1.32 g/cm^3. The samples were saturated with distilled water and saline solutions of NaCl with a concentration 0.1 and 0.77 S/m. The specific electrical conductivity (SEC) of the solution at direct current was measured with an S700-K conductometer with the InLab 731-ISM sensor from Metter Toledo. The SEC of distilled water is measured with this sensor with a large error, it is estimated at a value less than 0.001 S/m. The CDP measurements were carried out with a gradual decrease in the amount of saline solution by centrifuging a sample in the measuring cell. This method allows you to maintain a constant dry density addition when changing the amount of the solution.

Measurements of CDP were made in the same coaxial cell with Rohde & Schwarz ZNB20 vector network analyzer (VNA) and Keysight Technologies E4990A Impedance Analyzer [9]. At frequencies above 0.1 GHz, the complex transmission coefficient (CTC) of cell was measured by using VNA. For measurements in range 0.3–100 MHz, the cell was included into the break of the central conductor of a coaxial line of a large cross section. Admittance of the cell was determined by measuring of the CTC of the coaxial line of the large cross section. At frequencies below 10 MHz, the cell admittance was measured by E4990A. The output voltage of the impedance analyzer was set to 0.5 V, the output power of the network analyzer was 10 dBm. The measurement error of the real and imaginary parts of the CDP at different frequencies varied from 0.2 to 4%. The temperature of the samples was maintained in a TH-ME-25 climatic chamber (Republic of Korea) equal to 25° C with accuracy of 0.3° C.

3. Experimental results

First, consider the effect of particle shape on CDP. Figure 2 shows sample spectra with the same particle size and close values of saturation solution fraction that is about 0.2 m^3/m^3. It can be seen that the real (ε') and imaginary (ε") parts of CDP of sample # 3 containing debris at frequencies below 2-3 MHz (for ε') and below 600 MHz (for ε") are significantly less than those of other samples. When the samples are moistened with distilled water, the differences in the values of ε' increase, and when saturated with a salt solution with the SEC of 0.77 S/m they decrease (Figure 3).

In the frequency range from 1-2 to 10-30 MHz, the ε' values of the sample # 2 are slightly higher than these values are in sample # 1. This is due to the greater intensity of the high-frequency relaxation process in the sand having a higher specific surface area. The difference in the values ε' at frequencies below 1 MHz is explained by the difference in the low-frequency specific conductivity of the samples σ_0 (Figure 4). It is known that the real part of the CDP at frequencies of tens of kilohertz has a close correlation with σ_0 [11], [12].

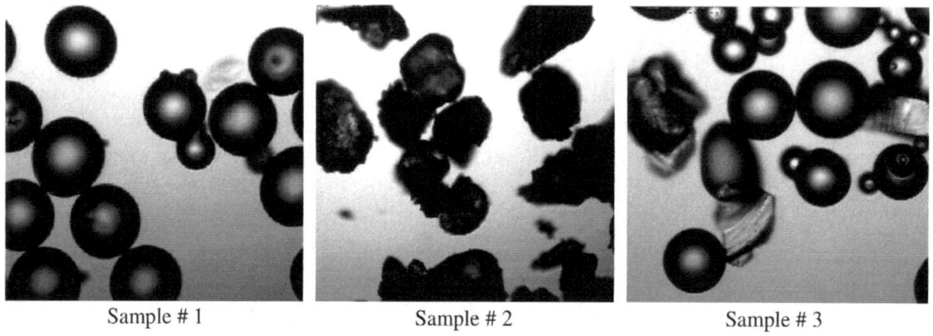

Figure 1. Form of the studied samples.

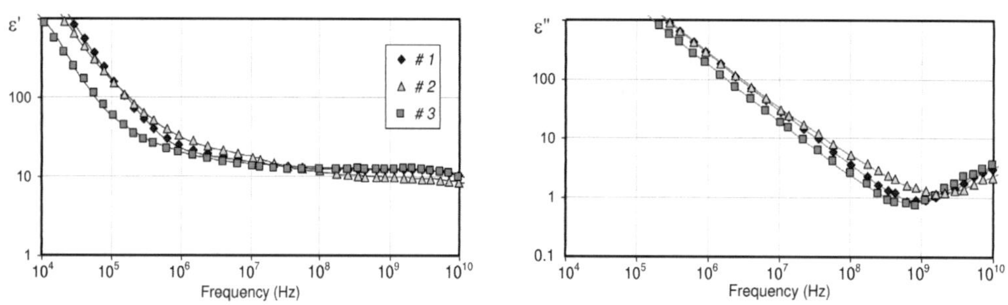

Figure 2. Frequency dependences of the CDP samples # 1, # 2 and # 3 with particle sizes 63-74 μm when saturated with salt solution with SEC of 0.1 S/m. Volume solution portion in the samples: # 1 – 0.197; # 2 – 0.191; # 3 – 0.201. Dry density in g/cm^3: # 1 - 1.47; # 2 - 1.29; # 3 - 1.47.

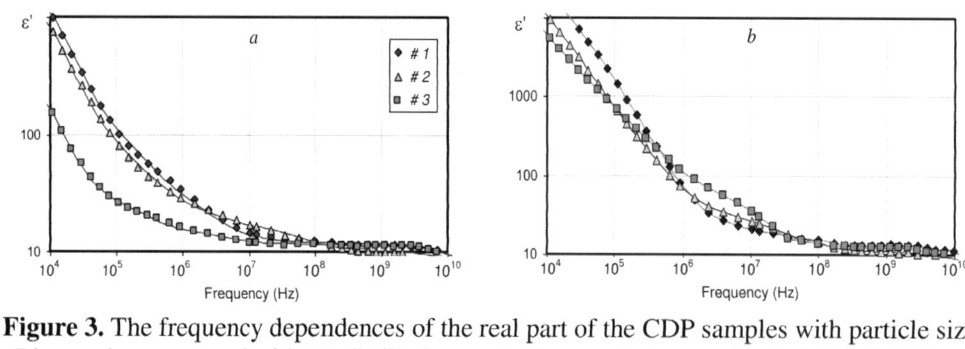

Figure 3. The frequency dependences of the real part of the CDP samples with particle sizes 63-74 μm when saturated with (a) distilled water (volume solution portion is 0.190-0.196) and (b) solution with SEC of 0.77 S/m (volume solution portion is 0.202-0.212).

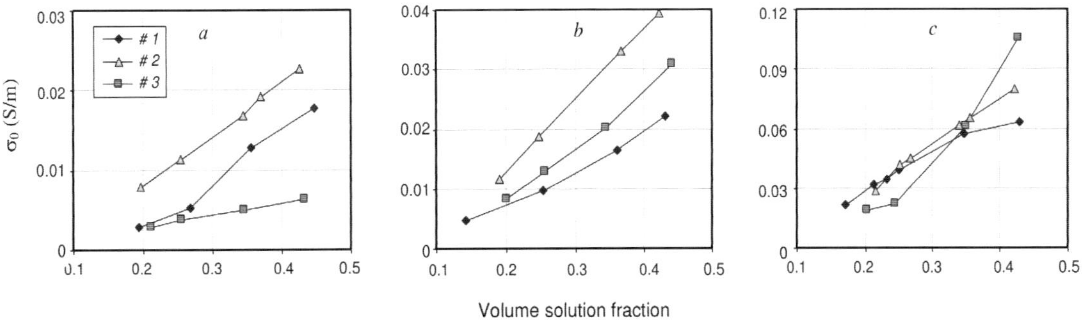

Figure 4. Specific electrical conductivity of the samples # 1, # 2, # 3 with particle sizes 63-74 μm when saturated with solution with SEC: (*a*) – 0.001 S/m (distilled water); (*b*) – 0.1 S/m; (*c*) – 0.77 S/m.

This dependence can be explained as follows. A significant contribution to the CDP at low frequencies is made by the polarization of the electric double layer (EDL), especially its inner part, called the Stern layer and also the conductivity of the samples is partially realized inside the EDL. Thus, the electrical properties of the EDL make a significant contribution to both the conductivity and dielectric relaxation. The correlation cannot be very high, since both the conductivity and dielectric relaxation are separately affected by other mechanisms that are different for these phenomena.

This behaviour of the SEC of the samples is due to the presence of different competing mechanisms of electrical conductivity: (i) surface conductivity - by the movement of ions inside the EDL; (ii) bulk conductivity - by the movement of ions in the volume of the solution.

The real part of CDP correlates mainly with surface conductivity, which makes the main contribution to the general conductivity at a low concentration of solution and its small quantity in the sample. In this case, samples with a different shape of particles have different superficial conductivity and different general conductivity (Figures 4*a* and 4*b*). With an increase in the amount of solution and its concentration, the overall conductivity of samples increases due to the bulk conductivity and the relative effect of surface conductivity is reduced. Therefore, the conductivity σ_0 of samples weakly depends on the form of particles (Figures 4*c*).

Thus, the shape of the particles significantly affects the CDP in the low-frequency band only at a low concentration of the solution and its small amount in the sample. Figures 5 and 6 show the results of measurements with almost complete filling with the solution. The higher the SEC of solution, the smaller the difference in CDP with different particle shapes. With the SEC of the solution, equal to 0.77 S/m differences in the real part of CDP are almost absent (Figure 6*b*). In this case, the conductivity σ_0 of all samples is almost the same (Figure 4*c*). Figure 7 shows the spectra of CDP powders of spherical globules and sand with particle sizes of 220-300 μm. At wetting with distilled water, large differences in ε' values are observed at frequencies below 2-3 MHz, and differences in ε" values at frequencies below 1 GHz. When saturated with saline, differences in ε' values do not exceed 20%, and in ε" values they are less than 10%.

Figure 8 shows the effect of particle sizes on the real part of the CDP of the sample when it was saturated with distilled water. There is a tendency of increasing ε' values with a decrease in particle size at frequencies below 10 MHz, especially noticeable with a smaller water fraction (Figure 8*b*). With a large water fraction, the ε' values are mainly affected by low-frequency relaxation processes. With a smaller water fraction, high-frequency relaxation process plays a significant role, raising the ε' values of the particles 2–9 μm in size at frequencies of 1-10 MHz almost an order comparable to other samples. Note that the ε' values of samples of close sizes 90-100 μm and 63-74 μm differ a little, and in sample 63-74 μm the values are even smaller than in sample of large sizes. The same situation is observed for the values of ε' in these samples. Thus, the relation of ε' and ε" values with particle sizes is statistical.

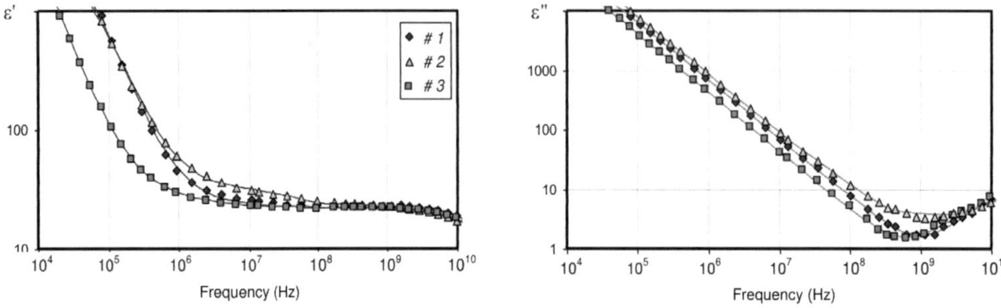

Figure 5. Frequency dependences of the CDP samples with particle sizes 63-74 μm when saturated with salt solution with SEC of 0.1 S/m. Volume solution portion in samples: # 1 - 0.356; # 2 - 0.368; # 3 - 0.367. Dry density in g/cm^3: # 1 - 1.47; # 2 - 1.29; # 3 - 1.47.

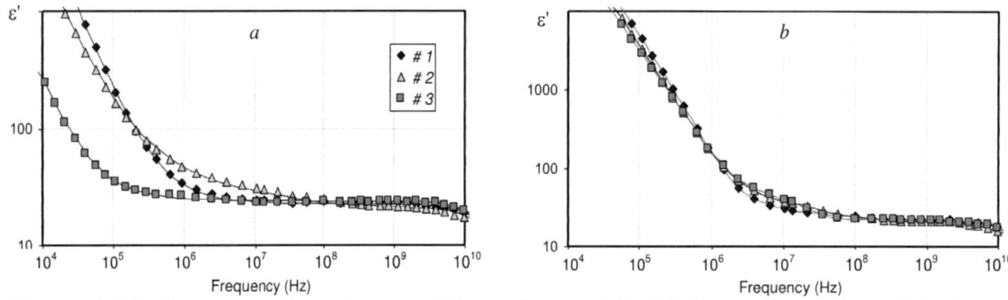

Figure 6. The frequency dependences of the real part of the CDP samples with particle sizes 63-74 μm when saturated with (a) distilled water and (b) salt solution with SEC of 0.77 S/m. Volume solution portion is 0.347-0.368.

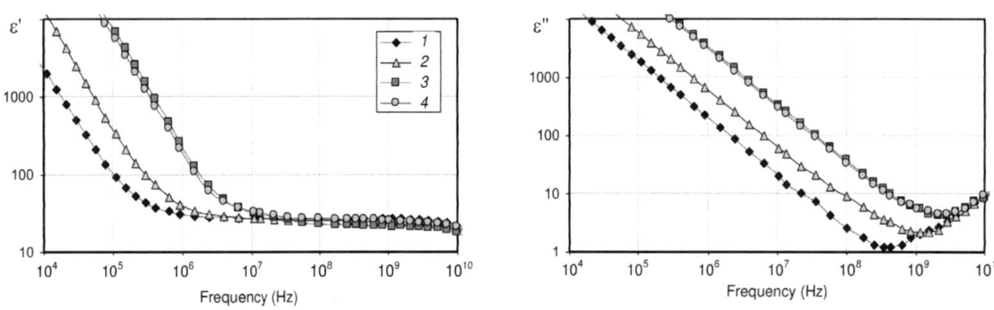

Figure 7. Frequency dependences of the CDP samples # 1 *(1, 3)* and # 2 *(2, 4)* with particle sizes 220-330 μm when saturated with distilled water *(1, 2)* and solution with SEC of 0.77 S/m *(3, 4)*. Volume solution and water portion is 0.371-0.397.

Figure 9 shows the frequency dependencies of CDP at saline saturation with SEC conductivity equal to 0.77 S/m. At near-full saturation, there is a slight decrease in ε' values at frequencies below 1-2 MHz with a decrease in particle size. With a smaller portion of solution, the ε' values of the samples with large particles decrease over the entire frequency range. The spectrum of the ε' values of the sample with the smallest particles differs sharply from the spectra of other samples. At frequencies of 0.5-100 MHz, ε' values are higher than in other samples, and at frequencies below 0.5 MHz these values are less.

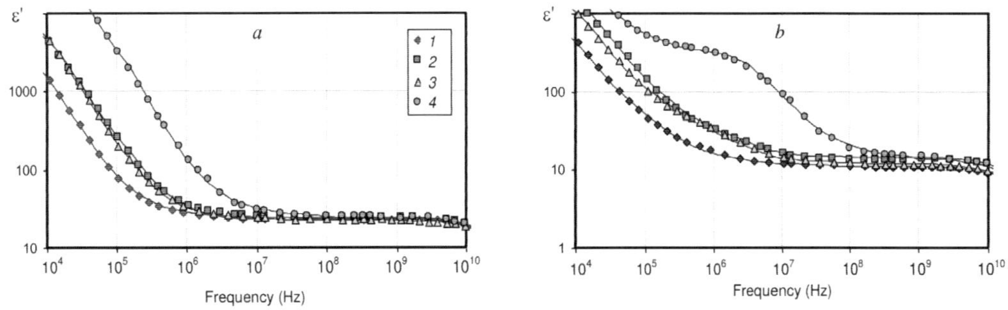

Figure 8. The frequency dependences of the real part of the CDP sample # 1 with different sizes of particles when saturated with distilled water. Volume water portion is (a) 0.347-0.359 and (b) 0.192-0.206. Particle sizes in μm: *1* - 150-180; *2* - 90-100; *3* - 63-74; *4* - 2-9.

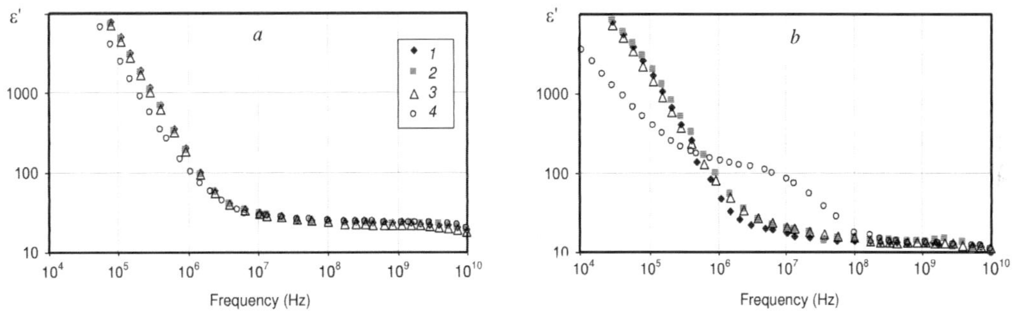

Figure 9. The frequency dependences of the real part of the CDP sample # 1 with different sizes of particles when saturated with salt solution with SEC of 0.77 S/m. Volume solution portion is (a) 0.345-0.354 and (b) 0.212-0.217. Particle sizes in μm: *1* - 150-180; *2* - 90-100; *3* - 63-74; *4* - 2-9.

Consideration of the middle part of the dielectric spectrum (frequency from units to tens megahertz) outlines a relaxation process which appears at a partial filling of pores and leads to an increase in the real part of CDP, as it is shown in [13]. The strength of this process is maximal with a particle size of several micrometers and the conductivity of the solution is in the tenth shares S/m. It can be noted that the increase in the real part of the CDP is observed not only in the sample with the smallest particles. A slight exceeding ε' values is observed in sample # 2 (Figures 2, 3, 5, 6), particles of which have irregular shape, as well as in sample # 3 containing a small part of small particles (Figure 3*b*).

In the high-frequency range, the effect of the shape and particle sizes is very weak. With the same portion of the solution in samples, regardless of its SEC, the ε' values at frequencies above 1 GHz, and the ε" values at frequencies above 3 GHz almost do not differ. In some cases (with a small portion of the solution and the low concentration), the values of ε' and ε" in the sand are slightly smaller than in other samples (Figures 2, 3*a*, 6*a*). This is explained by the fact that in the sand due to its higher specific surface, a small part of the water is in the bound state.

4. Conclusion

In this study, we used the method of broadband dielectric measurements to identify the effects of the size and shape of sand particles and quartz granules on dielectric spectra. We investigated three types of samples with a different shape of particles and with narrow distributions of particles in size. We showed that the shape and size of the particles significantly affect CDP in the low-frequency part of the spectrum, where CDP is determined by the polarization of the double electric layer. The properties of the double layer also determine the surface conductivity carried out inside it. With a small amount of solution in the sample and low concentration, the surface conductivity carried out in a double layer makes a major contribution to general conductivity. Therefore, the differences in surface conductivity in samples with different sizes and the shape are manifested in differences in CDP. With a large amount of solution and its high concentration, the conductivity is carried out in the volume of the solution and the contribution of the surface conductivity is negligible, therefore, the shape and size of the particles have little effect on CDP. In the frequency range from units to dozen megahertz, the differences in CDP are caused by the relaxation process arising in samples with small particles. The strength of this process is maximum at a medium moisture, when the amount of the air and the solution are approximately equal. On the high-frequency edge of the range, the effect of the shape and particle sizes is manifested very weakly.

Acknowledgments
This research was supported by the Russian Foundation for Basic Research, project 19-05-00395.

References

[1] Barreto A N, Dias C A 2014 *Geophysics* **79** D333–D347
[2] Robinson J *et al.* 2018 *Water Resources Research* **54** 3436–3452
[3] Garrouch A A, Sharma M M 1994 *Geophysics* **59** 909-917
[4] Breede K *et al.* 2012 *Near Surf. Geophysics* **10** 479–489
[5] Chen Y, Or D 2006 *Water Resources Research* **42** W06424
[6] Arcone S A, Boitnott G E 2012 *Journal of Applied Geophysics* **81** 97–105
[7] Loewer M *et al.* 2017 *Geophysical Journal International* **210** 1360–1373
[8] Datsios Z G, Mikropoulos P N 2019 *IEEE Trans. Dielectr. Electr. Insul.* **26** 904 – 912
[9] Bobrov P P *et al* 2015 *IEEE Trans. Geosci. Remote Sens.* **53** 2366–2372
[10] Bobrov P P *et al.* 2018 *Journal of Applied Geophysics* **159** 616–620
[11] Revil A, Skold M 2011 *Geophysical Journal International* **187** 813–824
[12] Bobrov P P *et al* 2020 *Radio communication technology* **46** 85–94
[13] Bobrov P P *et al* 2015 *Proceeding of the Progress in Electromagnetics Research Symposium (PIERS)* 1877–1879

Numerical simulation of a multi-beam cylindrical antenna in the GSM-900 frequency range

E V Balzovsky[1], **Yu I Buyanov**[1] **and T I Churikova**[2]

[1]Institute of High Current Electronics, Siberian Branch, Russian Academy of Sciences (IHCE SB RAS), Tomsk, Russia
[2]Tomsk State University, Tomsk, Russia

E-mail: bev@lhfe.hcei.tsc.ru

Abstract. The results of numerical simulation of a linearly polarized cylindrical multi-beam antenna with six ports are presented. The frequency dependence of the port coupling coefficient is considered when one antenna port is excited. The characteristics of the antenna directivity in the frequency domain are given.

1. Introduction

A multi-beam antenna (MA) is considered to be an antenna with several inputs, each of which corresponds to its own expressed maximum of the pattern often called the pattern beam. Usually, the MA consists of antenna elements (the radiating part) and a diagram-forming circuit (DFC). In MA, the number of DFC inputs corresponds to the number of generated beams, and antenna elements are connected to the DFC outputs, that is, the number of outputs is equal to the number of antenna array elements [1, 2]. Multi-beam antennas can be used as both receiving and transmitting ones in digital communication systems, radio monitoring, as well as in radar stations [3, 4].

The goal of the work is to develop and study a numerical model of a cylindrical receiving-transmitting MA operating in a given frequency range of 862...870 MHz and 902...928 MHz, with a gain of at least 11 dB. The paper presents the simulation results of a six-beam receiving and transmitting antenna with linear polarization.

2. A model of a multi-beam antenna array with linear polarization

Figure 1 shows a cylindrical MA model with a cone-shaped radiating part. The radiating part is divided into six emitters, each of which is connected to its own feed port. The z axis is parallel to the bisector of the angle between the vertical walls forming a sector with port 1. The antenna height is 460 mm, the diameter is 828 mm, and the gap between the truncated vertices of the cones is equal to 20 mm.

Content from this work may be used under the terms of the Creative Commons Attribution 3.0 licence. Any further distribution of this work must maintain attribution to the author(s) and the title of the work, journal citation and DOI.
Published under licence by IOP Publishing Ltd

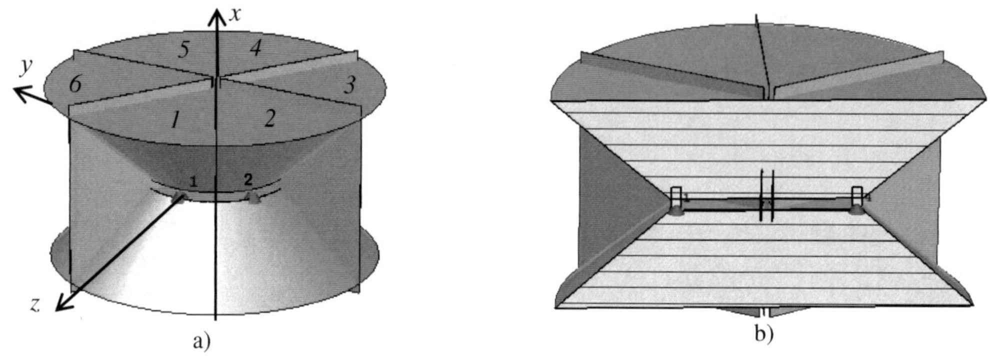

Figure 1. Model of a multi-beam antenna.

3. Results of numerical simulation

The antenna simulation was carried out in the licensed version of the CST Studio Suite program. Figure 2 shows the frequency dependence of the voltage standing wave ratio (VSWR) on the frequency. The minimum VSWR of the antenna has a value of 1.8 at a frequency of 900 MHz. For the frequency band of 845...960 MHz, the VSWR does not exceed the value of 2.2.

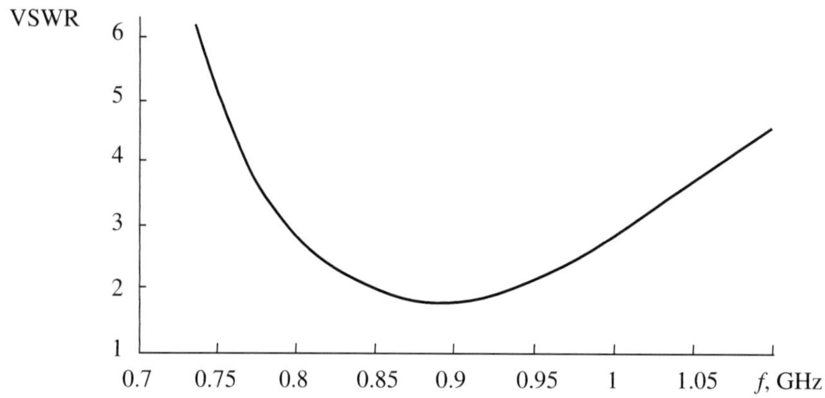

Figure 2. VSWR versus frequency.

Figure 3 shows the frequency dependence of the port coupling coefficient when power is supplied to port number 1. The location of the ports is shown in Figure 1*a*. The calculated decoupling for three pairs of ports 1-2, 1-3 and 1-4 in the form of a transmission coefficient module $|S_{x1}|$, where x is the number of the "receiving" port, is shown in Figure 3. For port 5, the result is similar to the result for port 3, and port 6 is similar to port 2. In the frequency range of 0.8...1 GHz, the transmission coefficient from port to port does not exceed -48 dB.

4. Characteristics of the antenna directivity

Figure 4 shows the results of calculating the pattern in the *E*-plane when the antenna is excited by port number 1. In the diagram presented, the normalized pattern of the antenna is multiplied by the maximum value of the antenna gain factor *G*.

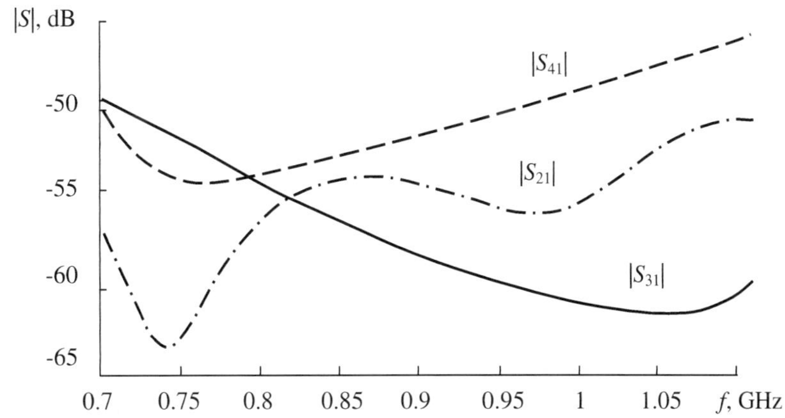

Figure 3. Frequency dependence of the transmission coefficient from port to port.

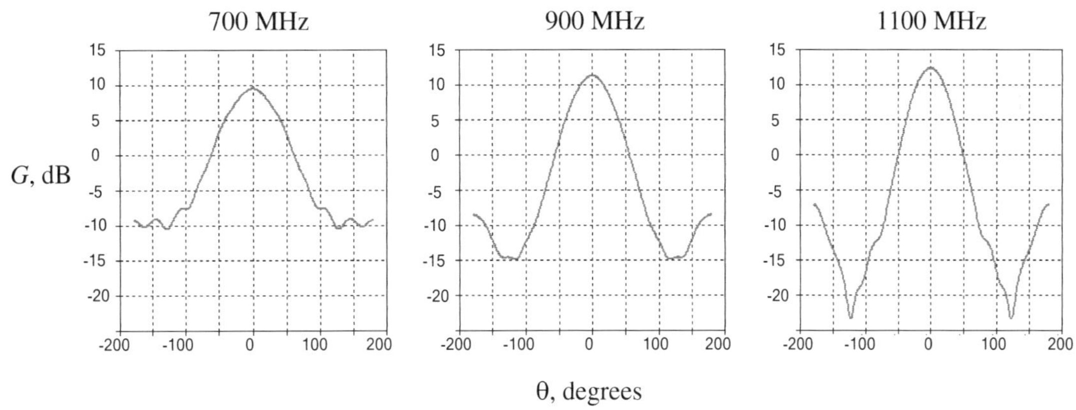

Figure 4. Antenna pattern when port 1 is excited.

Figure 5 shows the frequency dependence of the maximum gain values for a wider frequency band. The analysis of the obtained characteristics allows concluding that with increasing frequency, the gain value increases, the width of the bottom decreases, but at frequencies above 1.5 GHz, the level of the side lobes increases noticeably.

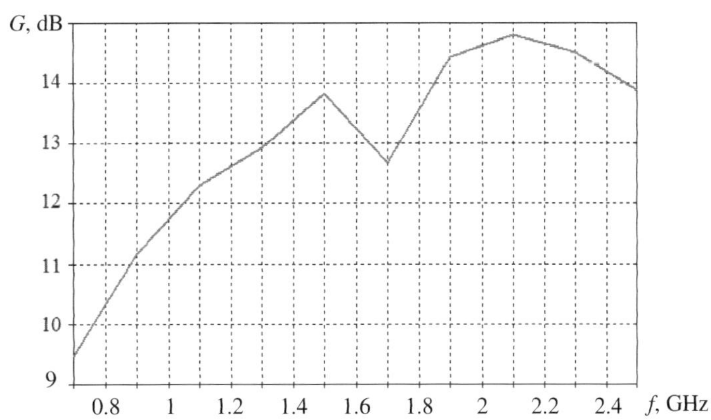

Figure 5. The antenna gain versus the frequency.

The results of calculating the pattern at a frequency of 900 MHz are shown in Figure 6 (in the *E*- and *H*-planes). In the *E*-plane, the width of the pattern is $\theta_{0.5} = 54.4$ degrees, the level of the side lobes is 19.8 dB. For the *H*-plane, the width of the pattern is $\theta_{0.5} = 91$ degrees, the level of the side lobes is 13.8 dB. The gain is equal G = 11.2 dB. The beams of the antenna pattern intersect at the level of 7.6 dB (Figure 7).

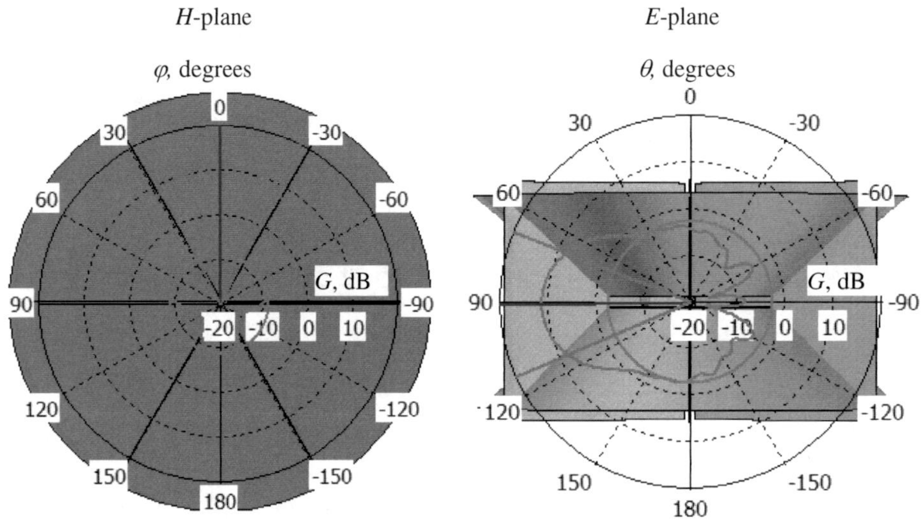

Figure 6. Antenna pattern in two planes at a frequency of 900 MHz

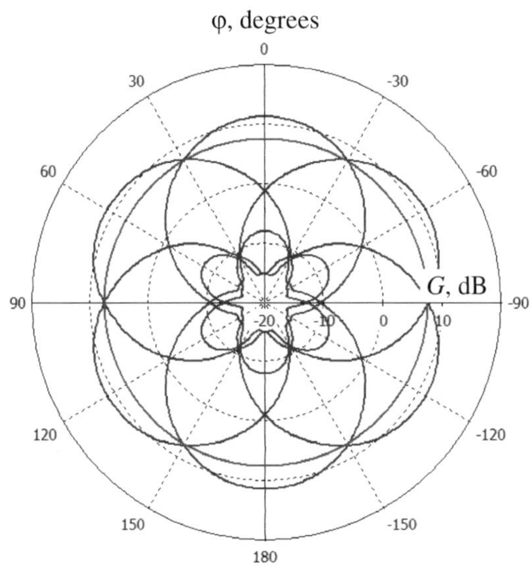

Figure 7. Antenna pattern with six beams in *H*-plane at a frequency of 900 MHz.

The antenna radiation is linearly polarized. The vector of the electric field strength of the main polarization lies in the *xz* plane. The ellipticity coefficient in the direction of the main maximum does not exceed the value of 60 dB. In the direction of 30 degrees in the horizontal and vertical planes from the direction of the main maximum, that is, along the metal walls separating the antenna sector, the coefficient of ellipticity reaches a maximum but has a value of no more than -18 dB.

5. Conclusion

A model of a six-beam receiving and transmitting antenna is proposed and its characteristics are numerically investigated. The antenna matching band at VSWR $\leq$ 2 is 850...940 MHz. The antenna radiation is linearly polarized. The antenna pattern width is 54.4 degrees in the E-plane and 91 degrees in the H-plane with a gain of 11.2 dB. The results obtained correspond to the goal of the research. Further work will be aimed both at improving the characteristics of the antenna discussed above, and at creating a multy-beam antenna with circular polarization.

Acknowledgments

The work was performed under State Assignment of the Ministry of Science and Higher Education of the Russian Federation (project No. FWRM-2021-0002).

The study was carried out using the software product CST Studio Suite® developed by Dassault Systemes on the basis of a valid license LN00031351538 dated July 21, 2020, provided by the IHCE SB RAS, and a valid license LN00029178508 from 10.12.2019, provided by Tomsk State University. All the conditions for the granting of such licenses when these organizations carry out joint research and publish their results are met.

References

[1] Dahlman E 2018 *5G NR: The Next Generation Wireless Access Technology* (London: Academic Press) 441 p.

[2] Osseiran A 2016 *5G Mobile and wireless communications technology* (Cambridge: Cambridge University Press) 406 p.

[3] Weiss S 2010 Wideband Beamforming Concepts and Techniques (Chichester, UK: John Wiley & Sons, Ltd) 304 p.

[4] Toit C 2003 *IEEE International Symposium on Phased Array Systems and Technology* 146–151 doi:10.1109/PAST.2003.1256972.

Journal of Physics: Conference Series **2140** (2021) 012006 doi:10.1088/1742-6596/2140/1/012006

Model of a single-electron transistor based on endohedral fullerene $(Sc_3N)@C_{80}$

A G Duisenova[1] and D M Sergeyev[2]

[1] K. Zhubanov Aktobe Regional University, Aktobe, Republic of Kazakhstan
[2] T. Begeldinov Military Institute of Air Defense Forces, Aktobe, Republic of Kazakhstan

E-mail: ainura_duisenova@mail.ru

Abstract. In this work within the framework of the density functional theory and the method of nonequilibrium Green's functions the dependences of the total energy of molecules C_{80}-SET and $(Sc_3N)@C_{80}$ - SET on their total charge, the dependence of the total energy from the gate voltage and the stability diagram of the single-electron transistor have been determined. It is noted that for transition to switch to on mode $(Sc_3N)@C_{80}$-SET it is necessary to apply the gate voltage in the range from $0.019 \le V_G \le 5.940$ with the bias voltage $-2.040 \le V_{SD} \le 2.155$ V. Considering that at values of bias voltage equal to $-0.381 \le V_{SD} \le 0.533$ V there is no voltage on the substrate ($V_G = 0$ V) and electric current does not flow. It is shown that the total energy at negative values of charge is higher than at positive charges and that the area of the Coulomb rhombus in fullerene with scandium nitride is 5.3 times larger than in "pure" fullerene.

1. Introduction

One of the promising nanoelectronic devices is a single-electron transistor (SET), which is based on the effects of correlated electron tunnelling and Coulomb blockade. The SET consists of a main element of the device – an island, which is reliably separated by tunneling electrode junctions (drain – source), and a control electrode (gate), which influences on this island solely through capacitive coupling. To date, there are many experimental works where there are various options for the implementation of single-electron devices based on single molecules or atoms [1–3].

According to experimental works [4–7], fullerenes have been shown as effective nanomaterials to create SETs because of their stability to charge variation. The unique construction of fullerenes lies in the molecular cavity that can contain an atom, a group of atoms or even small molecules. In consequence, new compounds are formed, known as endohedral fullerenes with unusual electronic properties. One of the most distinctive features of endohedral fullerenes is that they can be considered as consisting of a positively charged metallic core and a negatively charged carbon shell.

In previous work [8] we considered the SETs based on endohedral fullerenes $Me@C_{60}$ (Me = Na, Li, K), where was shown a noticeable effect of encapsulated atoms on the energy characteristics. A new method of synthesis was developed in work [9,10], where a new metallofullerene $(Sc_3N)@C_{80}$ was obtained. It has lower thermal reactivity due to the difference in energy level and spatial distribution of the higher occupied molecular orbital (HOMO) between the two endohedral fullerenes [11–12]. Metallic nitride fullerenes provide promising replacement materials for new types of electronic devices.

Content from this work may be used under the terms of the Creative Commons Attribution 3.0 licence. Any further distribution of this work must maintain attribution to the author(s) and the title of the work, journal citation and DOI.
Published under licence by IOP Publishing Ltd

In this work, within the framework of density functional theory (DFT) and nonequilibrium Green's function (NEGF) method, determined the values of total energy, and stability diagram of SET based on endohedral fullerene $(Sc_3N)@C_{80}$ [13,14].

2. Geometry

The geometry of the investigated SET models based on C_{80} is shown in Figure 1. The drain, source and gate electrodes are formed from gold. The width of the electrodes is 14 Å. Geometric dimensions of the SETs are shown in Figure 1.

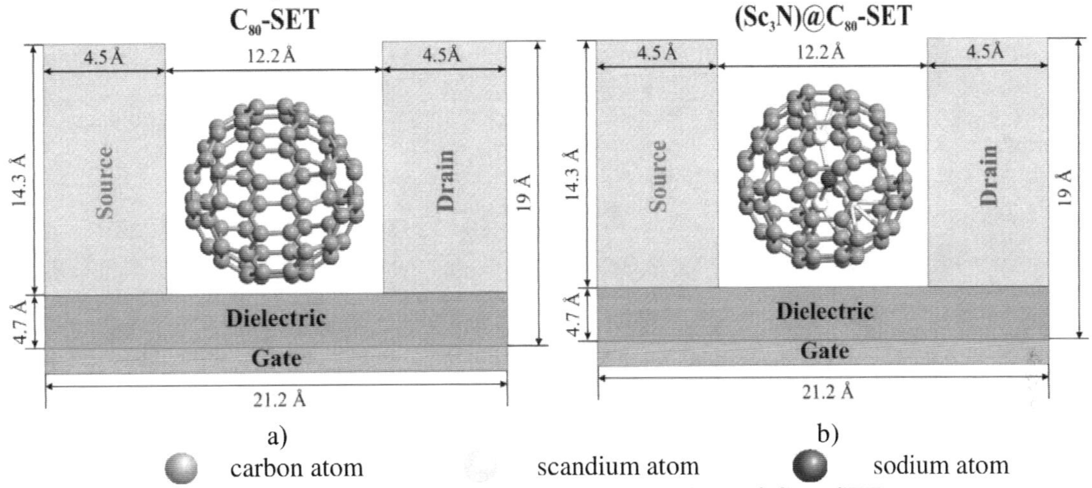

Figure 1. SET geometry: a) C_{80} – SET, b) $(Sc_3N)@C_{80}$ – SET.

The optimization of the geometric structures was carried out within the framework of DFT, the generalized gradient approximation PBE was used as an exchange-correlation functional. To calculate the electro-transport characteristics, DFT in the local density approximation was used. During the modeling process the electric charges (Q = -2, -1, 0, 1, 2) were chosen for each molecule and quantum-chemical properties of the geometrical structure of the molecule were calculated which matched the stable ground state of the molecule at each value of the chosen charge [15]. The selection of charge values from -2e to 2e is linked to the fact that the state of small molecules with more than two extra electrons becomes unstable.

The calculations of this work are realized in the program Atomistix ToolKit Virtual NanoLab [16, 17]. When fullerene C_{80} is optimized, the bond length between nearby C - C carbon atoms are 1.41 Å and the diameter of fullerene is 8.236 Å. The distance from the fullerene to the electrodes is 1.982 Å. In the endohedral fullerene $(Sc_3N)@C_{80}$, the C - C bond length ranges from 1.41 to 1.439 Å, and the distance between the Sc - C and N - Sc atoms is 2.306 Å and 2.054 Å, respectively. When Sc_3N is introduced into the C_{80} cavity, a minor change in its geometry is observed. The diameter of fullerene $(Sc_3N)@C_{80}$ is 8.19 Å and the distance from the endohedral fullerene to the electrodes is 2.005 Å.

3. Results

In Figure 2 is shown the dependence of the total energy of fullerene molecules from the charge state. Based on the results of the total energy calculations, it is shown that the total energy is higher when the charge is negative than when it is positive. The total energy of fullerene at neutral charge equals -12607 eV, while for endohedral fullerene $(Sc_3N)@C_{80}$ -13614 eV.

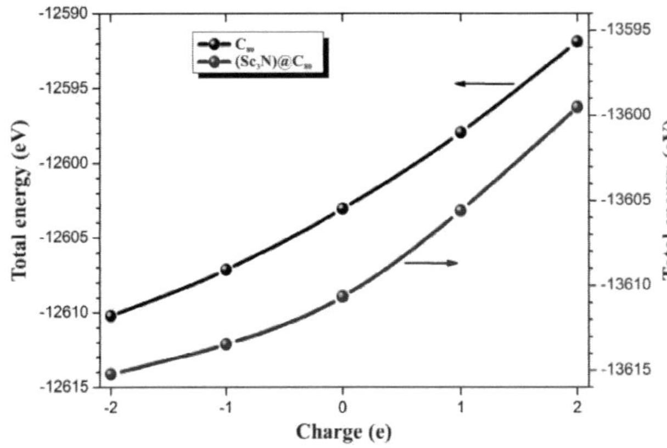

Figure 2. The dependency of the total energy of molecules C_{80} – SET and $(Sc_3N)@C_{80}$ – SET from their total charge.

Figure 3 shows the dependence of total energy on gate voltage (V_G) for C_{80} and $(Sc_3N)@C_{80}$ based SETs in different charge states. As can be seen at $V_G > 0$ the negative charge states (Q = -1, -2) show a significant decrease in the total energy of SETs, making them stable at positive values of V_G. At $V_G < 0$, the positive charge state (Q = +1, +2) is more stable relative to the negative state, due to the shift of HOMO and lowest unoccupied molecular orbital (LUMO) levels with gate voltage.

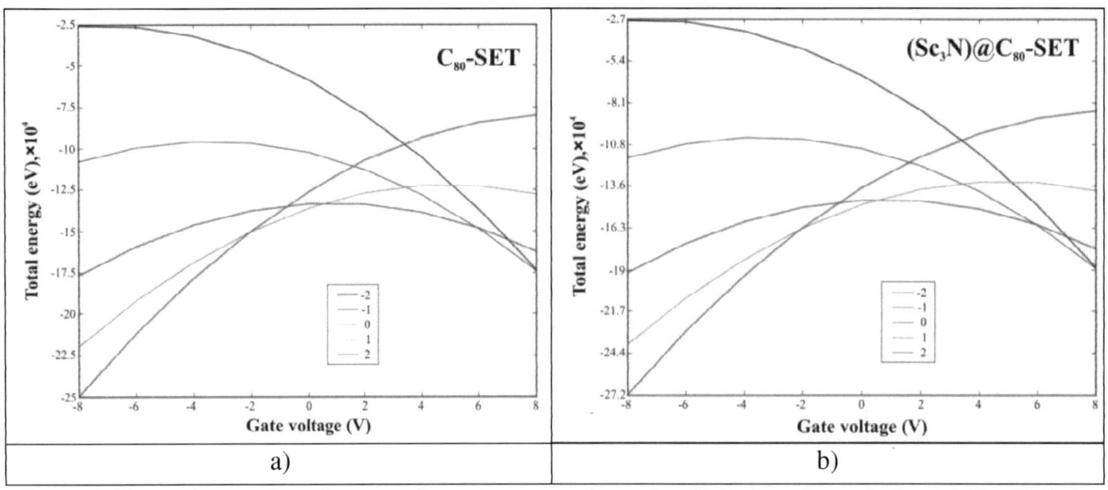

Figure 3. Dependence of total energy from the gate voltage for SETs based on: a) C_{80}, b) $(Sc_3N)@C_{80}$. (Different charged states of the molecule are shown with different colored curves: blue (-2), green (-1), red (0), turquoise (+1), violet (+2)).

The charge transfer in SET happens when it overcomes the Coulomb blockade, but to overcome this, it is necessary to supply sufficient charge energy by adjusting the gate voltage and the voltage at the electrodes. For the given values of gate voltage and bias voltage (V_{SD}) the number of molecular energy levels inside the bias window voltage is given by the color codes in Figure 4*a,b*.

In Figure 4 by controlling the gate electrode it is possible to transfer the Coulomb Island from the conductive mode to the tunnelling current blockade mode. Within the bias voltage interval $-0.896 \le V_{SD} \le 0.972$ V by adjusting the gate voltage in the interval $0.524 \le V_G \le 3.010$ V, the C_{80} - SET can be transferred to the blockade mode. Considering that in the absence of gate voltage

($V_G = 0\,\text{V}$), Coulomb blockade occurs at a bias voltage of $-0.457 \le V_{SD} \le 0.533\,\text{V}$. For transition to $(Sc_3N)@C_{80}$ - SET switch to on mode, the gate voltage must be applied in the range from $V_{G\min} = 0.019\,\text{V}$ to $V_{G\max} = 5.940\,\text{V}$ with a bias voltage established as $-2.040 \le V_{SD} \le 0.533\,\text{V}$. Considering that at bias voltage values equal to $V_{SD\min} = -0.381\,\text{V}$ and $V_{SD\max} = 0.533\,\text{V}$ the gate voltage is absent ($V_G = 0\,\text{V}$) and the electric current does not pass through.

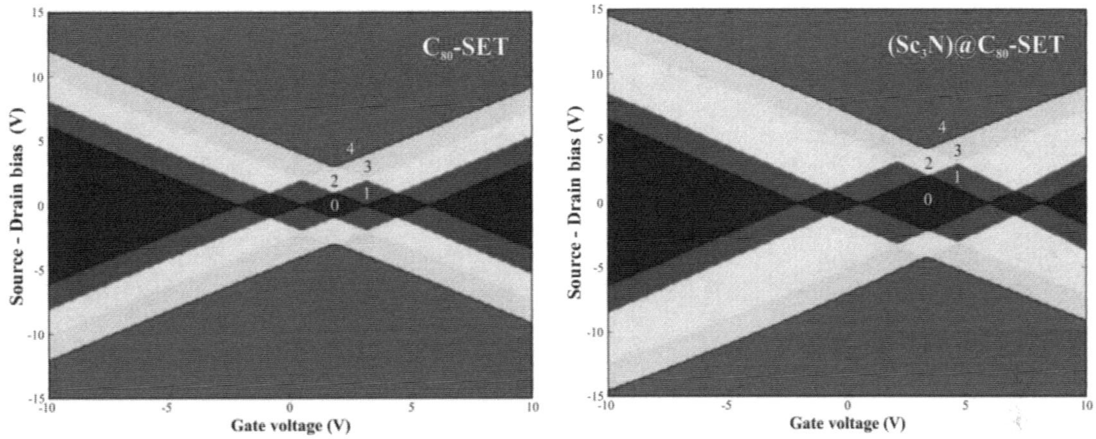

a) b)

Figure 4. Charge stability diagrams for SETs - (a) C_{80}, b) $(Sc_3N)@C_{80}$. (The number of charge states for given bias potentials are represented by the following colors: blue (0), light blue (1), green (2), yellow (3) and red (4)).

Table 1 shows that the Coulomb diamond area in fullerene with scandium nitride is significantly larger than in "pure" fullerene. The change of Coulomb diamond area is related to the electronic structure of the encapsulated molecule.

Table 1. The main parameters obtained from the central Coulomb diamond from the stability diagrams shown in Figure 4.

	$V_{SD\min}$	$V_{SD\max}$	ΔV_{SD}	$V_{G\min}$	$V_{G\max}$	ΔV_G	Diamond area
C_{80}-SET	-0.896	0.972	1.868	0.524	3.010	2.486	2.321
$(Sc_3N)@C_{80}$-SET	-2.040	2.155	4.195	0.019	5.940	5.959	12.499

4. Conclusion

In summary, the geometrical structures of C_{80} and $(Sc_3N)@C_{80}$ fullerenes were optimized within the framework of DFT + NEGF and on their basis the dependences of the total energy of molecules C_{80} - SET and $(Sc_3N)@C_{80}$ - SET from their total charge, the dependence of total energy from the gate voltage and the stability diagram of the SET were determined. This shows a significant influence of the encapsulated Sc_3N molecule on the electronic properties of the fullerene C_{80}. The results will be useful in the development of one-electronic components.

Acknowledgments

The research has been supported by the grant of the Ministry of Education and Science of the Republic of Kazakhstan AP08052562.

References

[1] Koenraad P M and Flatte M E 2011 *Nature materials* **10**(2) 91–100 doi:10.1038/nmat2940

[2] Zhang X, Li W, Feng L, Chen X, Hansen A, Grimme S, Chen N *et al* 2018 *Nature Communications* **1**(9) 2753 doi: 10.1038/s41467-018-05210-8
[3] Sergeyev D, Ashikov N and Zhanturina N 2021 *International Journal of Nanoscience* **20**(1) 2150007 doi: 10.1142/S0219581X21500071
[4] Popov A A 2017 *Endohedral Fullerenes: Electron Transfer and Spin* (Switzerland: Springer International Publishing)
[5] Konarev D, Popov A A, Zorina L V, Khasanov S S K and Lyubovskaya R N 2019 *Chem.-Eur. J.* **25** 14858–14869 doi: 10.1002/chem.201902782
[6] Plonska-Brzezinska M E, Athans A J, Phillips J P, Stevenson S and Echegoyen L 2008 *Journal of Electroanalytical Chemistry* **614** 171–174 doi: 10.1016/j.jelechem.2007.11.013
[7] Takase T, Sakaino M, Kirimoto K and Sun Y 2013 *Applied Physics A* **112**(4) 927–931 doi: 10.1007/s00339-012-7449-6
[8] Sergeyev D M 2020 *J. Nano- Electron. Phys.* **12**(3) 03017 doi: 10.21272/jnep.12(3).03017
[9] Krause M and Dunsch L 2004 *Chem Phys Chem* **5**(9) 1445–1449 doi: 10.1002/cphc.200400085
[10] Iezzi E B, Duchamp J C, Harich K, Glass T E *et al* 2002 *J. Am. Chem. Soc.* **124**(4) 524–525 doi: 10.1021/ja0171005
[11] Stevenson S, Rice G, Glass T, Harich K, Balch A L, Dorn H C *et al* 1999 *Nature* **401**(6748) 55–57 doi: 10.1038/43415
[12] Pykhova A D, Semivrazhskaya O O, Samoylova N A, Rybalchenko A V *et al* 2020 *Dalton Transactions* **49** 9137–9147 doi: 10.1039/D0DT01513J
[13] Ferre N, Filatov M and Huix-Rotllant M 2016 *Density functional methods for excited states* (Vol. 368) (Switzerland: Springer International Publishing) p 487
[14] Smidstrup S, Stradi D, Wellendorff J, Khomyakov P A *et al* 2017 *Physical Review B.* **96** 195309 doi: 10.1103/PhysRevB.96.195309
[15] Xing-Xing Y, Bairu L, Hao-Sheng L, Fei J *et al* 2020 *Research* **21** 2059190 doi: 10.34133/2020/2059190
[16] Brandbyge M, Mozos J L, Ordejon P, Taylor J and Stokbro K 2002 *Phys. Rev. B* **65** 165401 doi: 10.1103/PhysRevB.65.165401
[17] Stradi D, Martinez U, Blom A, Brandbyge M and Stokbro K 2016 *Phys. Rev. B* **93** 155302 doi: 10.1103/PhysRevB.93.155302

Journal of Physics: Conference Series **2140** (2021) 012007 doi:10.1088/1742-6596/2140/1/012007

Active positioning system based on hall-effect sensors

G F Karlova, B I Avdochenko and E A Belozyorova

Tomsk State University of Control System and Radioelectronics, Tomsk, Russia

E-mail: ekaterina.a.belozerova@tusur.ru

Abstract. The paper investigates the behavior of Hall-effect sensors (HES) operating in microwave frequency range. In order to evaluate the capabilities of small radars based on HES's, the calculation of HES output voltage dependence on microwave intensity is carried out. It is shown that at power flow density $I(t)$ = 10 μW/cm^2 (a permissible value for a long-term human body exposure) the single GaAs sensor output Hall voltage U_H ~ 0.03 μV.
It is also demonstrated that integration of signals with due regard for their delays (as in the case of range-difference radar systems) gives an opportunity to obtain electromagnetic image of any point within the antenna coverage area with centimeter-level accuracy. The paper describes the experimental assembly and estimates the distances to the detected contrasting objects.

1. Introduction

Using HES's to receive microwave radiation is an advanced technology for various devices design. As it is shown in [1], the simultaneous impact of the mutually perpendicular electric and magnetic fields with given orientation on a HES can be used to receive microwave radiation with the frequency of up to 30 GHz. It is also demonstrated in [1] that detector higher operating frequency is limited mainly by charge carrier relaxation time τ. The relaxation time is estimated in [2], and for gallium arsenide epitaxial material it is equal to tens of picoseconds that corresponds to the data presented in [3]. Therefore, the investigation of Hall-effect sensor output voltage dependence on electromagnetic microwave intensity is of great interest because it enables estimation of the possibility to build up small HES radars with the resolution of centimeters.

The research conducted in [4] reveals sufficiently high sensitivity of gallium arsenide Hall-effect sensors. It is up-to-date to study their capability to receive pulsed magnetic fields of nanosecond range and react to electromagnetic field.

2. Theory

The work [1] considers the charge carrier behavior in sensor surrounded by altering electromagnetic field. The field strength is obtained from the motion equation for charge carriers e moving inside the isotropic conductor with dielectric permittivity ε, in alternating harmonic fields with frequency ω, at current density $j_x = 0$, neglecting relaxation phenomena, i. e. $t >> \tau$.

$$E_H = \frac{u}{2} \cdot \frac{\mu_0 \mu E_{0y} H_{0z}}{1+(\omega\tau)^2}, \tag{1}$$

where $u = \frac{e\tau}{m}$ is charge carrier mobility, μ is medium relative magnetic permeability, μ_0 is the magnetic constant.

Content from this work may be used under the terms of the Creative Commons Attribution 3.0 licence. Any further distribution of this work must maintain attribution to the author(s) and the title of the work, journal citation and DOI.
Published under licence by IOP Publishing Ltd

A special assembly has been designed to estimate the value of Hall-effect sensor output signal voltage under the impact of pulsed magnetic field created by a single coil loop with the diameter $d = 3$ cm (or radius $r = 1.5$ cm) and measured at the distance l from the coil center. The pulse duration $\tau_p = 5$ ns, the pulse current $I = 1000$ A. The assembly generates pulses with steep leading and trailing edges [4] that enables using formulae for static fields. In this case, in SI system, the magnetic flux density at the point situated on the axis of the ring with the radius r at a distance l from its center is found with the use of the following expression:

$$B = \frac{\mu_0 \cdot I \cdot r^2}{2(r^2 + l^2)^{3/2}}, \tag{2}$$

where μ_0 is equal to $4\pi \cdot 10^{-7}$ H/m = $4\pi \cdot 10^{-7}$ Ω·cm/m.

This field gives rise to a Hall-effect sensor signal:

$$U_H = \gamma \cdot \frac{\mu_0 \cdot I \cdot r^2}{2(r^2 + l^2)^{\frac{3}{2}}}, \tag{3}$$

where γ is sensor magnetic field sensitivity.

The signals produced by Hall-effect sensor under the influence of the fields created by the coil loops of various diameter and at various distances from the loop center have been estimated. The GaAs Hall-effect sensor sensitivity to magnetic flux density [4] was equal to 1 V/T. The results are gathered together in Table 1 where the data for commercial InSb samples are used to compare the materials.

Table 1. The Hall voltage dependence on the distance from the coil center l and loop radius r for GaAs and InSb samples.

***r*, cm**	***l*, cm**	***B*, T**	**U_H, μV for GaAs (*I*=1 mA)**	**U_H, μV for InSb**	
				***I*=1 mA**	**I=5 mA**
1.5	20	$17.592 \cdot 10^{-6}$	17.59	87.95	437.75
1.5	100	$0.138 \cdot 10^{-6}$	0.138	0.690	3.45
5	20	$1.96 \cdot 10^{-4}$	196	980	4900
5	100	$1.57\ 10^{-6}$	1.57	7.85	39.35

The obtained data are used below to estimate the distances to the explored objects.

As it follows from the expression (1), the sensor responds to field magnetic component variations. This means that it can register inhomogeneities while exploring objects. Let's consider the relation between microwave frequency wave intensity and Hall-effect sensor output voltage. The power flux density (PFD) average value $P_{av} = \frac{E_{0y} \cdot H_{0z}}{2}$. Then, the Hall field strength average value:

$$E_{Hav} = \frac{u \cdot \mu_0\ \mu \cdot P_{av}}{1 + (\omega\tau)^2}, \tag{4}$$

where ω is oscillation frequency equal to 10 GHz. Substitution of the values for GaAs—$u = 8000$ cm²/V·s, $P_{av} = 0.1$ W/m² (a permissible value for human body exposure), $\mu_0 = 4\pi \cdot 10^{-7}$ Ω·s/m, $\mu = 1$, $\omega = 10^{10}$ Hz, $\tau = 10^{-8}$ s—gives the GaAs Hall-effect sensor die output voltage U_H ~0.03 μV.

To obtain $P_{av} = 0.1$ W/m² at the area of the sphere (with the radius $r = 1$ m = 100 cm) $S = 4\pi \cdot 100^2$ =250,000 cm² = $25 \cdot 10^4$ cm² and transmitting antenna circular pattern (CP) it is necessary to have the following value of the transmitter power:

$$P_{cr} = P_{av} \cdot S = 10 \cdot 25 \cdot 10^4 = 250 \cdot 10^4\ \mu\text{W} = 2.5\ \text{W}. \tag{5}$$

In order to use the sensor it is important to be aware of its threshold sensitivity which is defined by the noise. The investigation [5] shows that in microwave frequency broadband applications the main noise component is a thermal noise. In the case of thermal noise, the mean-square noise voltage is determined by Nyquist formula [6]:

$$U_n^2 = 4kTR\Delta f, \quad (6)$$

where R is an active resistance of noise source, Δf is frequency band, $K = 1.23 \cdot 10^{-23}$ is the Boltzmann constant. The noise spectral density S_n at the HES resistance $R = 100\ \Omega$ and the temperature $T = 300$ K is equal to $3 \cdot 10^{-16}$ V^2/Hz. At this value of the noise density the receiver should be sufficiently narrowband—at signal/noise ratio of 1 it should have bandwidth $\Delta f \leq U^2_n/S_n = (0.03 \cdot 10^{-6})^2/3 \cdot 10^{-16} = 3$ Hz.

3. Experiment

Figure 1 shows the measurement scheme for the Hall-effect sensor response to electromagnetic pulse. The pulsed magnetic field former is described in [2]. The pulses with the duration of 10 ns and the front time less than 1 ns are formed by pulsed magnetic field generator. One of the main Hall-effect sensor characteristics is its threshold sensitivity. To increase it, the output voltage is amplified with the use of fast amplifier AD8000 with the higher frequency of 1 GHz.

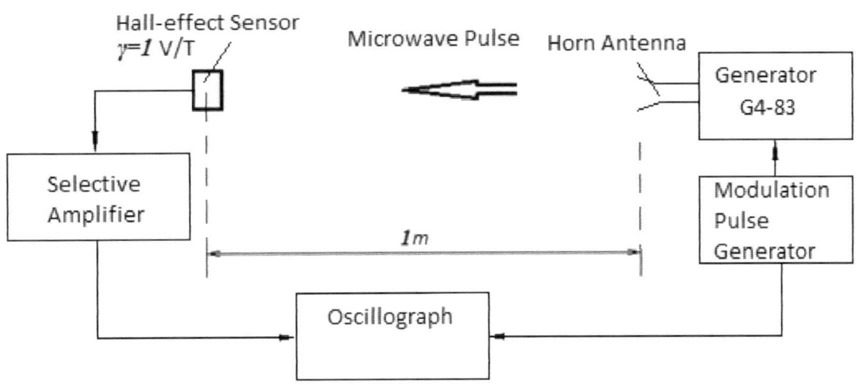

Figure 1. The measurement scheme for Hall-effect sensor response to electromagnetic pulse.

In the course of experiment the planar gallium arsenide high-sensitivity magnetic field sensors (designed in the Research Institute of Semiconductor Devices JSC—RISD JSC) and indium antimonide sensors. As Hall voltage is inversely proportional to the thickness, the main condition for designing the high-sensitivity Hall-effect sensors is manufacturing thin epitaxial films.

The sensors designed in RISD JSD are manufactured from GaAs layers formed on semi-insulating substrate by means of vapor-phase epitaxy methods. The vapor phase epitaxial structures consist of 4 layers: n^+, n, n_{bi}, and n_i (contact layer, active layer, buffer high-resistivity layer, and semi-insulating substrate, correspondingly). The technological process includes photolithography, metal deposition, chemical etching and precipitation, ohmic contacts forming, thermal-compression wire bonding and packaging. The sulfur-doped layers are grown by means of chloride vapor-phase epitaxy. The charge carrier concentration in 1 μm n-layer is equal to $(1.5–2) \cdot 10^{16}$ cm^{-3}. The contact n^+-layer thickness is equal to 0.25 μm, concentration $n^+ = 1 \cdot 10^{18}$ cm^{-3}. The substrate n_i is made of AGChP-9 ingot, the buffer layer (n_b^-) thickness is equal to 1–5 μm. The volt-ampere characteristics (VACs) of HES manufactured on the basis of GaAs epitaxial structures are linear just up to certain U_t value. At $U > U_t$, VACs usually become saturated. Moreover, the smaller ($n \cdot d$) product, the earlier saturation region starts. This can be explained by the influence of the reverse control effect through the substrate. The VAC linear region is the working one.

The InSb sensors are manufactured on the basis of the epitaxial material too.

The measurement is conducted at frequency of 10 GHz applied to the horn antenna. The modulation is provided by the external rectangular pulses with the pulse ratio of 2 and frequency of 20 kHz formed by TTL generator. The same pulses are applied to oscillograph synchronization input. The selective amplifier is represented by Wien-bridge small-noise operational amplifier (OA) with the gain of 60 dB, resonant frequency of 20 kHz, and bandwidth of 30 GHz. The calculated OA noise level is no more than 6 nV. The pulse sequence from the amplifier output is applied to measuring oscillograph input. The microwave pulses detection threshold is defined by eye when they become noticeable in contrast to the noise. The G4-83 generator internal calibrated attenuator is used to measure the microwave radiation level. The observed signal with the level of 20 μV is shown in Figure 2. The voltage applied to HES amplifier input is equal to 20 nV. At the observed signal/noise ratio of 2 the voltage detected on HES is equal to 10 nV. The obtained value well corresponds to the value calculated with allowance for bandwidth widening up to 30 Hz.

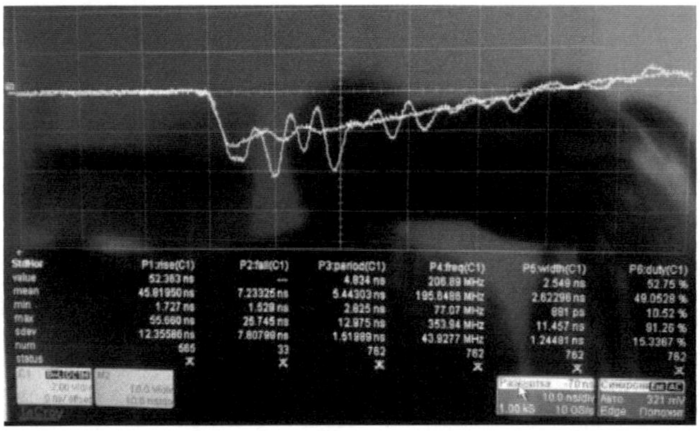

Figure 2. The observed signal.

4. Active positioning system based on Hall-effect sensors

Let's consider the possibility to detect a metal ball with the 10 cm^2 reflecting surface. If the distance to the transmitter is 1 m, the ball obtains the power 10 times greater than at PFD of 10 $\mu W/cm^2$, i. e. 150/25,000 = 0.006 of transmitter power. At the microwave transmitter power of 1 W it is equal to 6 mW (-26 dBm).

Accordingly at the distance of 1 m between the receiver and the HES (i. e. in the transmitter range), in the condition of circular reflection from the target, the target reflection PFD is $6/25\cdot10^4=0.024$ $\mu W/cm^2$. So the voltage at HES $U_{output2} = 0.07$ nV.

In order to detect such a signal it is necessary to reduce the receiver noise level by means of narrowing receiver bandwidth: $\Delta f \leq U^2_{output2}/S_n = (0.07\cdot10^{-9})^2/3\cdot10^{-16} = 0.15$ Hz. This can be achieved with the use of synchronous reception.

The high-intensity focused electromagnetic video pulse with the spectrum of 0.1÷10 GHz is radiated. The signal reflected from electromagnetic inhomogeneities (not only from ferromagnetic and metal materials but also from dielectric ones) is applied to the spaced HES's (Figure 3).

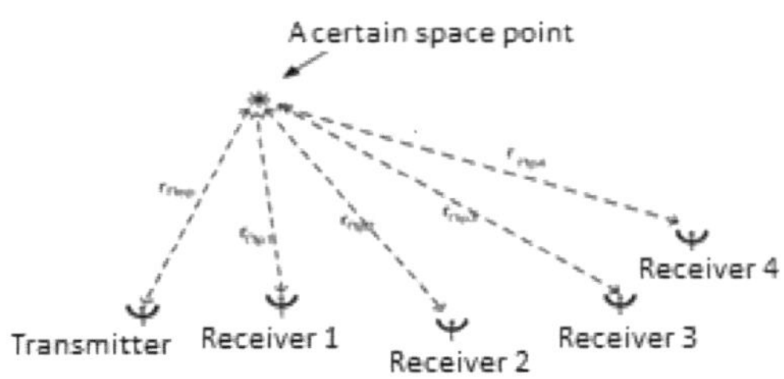

Figure 3. A simplified measurement scheme: the delay time is defined by the distance to the analyzed point R_1, R_2, R_3, R_4.

Picosecond signals integration with allowance for their delays (analogously to range-difference location systems) enables obtaining electromagnetic image of each space point in the antenna operational range (Figure 4).

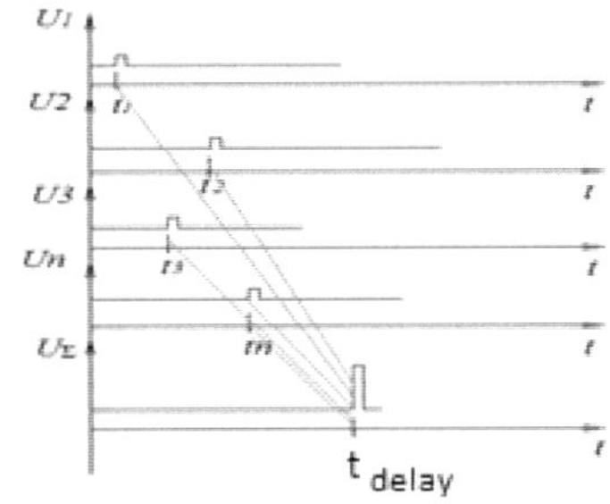

Figure 4. Obtaining the image of the contrasting space point.

5. Conclusions

The carried out investigation results in conclusions stated below.

- The paper describes the experimental assembly designed for GaAs magnetic field sensor output voltage measurement in pulsed microwave field, and presents the obtained results.
- The paper contains the calculation of output voltage for the HES surrounded by electromagnetic field with the intensity of maximum permissible level for human body exposure.
- To increase a HES threshold sensitivity which is dependent on noise level a Wien-bridge selective amplifier is used.
- It is shown that there is a possibility to build up Hall-effect magnetic locators. The distances to the detected objects and their surface areas are estimated. This information can be used for medical purposes.

References

[1] Antropov V A, Antropova L H 2011 *Using Galvanomagnetic Phenomena in Semiconductors for Microwave Device Designing* (Penza: PGU) pp 80–83

[2] Karlova G F, Avdochenko B I, Dedkova O A, Shandrikov M V, Yurchenko V I 2015 *Izvestiya Vuzov. Fizika* **58** 129–131

[3] Trokhimchuk P P 2015 *Advances in Applied Physics* **3** 325–341

[4] Karlova G F, Avdochenko B I, Belozyorova E I 2019 *Proceedings of the Advanced Topics in Rediophysics (APR 2019)* 244–249

[5] Karlova G F, Gradoboev A V 2006 *Proceedings of the 9th Conference Gallium Arsenide and III-V Compound Semiconductors* 22–26

[6] Van der Ziel A V 1973 *Noise* (Moscow: Sov. Radio) 227p

Journal of Physics: Conference Series **2140** (2021) 012008 doi:10.1088/1742-6596/2140/1/012008

Investigation of the effect of the summation of monopolar pulses of individual generators for a bipolar UWB pulse formation

E V Balzovsky[1] and P M Lysykh[2]

[1]Institute of High Current Electronics, Siberian Branch, Russian Academy of Sciences (IHCE SB RAS), Tomsk, Russia
[2]Tomsk State University, Tomsk, Russia

E-mail: bev@lhfe.hcei.tsc.ru

Abstract. A technique for the formation of ultra-wideband bipolar pulses using two synchronized generators of monopolar pulses of different polarities and a broadband ring summing unit is considered. In contrast to the traditional circuit on short-circuited segments of line, the proposed approach allows varying the duration of bipolar pulses by changing the time delay between the generators

1. Introduction

Ultra-wideband (UWB) pulses are used in short-range radar [1] and subsurface sensing [2] due to their wide spectrum and high spatial resolution. To form short UWB pulses, gas-discharge switches used in high-power high-voltage radiation sources are applied [3]. However, semiconductor generators have become widely used, primarily due to the high pulse repetition rate up to hundreds of kilohertz [4]. There are two types of generators in which a semiconductor element is used either in the mode of a sharp transition from a non-conducting state to a conducting one, connecting an energy storage device to a load, or in the mode of transition from a conducting state to a state with high resistance. In the latter case, step recovery diode (SRD) are used, and the generator circuit consists of a voltage source, inductance and a diode connected in series. Such a simple scheme allows forming a monopolar voltage pulse of positive or negative polarity. Depending on the type of SRD, the pulse duration ranges from tens of picoseconds to several nanoseconds. The peak voltage ranges reaches of tens of volts. It is of interest to use SRD generators as inexpensive sources of short UWB pulses for practical application.

The spectrum of a monopolar pulse is rich in low-frequency components, so it is energetically unprofitable to excite antennas with monopolar pulses. Most of the energy is reflected from the antenna input due to the limited matching band of the antenna and does not participate in the formation of radiated wave. It is more effective to excite the antenna by bipolar pulses. If the areas of the positive and negative half-waves are equal, there is a minimum energy at zero frequency in the spectrum of such pulses. It denote that just a small part of the energy is contained in the low frequency region. The maximum energy in the spectrum is concentrated near the frequency inversely proportional to the pulse duration. For the formation of bipolar pulses from monopolar ones, short-circuited segments of line are traditionally used [5], however, the reflection from the place of inclusion of the segments of lines is large.

Content from this work may be used under the terms of the Creative Commons Attribution 3.0 licence. Any further distribution of this work must maintain attribution to the author(s) and the title of the work, journal citation and DOI.
Published under licence by IOP Publishing Ltd

Alternatively, you can use two generators of monopolar pulses of positive and negative polarity, shifted in time by a fixed delay, and then summ the two pulses in the load using a wideband summation unit. This approach will not only increase the voltage range of the output pulse, but also allows you to control the duration of the bipolar pulse by changing the delay time. In this paper, it is proposed to evaluate the feasibility of such a method and the prospects of its application to obtain pulses of different waveform.

2. Analysis of the scheme based on the segments of transmission lines

Methods of forming bipolar pulses using transmission lines are widely used. One of them is the scheme shown in Figure 1. A generator with an internal impedance Z_g is connected through a line of length L_1 and wave impedance Z_1 to the point of connection of two segments of line of length L_2 and L_3 with wave impedance Z_2 and Z_3, respectively, which are shorted at the ends. A line of length L_4 with a wave impedance Z_4 is connected between connection point and the load impedance Z_l. To analyze the circuit, we take $Z_g = Z_1 = Z_4 = Z_l = 50$ Ohms.

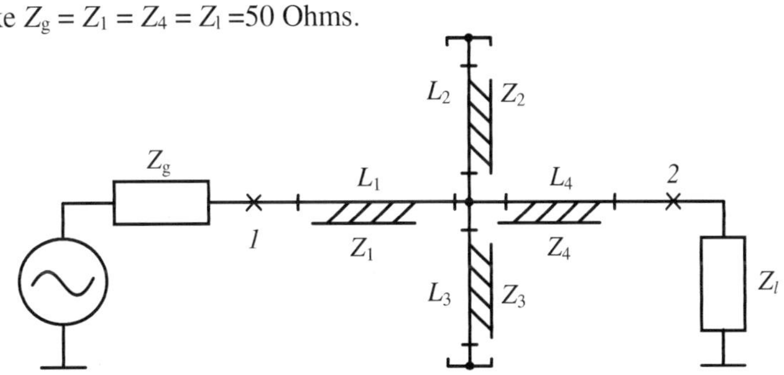

Figure 1. The scheme of pulse formation on short-circuited segments of line.

A monopolar pulse with a half-height duration of 0.1 ns is fed to the input of the circuit. The waveform form of pulse is shown in Figure 2*a* (region A). After an interval of time determined by the time of the double path of the wave, a reflected signal appears in the L_1 line (Figure 2, region B) from the point of connection of the segments of lines with $Z_2 = Z_3 = 50$ Ohms, caused by an impedance mismatch. The signal at the output of the circuit is shown in Figure 2*b*. This is the sum of the past monopolar pulse and the second pulse reflected from the short-circuited ends of the segments of lines L2 and L3, which changed the polarity. Depending on the length of the segments of the line $L_2 = L_3$, the duration of the bipolar pulses at the output may vary. The pulses for several length values are shown in Figure 2*b*.

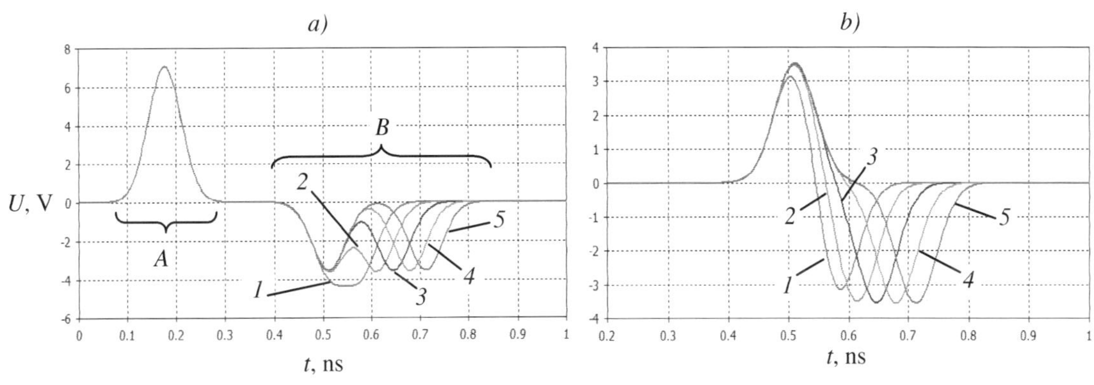

Figure 2. Pulses at the input (a) and output (b) of the pulse former with the line segments of length $L_2 = L_3 = 10$, 15, 20, 25 and 30 mm (curves 1-5, respectively). Area A is the incident pulse, and B is the reflected pulse.

Journal of Physics: Conference Series **2140** (2021) 012008 doi:10.1088/1742-6596/2140/1/012008

Despite the simplicity of the scheme, it has two significant drawbacks. The first is the need to mechanically change the length of the short-circuited segments of lines for changing the duration of the bipolar pulses. The second disadvantage is associated with the loss of energy for reflection from the switching point of the line segments. The peak voltage value of the reflected pulses reaches half of the peak value of the incident pulse. A decrease in Z_2 and Z_3 leads to an even greater increase in reflection (Figure 3*a*). Changing the values of Z_2 and Z_3 results in a distortion of the shape of the formed bipolar pulse, Figure 3*b*. An increase to Z_2 and Z_3 to 100 ohms allows reducing the proportion of reflected energy, but at the same time there are significant distortions in the shape of the output pulse.

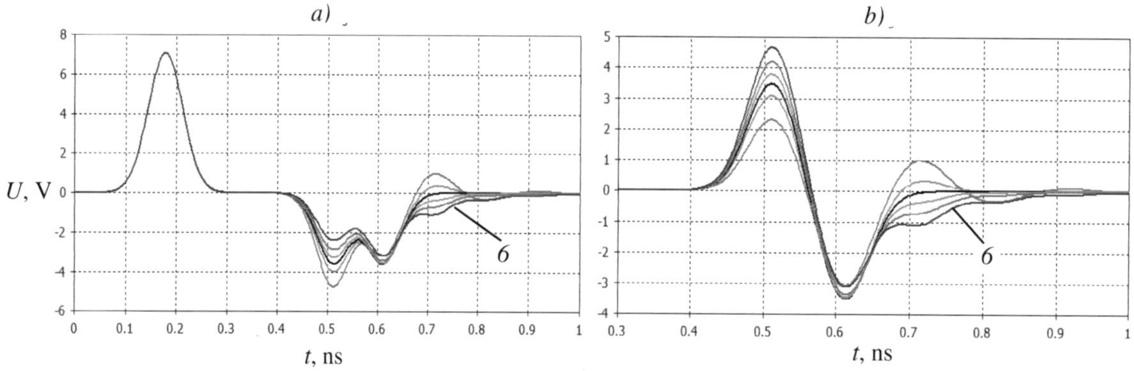

Figure 3. Pulses at the input (a) and output (b) of the pulse former with segments of line with wave impedance $Z_2 = Z_3 = 25, 40, 50, 60, 75$ and 100 Ohms (curves 1-6, respectively).

3. The summation of the pulses of two generators

A wideband ring summing unit (Wilkinson power divider) is used to combine two pulses of opposite polarities into a single bipolar pulse. The inputs of the summing unit receive pulses shifted in time from two generators, as shown in Figure 4. In order to ensure a time shift τ of the signals relative to each other, you can either increase the propagation time of one of the pulses from the output of the circuit to the summing unit, for eample, by changing the length of the line segment, or control the moment of operation of one of the generators. This method of formation allows achieving the effect of compensation of positive and negative half-waves by reducing the frequency and decreasing the duration of the output bipolar pulse with a decrease in the amplitude.

Combining of the pulses of a larger number of generators allow synthesizing a pulse of a more complex waveform. An example of such pulse is shown in Figure 5. The position of the time lobes inside pulse can be changed, thereby encoding information. However, the synchronization of generators and the preservation of the specified time shifts between the initial monopolar pulses is a separate task.

The CST Studio Suite software product was used to the numerical simulation of the bipolar pulse former scheme with a ring summing unit. To expand the bandwidth, the summing unit is made multi-stage. According to the recommendations in [6], an increase in the number of stages should result in the increase of the wave impedance of the half-circular line closest to the output. However, the implementation of a strip line with a high wave resistance (over 100 Ohms) on a microwave dielectric is associated with the technological limitations of the printing board technology, which does not allow performing too narrow conductors.

In this paper, it is proposed to use a three-stage summing unit made on a foil-coated microwave dielectric FLAN with a thickness of 1 mm and a relative permittivity equal to 5. The choice of such a material is due to the balance between the ability to perform small-width microstrip line segments with the necessary accuracy and to reduce the physical dimensions of these segments while maintaining their electrical length.

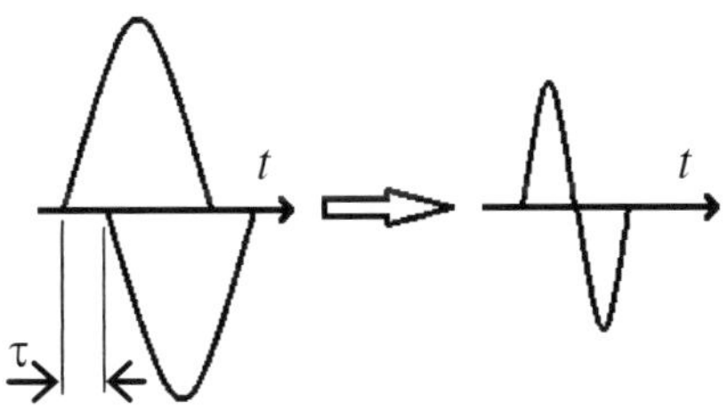

Figure 4. Decreasing of a bipolar pulse duration with a shift of monopolar pulses.

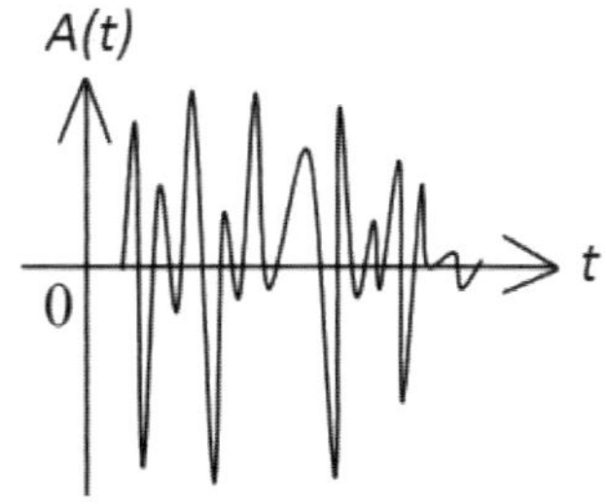

Figure 5. UWB pulse of complex waveform.

The model of a three stage ring summing unit is shown in Figure 6. Input ports are connected to microstrip lines 1 and 2, and output port is connected to the line 3. Lines 1 and 2 are connected to three consecutive segments of transmission lines 4 in the shape of half-rings with wave impedance of 58, 71 and 88 Ohms, respectively. In the gaps between the half-rings, chip resistors 5 with nominal values of 100, 200 and 390 ohms are included. The total bandwidth of the summing unit is limited by the matching bandwidth and is equal to 0.4...3.9 GHz.

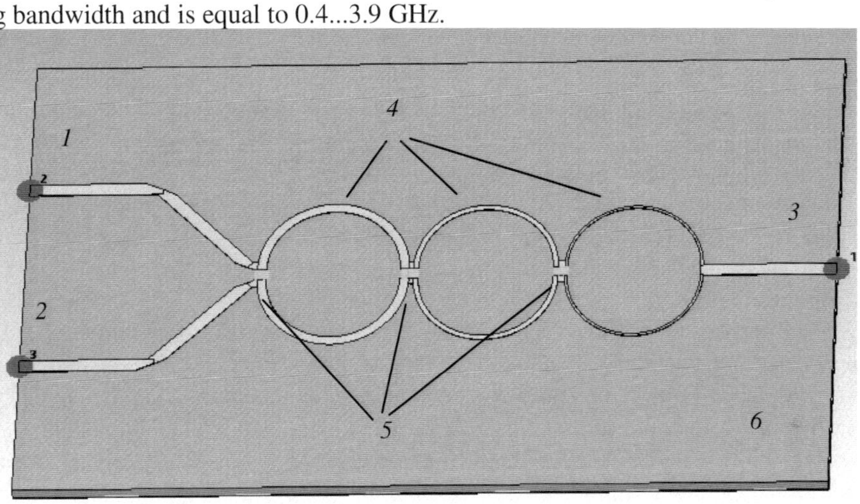

Figure 6. Model of a three-stage ring summing unit.

A rectangular monopolar pulse with a rise time and fall time of 0.05 ns was applied to the input port 1 of the summing unit. The total pulse duration was 0.2 ns. The input port 2 was supplied with a pulse of the same waveform, but having reverse polarity and shifted by the time τ. The waveform of the input monopolar pulses is shown in Figure 7*a*. Distortions in the form of an uneven vertex and post-pulse oscillations are associated with reflection from the summing unit due to the imperfection of its matching with the feeder. The waveform of the output bipolar pulses is shown in Figure 7*b* for several values of τ. On can see that it is possible to change the duration of the output bipolar pulses within a wide range with the help of the relative shift of the monopolar pulses. The change in the shape of the half-waves is due to the limited bandwidth of the summing unit, which can be increased by using a larger number of stages.

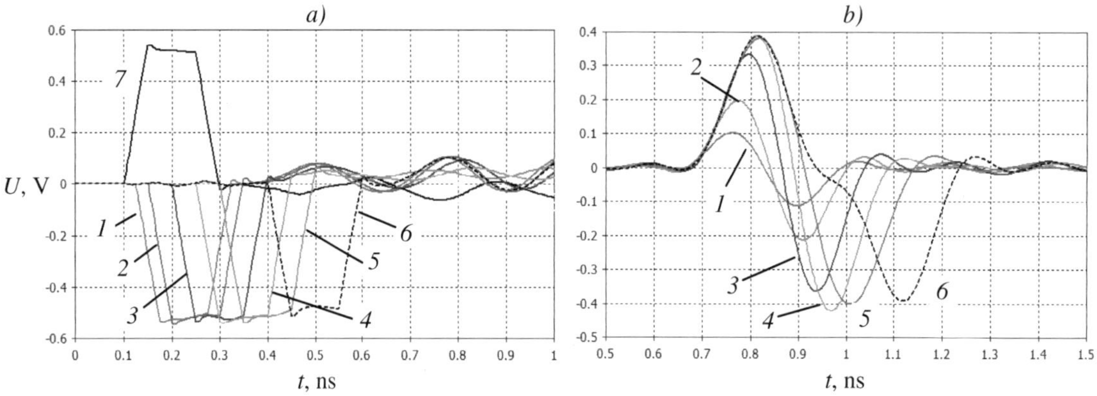

Figure 7. The waveform of the pulses at input port 1 of the summing unit (curves 1-6) and the input port 2 of the summing unit (curve 7) (a) and at the output port 3 of the summing unit (b). Curves 1-6 correspond to the delay time of τ = 25, 50, 100, 150, 200 and 300 ps between the pulses at the input ports 1 and 2.

4. Conclusion

The paper considers the method of forming bipolar pulses by summing of monopolar pulses from two generators of opposite polarity. The possibility of controlling the waveform of the output bipolar pulses by changing the time shift between the pulses of the generators is shown. Further work will be aimed at the practical realization of low-power generators of monopolar pulses and their synchronization for the synthesis of bipolar pulses.

Acknowledgments

The work was performed under State Assignment of the Ministry of Science and Higher Education of the Russian Federation (project No. FWRM-2021-0002).

The study was carried out using the software product CST Studio Suite® developed by Dassault Systemes on the basis of a valid license LN00031351538 dated July 21, 2020, provided by the IHCE SB RAS, and a valid license LN00029178508 from 10.12.2019, provided by Tomsk State University. All the conditions for the granting of such licenses when these organizations carry out joint research and publish their results are met.

References

[1] Amin M 2011 *Through-the-Wall Radar Imaging* (Boca Raton: CRC Press) p 450
[2] Daniels D J 2004 *Ground-penetrating radar* (London: The Institution of electrical engineers) p 734
[3] Mesyats G A, Pegel I V 2009 *Introduction to nanosecond pulsed power engineering and electronics* (Moscow: FIAN) p 240 (in rus.)
[4] UWB generators 2021 Available at: http://www.fidtechnology.com
[5] Pichugina M T 2011 *High-voltage electrical engineering* (Tomsk: TPU Publishing House) p 120 (in rus.)
[6] Maloratsky L G 2004 *RF and Microwave Integrated Circuits: Passive Components and Control Devices* (Oxford, UK: Elsevier Inc.) p 384

Communication based on scattered laser radiation in a medium with high turbidity or in the presence of a noise source

E S Poznakharev, V V Belov, M V Tarasenkov, A V Fedosov and V N Abramochkin

V.E. Zuev Institute of Atmospheric Optics SB RAS, Tomsk, Russia

E-mail: 724_pes1992@iao.ru

Abstract. The results of laboratory studies of an optical communication channel based on scattered radiation in the Big Aerosol Chamber of IAO SB RAS in a clean atmospheric environment, in the case of the chamber filled with vapor of the water-glycerin mixture, and in the presence of an noise laser source are analyzed. It is found that with the non-coplanar communication geometry in the chamber filled with the water-glycerin mixture, stable data transmission is possible with the azimuthal angle of orientation of the detector optical axis up to 5°. The error probability in the communication channel increases more slowly with an increase of the detector elevation angle in the chamber filled with the water-glycerin mixture than that in the chamber without this mixture does. The presence of an interfering laser radiation at a wavelength $\lambda = 510$ nm in the communication channel affects the communication quality. When the power of the noise laser source achieves 70 mW, the maximal error probability corresponds to 0.02.

1. Introduction

Due to the intense development of lasers and photodetectors over the past decade, it has become possible creating new communication systems that operate in the optical range through an open atmospheric channel. The advantage of optical communication over radio and acoustic communication is the feasibility to transmit information (1) at higher rates (as compared to acoustic communication), (2) without licensing the frequency range, and (3) in the presence of radio and acoustic opaque obstacles between a receiver and a source.

Even though information can be transmitted through the atmospheric optical communication channel at a high rate over rather short distances within the line of sight, turbulence can lead to moving of the laser beam over the receiver aperture and, in some cases, even its escape from the aperture. The presence of external objects in the communication line can also interrupt communication. Non-line-of-sight (NLOS) communication systems are mostly free of these shortcomings. A feature of these systems is that the useful signal in them is represented by the scattered or reflected radiation. Communication systems on reflected radiation in the atmosphere allow the information transmission even in the presence of foreign objects in the line-of-sight (LOS). At short distances, high-rate information transfer is possible according to [1]. Optical communication based on scattered radiation is much more promising from the versatility point of view. Communication of this kind allows information to be transmitted to many subscribers located at

Content from this work may be used under the terms of the Creative Commons Attribution 3.0 licence. Any further distribution of this work must maintain attribution to the author(s) and the title of the work, journal citation and DOI.
Published under licence by IOP Publishing Ltd

baseline distances from hundreds of meters to tens of kilometers depending on the meteorological conditions of the atmospheric environment and the parameters of the transceiver equipment.

Currently, the interest to theoretical and experimental studies of NLOS communication systems is significantly growing [1–16]. In particular, the number of publications dealing with visible light communication (VLC) based on diffusive radiation has increased considerably (for example, [1-5]). A method for increasing the accuracy of localization of objects and an algorithm for orienting a transceiver system in a multicast indoor network are considered in [2,3]. The hardware implementation of the transceiver system for indoor localization of objects in 3D space and communication at baseline distances up to 6 m are presented in [4,5]. A transceiver communication system with the bitrate up to 40 Gbit/s at the 25-cm communication line is tested in [1].

The idea of optical communication systems based on scattered radiation was developed in [6–14]. Reviews of the current state of the art can be found, for example, in [6,7]. Communication channels based on the scattered radiation are now being studied for both coplanar [8–10] and non-coplanar [11–14] geometry in the UV range. Thus, a model of NLOS UV communication for single-scattered radiation was proposed, the path loss for this model was estimated, and the obtained theoretical results were compared with the experimental ones in [8]. A communication channel employing devices for retranslation to eliminate signal distortions caused by turbulence of the propagation medium is studied in [9]. In [10], an above-water channel for communication with a ship located at a distance of 164 m was considered, and losses in the communication channel were estimated by the Monte Carlo method.

Among papers considering the non-coplanar geometry, paper [11] presents a new spatial division multiplexing method for non-coplanar schemes and considers the effect of spatial division of communication channels on inter-channel interference (ICI). Security of LOS and NLOS communication lines is analyzed in [12]. A model of NLOS optical communication is considered in [13] based on the Monte Carlo method for different sources (laser, LED), and the calculated results are compared with those obtained by analogous methods. The Wavelet algorithm allowing reduction of noise in the received signal is presented and tested in [14].

The analysis of publications shows that NLOS systems based on reflected light in atmospheric communication channels are under development. Currently, such systems are being developed for indoor communication and object positioning. NLOS communication channels based on radiation scattered in the atmospheric environment are investigated in the UV range for the coplanar and non-coplanar geometry, but experimental results are much less discussed than theoretical ones.

Earlier, it was shown in our papers [15,16] that atmospheric optical communication based on scattered radiation is possible in the visible and ultraviolet (UV) ranges. However, the issues of estimation the influence of high turbidity of the propagation medium on the quality of data transmission and the noise immunity of the communication system to external illumination sources were not touched. That is why in this paper we estimate the effect of high turbidity of the propagation medium on the communication quality in the visible range at a wavelength λ=450 nm for the coplanar and non-coplanar communication geometry, as well as the noise immunity of the communication system in the presence of external laser radiation at another wavelength. The experiments were carried out in the Big Aerosol Chamber (BAC) of IAO SB RAS.

2. Instrumentation and geometry of optical communication based on scattered radiation

The optical communication channels were studied using the transceiver system comprising optoelectronic and optical units. Information was transmitted with a B 2000 laser system converted to the pulsed operation mode at a wavelength λ=450 nm. The laser system and the modulating device were made as a single unit. Information was received by an UFK 4G-2 photomultiplier tube (PMT) with a Semrock bandpass filter (centered at a wavelength of 450 nm) installed on the input window. PMT was placed in a lens refractor so that the received radiation completely fell on the entire working area of the photocathode. The demodulator as a part of the receiving system was made as a separate unit. To model the turbid medium, the LEXOR LM50003 fogging machine was used to fill the chamber with the water-glycerin mixture. The noise radiation was generated by a laser operating at a wavelength λ= 510 nm.

The experiments were conducted in two series. The first series was carried out on December 20 of 2020. It was aimed at estimating the effect of external light on the communication channel in the geometry shown in Figure 1*a,b*. The second series took place on February 17 of 2021. In this series, we estimated the effect of the high concentration of aerosol particles on the optical communication channel in the geometry shown in Figure 2*a,b*.

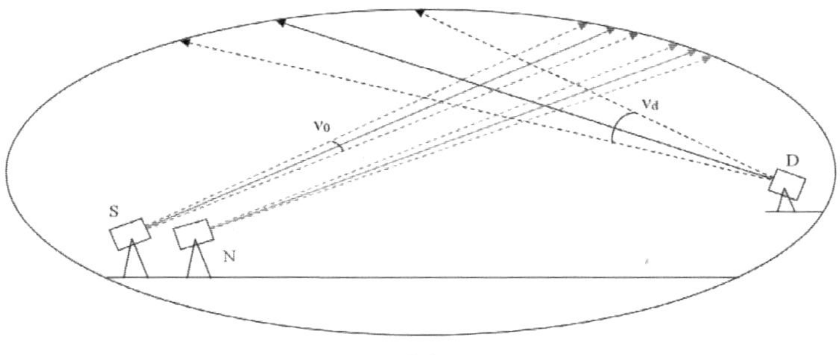

(a)

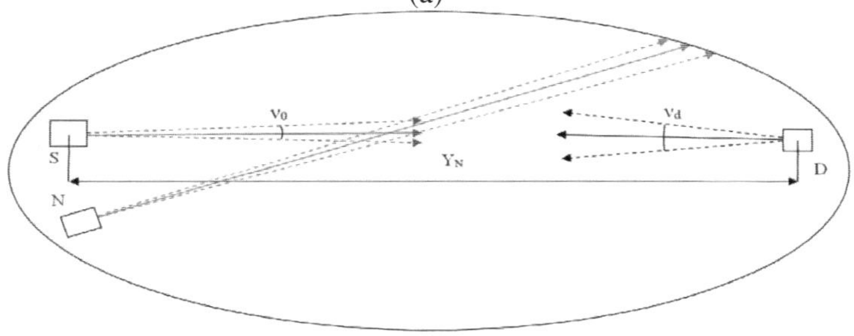

(b)

Figure 1. Geometry of optical communication channel based on scattered radiation in BAC in the presence of a noise source in the communication channel: (a) side view side view, (b) top view. Notations used: source S, detector D, baseline distance Y_N, noise laser source N.

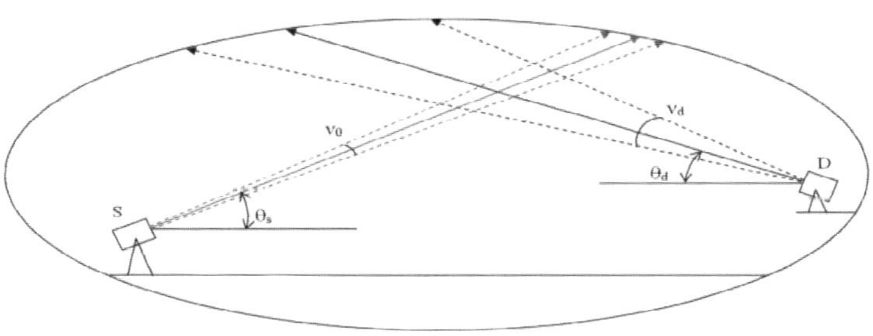

(a)

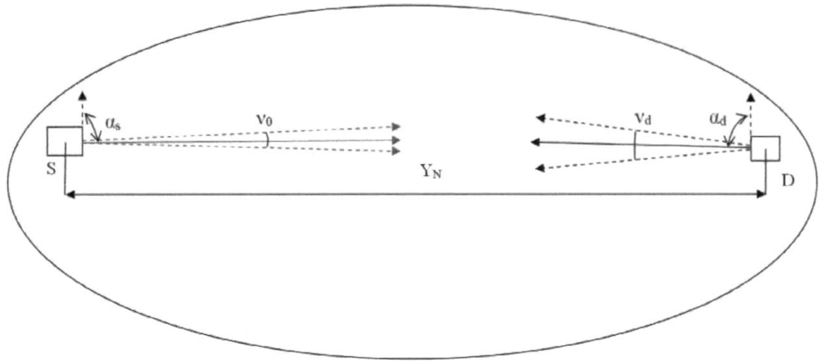

(b)

Figure 2. Geometry of optical communication channel based on scattered radiation in BAC in the presence of high concentration of particles of the water-glycerin mixture: (a) side view, (b) top view.
Notations used: source S, detector D; detector elevation angle θ_d, source elevation angle θ_s, basic distance Y_N, detector azimuthal angle α_d (measured from the direction to the source), source azimuthal angle α_s (measured from the direction to the detector).

Laboratory studies within the first experimental series were performed for the coplanar geometry of the system with the noise laser source operating at a wavelength λ= 510 nm and the chamber not filled by the model scattering medium. At the first stage, the power of the noise laser was set so that the noise in the channel was insignificant, and then the data transmission was recorded. At the second and third stages, the power of the noise laser was increased until the marked noise appeared in the communication channel, and the information transmission process was recorded similarly to the first stage.

The experiments within the second experimental series were carried out without the noise laser source. At the first stage, the chamber was not filled with the water-glycerin mixture. Then, the LEXOR LM50003 fogging machine generated the water-glycerin mixture for 30 minutes. Once the water-glycerin aerosol distributed over the chamber volume, the measurements similar to those at the first stage were conducted. For simpler reading of the detector azimuthal angles, the communication system was located in one plane according to the geometry shown in Figure 2. The zero value of the azimuthal angle corresponded to the direction of the receiving system to the source.

The experiments were organized in the following way. The modulated (time-pulse modulation) signal with the radiation divergence v_0=0.30° at a wavelength λ=450 nm was emitted from the semiconductor laser unit S at the elevation angle $\theta_s = 15°$ and the azimuthal angle $\alpha_s = 0°$ in the direction to the detector D. At the base line distance Y_N= 23 m, the receiving system D with the field-of-view angle v_d=1° recorded the scattered laser pulse at different elevation θ_d and azimuthal α_d angles. The noise laser source was oriented so that it formed the greatest noise in the communication channel at the optimal geometric arrangement of the transceiver equipment. To estimate the communication quality, the error probability in the communication channel $\overline{x}$ and its standard deviations $\overline{\sigma}$ were used. The values of these parameters were determined by equations (1) and (2) from [6]:

$$\overline{x} = \frac{1}{n}\sum_{i=1}^{n} x_i \tag{1}$$

$$\overline{\sigma} = \sqrt{\frac{1}{n}\sum_{i=1}^{n} x_i^2 - \overline{x}^2} \qquad (2)$$

where x_i is the error symbol of the i-th element of a burst; n is the total number of elements in a burst.

3. Results of the studies on December 09, 2020 (first series)

The experiments were conducted under the following conditions: the extinction coefficient in the empty (without fog particles) chamber σ_t =0.055 km^{-1}, the elevation angle of the detector optical axis θ_d= 0°, the azimuthal angle α_{d}= 0°, the mean power of source laser P = 1 mW, the mean power of the noise laser P= 44, 55, and 70 mW. The obtained values of the error probability and the standard deviation are shown in Figure 3.

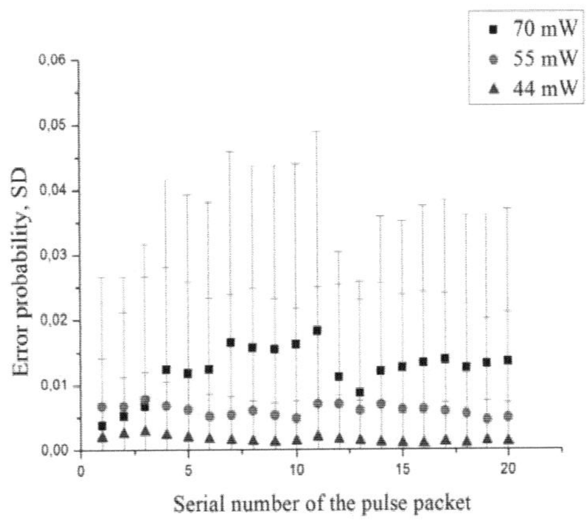

Figure 3. Error probabilities and standard deviation (SD) in the experiments on December 09, 2020.

The results of the first experimental series have shown that the laser radiation at a wavelength $\lambda = 510$ nm affects the quality of information transmission. As the interferring radiation power increases from 44 to 70 mW, the maximal value of the error probability grows from 0.004 up to 0.02. The presence of the noise radiation at a wavelength $\lambda = 510$ nm in the received signal is caused by the low power of the information signal, the bandwidth of the narrow-band filter, and the spectral range of the PMT sensitivity. An increase in the error probability at the noise source power of 70 mW is due to the unstable operation of the noise laser.

4. Results of the studies on February 17, 2021 (second series)

The experiments were carried out under the following conditions: the extinction coefficients σ_t – 0.055 km^{-1} in the empty (free of fog) chamber and σ_t = 393 km^{-1} in the chamber filled with the water-glycerin mixture, the elevation angle of the detector optical axis θ_d= 0°,15°, and 30°, the azimuthal angles α_d = 0° and 5°, the mean power of laser radiation P = 0.4 mW. The obtained error probabilities and their standard deviations (SD) are shown in Figure 4.

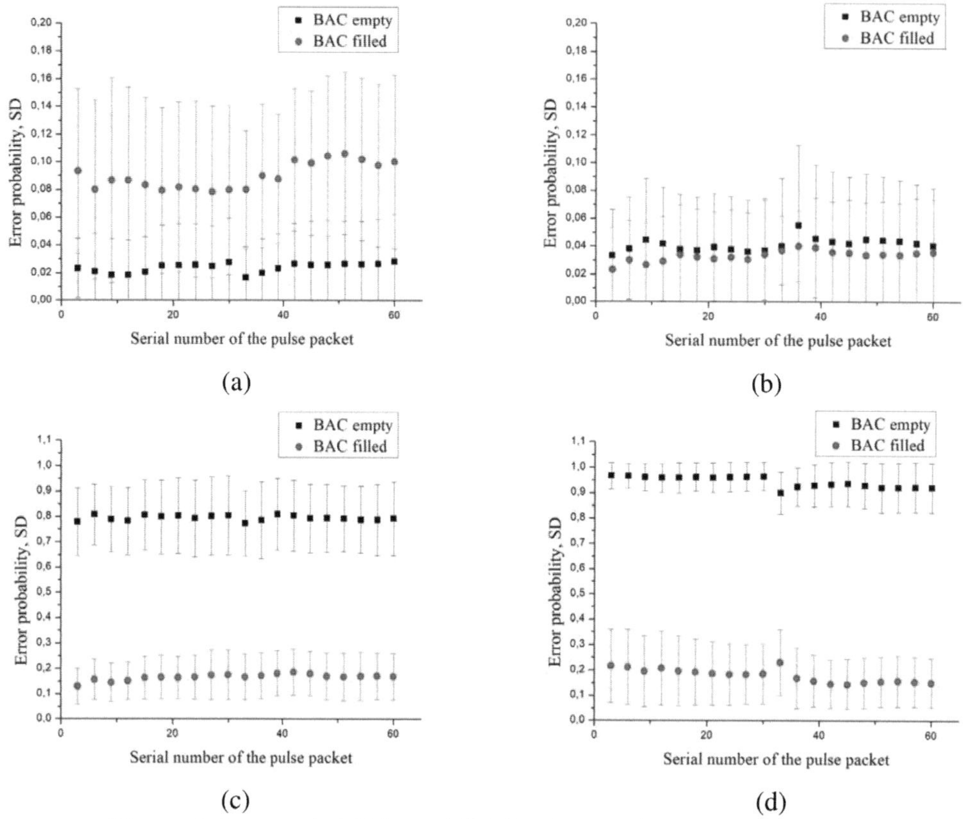

Figure 4. Error probabilities and SD for the experiments on February 17, 2021: (a) $\theta_d = 0°$ and $\alpha_d = 0°$; (b) $\theta_d = 15°$ and $\alpha_d = 0°$; (c) $\theta_d = 0°$ and $\alpha_d = 5°$; (d) $\theta_d = 30°$ and $\alpha_d = 0°$.

The analysis of results of the second experimental series has shown that at the angles of the optical axis of the receiving system $\alpha_d = 0°$ and $\theta_d = 0°$, the water-glycerin mixture attenuates the information signal due to scattering and absorption. In this connection, the mean error probability in the communication channel increases from 0.02 up to 0.09. As the angles of orientation of the receiving system change from $\alpha_d = 0°$ and $\theta_d = 0°$ to $\alpha_d = 5°$ and $\theta_d = 0°$, the mean error probability increases from 0.09 up to 0.15, because multiple scattering predominates over single scattering in the received signal in media with high extinction. In the medium with lower extinction, the data transmission degrades sharply, and the mean error probability in the communication channel increases from 0.02 up to 0.8. Significant deterioration of the communication quality is caused by the non-coplanar communication geometry and the predominance of the single scattered radiation over the multiple scattered one in the received signal when the chamber is not filled with the model medium (water-glycerin mixture). At the different arrangement of the receiving system and $\alpha_d = 0°$, $\theta_d = 0°$, 15°, and 30°, the mean error probability increases from 0.09 to 0.18 as the elevation angle of the detector optical axis increases under conditions of the chamber filled with the model medium. For the chamber without the model medium, the mean error probability increases from 0.03 to 0.94. Consequently, at small basic distances, the communication error probability increases more slowly in the medium with high turbidity than in the chamber without the model medium.

5. Summary

The effects of the model medium represented by the water-glycerin mixture and the interfering laser radiation at a wavelength $\lambda = 510$ nm on errors in the communication channel have been analyzed. It is shown that the laser radiation at a wavelength $\lambda = 510$ nm gives rise to noise in the communication channel, and an increase in the noise radiation power up to 70 mW leads to unstable communication with the error probability up to 0.02. The effect of the noise laser radiation on the received signal was primarily due to the fact that the receiving system recorded signals in a wide spectral range.

The studies of the communication channel in a turbid medium were carried for both the coplanar and non-coplanar geometry. The obtained experimental estimates show that in the turbid medium at a base line distance of 23 m the communication deteriorate due to the medium properties only at $\alpha_d = 0^\circ$ and $\theta_d = 0^\circ$. In other cases, due to predominance of multiple scattering, the medium with high turbidity allows the information transmission with the error probability not exceeding that in the chamber without the turbid medium. It should be noted that the significant deterioration of communication in the chamber not filled with the turbid medium for the non-coplanar geometry is caused by the predominance of single scattering in the propagation medium.

References

[1] Zhang X, Cao Z, Li J *et al.* 2020 *Journal of Lightwave technology* **38**(24) 6801–6806 doi: 10.1109/JLT.2020.3022329

[2] Shawky E, El-Shimy M , Mokhtar A *et al.* 2020 *Journal of the optical society of America B-optical physics* **37**(11) A130–A138 doi: 10.1364/JOSAB.395056

[3] Morales-Cespedes M, Haas H, Armada A G 2021 *IEEE communications* letters **25**(3) 921–925 doi: 10.1109/LCOMM.2020.3038027

[4] Mahmoud A A, Ahmad Z U, Haas O C *et al.* 2020 *IET Optoelectronics* **14**(6) 440–446 doi: 10.1049/iet-opt.2020.0046

[5] Seminara M, Meucci M, Tarani F et al. 2021 *Photonics Research* **9**(4) 548–557 doi: 10.1364/PRJ.414394

[6] Shubin D N, Varlamov V O, Lobov E M 2020 *Systems of Signal Synchronization, Generating and Processing in Telecommunications (SYNCHROINFO)* 19889713 doi: 10.1109/SYNCHROINFO49631.2020.9166129

[7] Vavoulas A, Sandalidis H G, Chatzidiamantis N D *et al.* 2019 *IEEE Communications Surveys and Tutorials* **21**(3) 2111–2133 doi: 10.1109/COMST.2019.2898946

[8] Ke X, Wu Y 2021 *Optical Engineering* **60**(3) 036104 doi: 10.1117/1.OE.60.3.036104

[9] Reeves A D, Torres E, Cerrillo-Marchan S *et al.* 2019 Proceedings of SPIE - The International Society for Optical Engineering **11133** UNSP 1113315 doi: 10.1117/12.2536618

[10] Garg K K, Bhatia V 2020 *Proceeding of the 26th National Conference on Communications* 19510242 doi: 10.1109/NCC48643.2020.9056074

[11] Li F, Zuo Y, Du Z *et al.* 2021 *IEEE Photonics Journal* **13**(3) 7300119 doi: 10.1109/JPHOT.2021.3073044

[12] Weisman M J, Dagefu F T, Moore T J *et al.* 2020 *Optics Express* **28**(16) 23640–23651 doi: 10.1364/OE.399196

[13] Zhang Q, Zhang X, Wang L *et al.* 2020 *Applied Sciences* **10**(3564) 1–14 doi: 10.3390/app10103564

[14] Song P, Tan Y, Geng X *et al.* 2020 *IEEE ACCESS* **8** 131626–131635 doi: 10.1109/ACCESS.2020.3009944

[15] Belov V V, Juwiler I, Blaunstein N *et al.* 2020 *Atmosphere* **11**(10) 1122 doi: 10.3390/atmos11101122

[16] Belov V V, Gridnev Y V, Kudryavtsev A N *et al.* 2018 *Proceedings of SPIE - The International Society for Optical Engineering* **10833** UNSP 1083311 doi: 10.1117/12.2503308

Journal of Physics: Conference Series **2140** (2021) 012010 doi:10.1088/1742-6596/2140/1/012010

Electromagnetic response from composite carbon-containing structures with technological inhomogeneities at EHF

E A Trofimov, G E Kuleshov, V D Moskalenko, A V Badin, K V Dorozhkin

National Research Tomsk State University, Tomsk, Russia

E-mail: grigorij-kge@sibmail.com

Abstract. The results of studies of the electromagnetic response from composite structures made of a carbon-containing polymer with the inclusion of spherical pores in the bulk of the material and with pyramidal corrugation on the surface of the material are presented. The results of modeling the frequency dependences of the transmission, reflection and absorption coefficients in the EHF range are shown. Samples of composite carbon-containing structures with technological inhomogeneities have been fabricated by 3D printing. Measurements of the electromagnetic response from experimental samples were carried out in the frequency range from 100 to 1000 GHz. At frequencies up to 250 GHz, the inclusion of air pores in the polymer volume reduces the transmission coefficient, practically does not affect the reflection, and increases the absorption. Pyramid corrugated material absorbs more than 99% of radiation in the frequency range from 200 to 635 GHz.

1. Introduction

The development and miniaturization of modern electronics contributes to the transition to ever higher operating frequencies [1, 2] and improvement of microwave technology [3]. Today, microwave equipment surrounds us everywhere and it is no longer possible to imagine a world without it. In this regard, the problem of monitoring and control of the electromagnetic environment is urgent. This is necessary to ensure electromagnetic compatibility and the normal functioning of high-frequency radio-electronic devices [4], to protect computer information processing systems [5], to protect biological objects from unwanted effects of electromagnetic fields [6]. One of the most widespread solutions to this problem is the shielding of electromagnetic radiation using special radioabsorbing materials and structures. As radioabsorbing materials in the HF and VHF ranges, spinel-structure ferrites have proven themselves best; in the microwave range - hexaferrites, carbonyl iron or carbon materials; in the EHF range, only different types of carbon-containing materials are used [7]. These materials are used directly and in composites.

The carbon materials are most often used as fillers in modern radioabsorbing composites [8, 9]. This is due to their unique structure, as a result of which there are whole classes of carbon materials (single-walled and multi-walled carbon nanotubes, graphenes, fullerenes, onion structures, nanodiamonds, etc.) with a wide variety of properties [10]. This allows the creation of lightweight, durable and broadband radio-absorbing materials.

In addition, the last decade has given a tremendous impetus to the development of additive technology. Nowadays, 3D printing is a very popular and most importantly effective tool for the manufacture of components for electronic equipment. Its use can significantly reduce the cost and speed up the production process of components for radio electronic devices, which is achieved due to

Content from this work may be used under the terms of the Creative Commons Attribution 3.0 licence. Any further distribution of this work must maintain attribution to the author(s) and the title of the work, journal citation and DOI.
Published under licence by IOP Publishing Ltd

the possibility of creating 3D models in CAD applications and subsequent printing with a composite filament of objects and structures of the required volume and shape [11].

The purpose of this work is a model and experimental study of the electromagnetic response from thin layers of complex geometry (with the inclusion of spherical pores in the volume of the material and with pyramidal corrugation on the surface of the material) and determination of the possibility of using the investigated structures as absorbers of electromagnetic radiation in the EHF range.

2. Models, materials and research methods

2.1. Models and method of the model experiment

The simulation was carried out in a specialized software package for electromagnetic analysis CST Studio Suite manufactured by DS Simulia Corp. The calculation was carried out in free space according to the "per pass" scheme for three models of samples (Figure 1). We studied the electromagnetic response from layer of a plane-parallel sample (Figure 1*a*), samples with pores of different radius and different periodicity incorporation into the material (Table 1, Figure 1*b*), samples with pyramidal corrugation of the surface of different heights and widths of the base of the pyramids (Table 2, Figure 1*c*). The frequency range of the simulation is limited by the parameters of the previously measured materials.

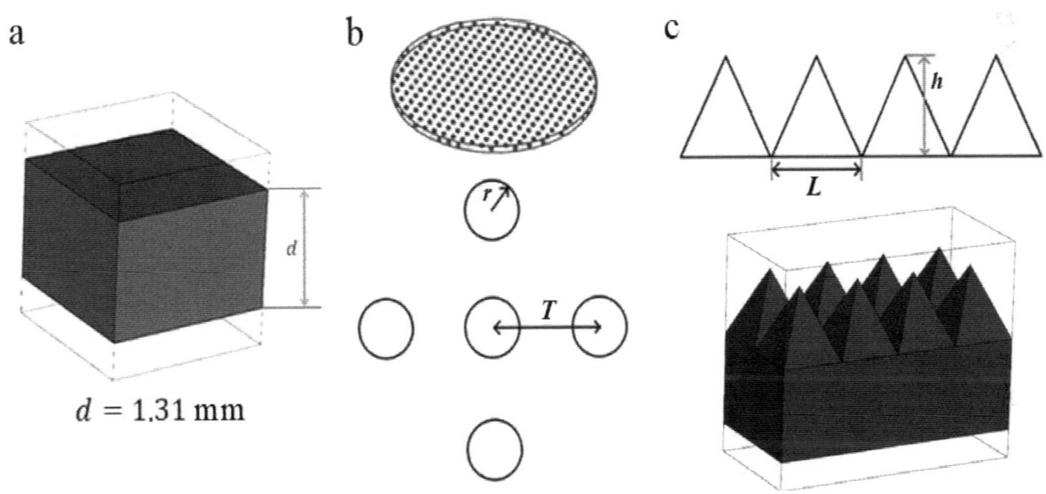

Figure 1. Models: solid sample (a), with pore inclusion (b), with pyramidal corrugation of the surface (c).

Table 1. Parameters of models with the inclusion of air pores in the material volume

Sample No	Pore radius r, mm	Pore repetition period T, mm
1	0.3	2.0
2	0.4	2.0
3	0.4	1.5
4	0.4	2.5
5	0.5	2.0
6	0.7	2.0
7	0.8	2.0

Table 2. Parameters of models with a pyramidal corrugation of the surface

Sample No	Height of pyramids *h*, mm	Pyramid base width *L*, mm
1	0.2	1.0
2	0.5	1.0
3	1.0	1.0
4	1.0	0.5
5	1.0	1.5

2.2. Materials and samples
As a material for modeling, plastic for 3D printing of the brand "ABS Conductive" of the company "Mellow" was chosen. It contains carbon fibers and has a specific electrical resistance of 10^3–10^5 Ohm/cm. Its electromagnetic parameters were previously measured on the STD-21 spectrometer in the frequency range from 115 to 258 GHz [11]. The real part of the complex permittivity is about 6.5 rel. units, and the imaginary part is 0.8 rel. units. They were used to create models of heterogeneous structures.

Also, on the basis of this material, using a 3D printer using FDM technology, two experimental samples were made in the form of washers with technological inhomogeneities. Spherical air pores with a diameter of 1.2 mm were added to the first sample (Figure 2*a*), and pyramids with a base width of 1.5 mm and a height of 1 mm were applied to one of the flat surfaces of the second sample (Figure 2*b*). It is worth noting that due to the inherent limitations of the layer-by-layer additive technology (the size of the extruder nozzle is 0.1 mm), the original geometry of the experimental samples has undergone some changes. The thickness of the base layer *d* was 1.38 mm instead of 1.31 mm. When printing, the air inclusions in the first sample became smaller in size and acquired a more stepped appearance, and the pyramids on the surface of the second sample also acquired a more stepped appearance.

Figure 2. Manufactured samples with air pores (a) and with pyramidal surface corrugation (b).

2.3. Measuring equipment
The electromagnetic response of experimental samples of composite structures was measured in the time domain using the fast Fourier transform on a T-Spec 1000 spectrometer. It uses a femtosecond laser, while the emitter and detector is a photoconductive antenna based on GaBiAs. The studies were conducted in the frequency range from 100 to 1000 GHz.

3. Model experiment results and discussion

3.1 Material with the inclusion of air pores
Figure 3 shows the frequency dependences of the transmission coefficient for sample models with different pore inclusions (Table 1).

The graphs show that a decrease in the distance between the air inclusions contributes to a decrease in the proportion of the passing wave. This is also facilitated by an increase in the size of the pores themselves, even if the pore diameter (2*r*=1.4 mm) slightly exceeds the thickness of the layer itself (*d*=1.31 mm). With a further increase in the diameter, the proportion of passing power begins to increase markedly, which is probably due to dimensional effects.

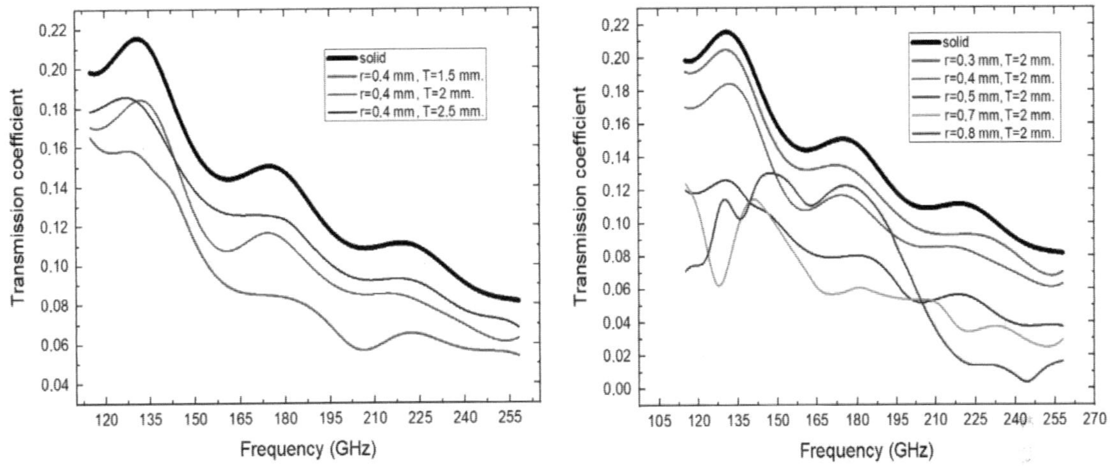

Figure 3. Frequency dependences of transmission coefficients of material with air pores.

The reflection coefficient is not significantly affected by the inclusion of pores in the material thickness with a constant plane-parallelism of the layer surface (Figure 4). This is due to the constant impedance at the interface of the media as long as the pore diameter does not exceed the thickness of the material.

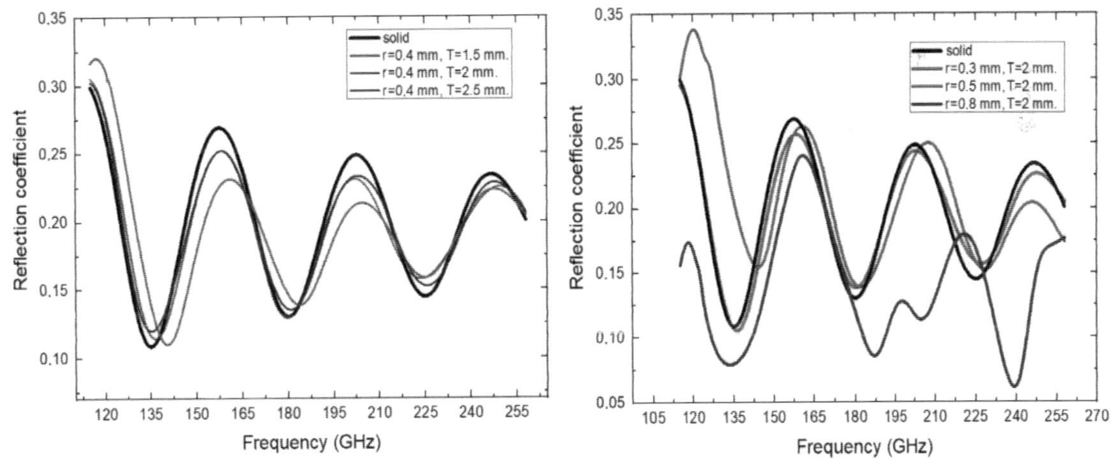

Figure 4. Frequency dependences of reflection coefficients of material with air pores.

Figure 5 shows the frequency dependences of the absorption coefficient calculated on the basis of the above reflection and transmission coefficients (*A*=1-*T*-*R*).

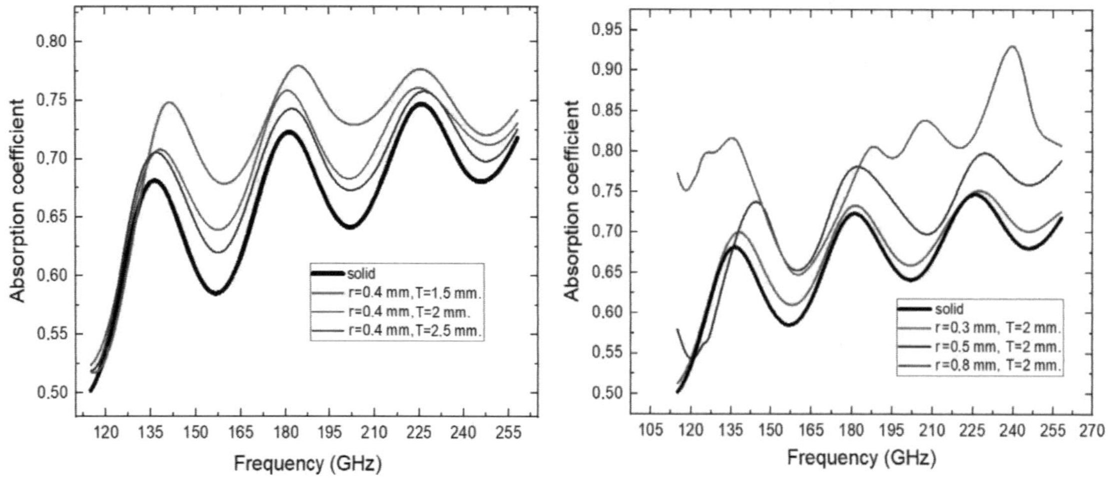

Figure 5. Frequency dependences of absorption coefficients of material with air pores.

Both as the pore size increases and as the distance between them decreases, the value of the absorption coefficient increases markedly. Moreover, even for a sample whose pore diameter ($2r$=1.6 mm) is noticeably greater than the thickness of the material layer (d=1.31 m), the curve of the absorption coefficient values lies significantly higher than the curve for a solid sample. This is probably due to the size effects in carbon-containing structures that occur when the radiation wavelength is comparable to the size of inhomogeneous inclusions.

3.2 Material with pyramidal surface fluting

Figures 6, 7 and 8 show the frequency dependences of the transmission, reflection and absorption coefficients, respectively, modeled and calculated for several samples with different parameters of the height and width of the base of the pyramids (Table 2).

The qualitative effect of changes in the geometry parameters of the samples (the height and width of the base of the pyramids) on the behavior of the curves in Figures 6 and 8 in this case is absolutely the same as in the previous paragraph. Only the reflection decreased noticeably. For the best absorption of the electromagnetic wave and to reduce the transmission and reflection, it is advantageous to increase the height of the pyramids and the width of their base.

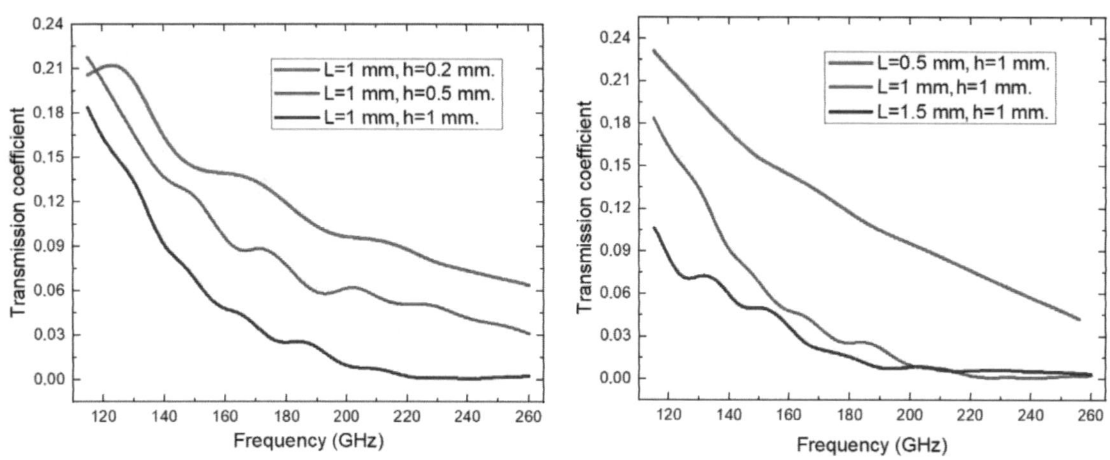

Figure 6. Frequency dependences of transmission coefficients of material with pyramidal fluting.

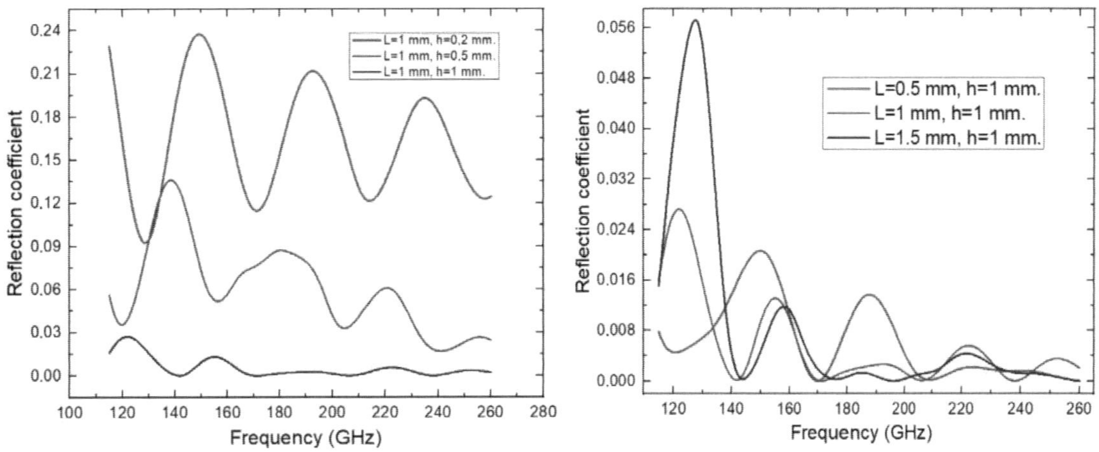

Figure 7. Frequency dependences of reflection coefficients of material with pyramidal fluting.

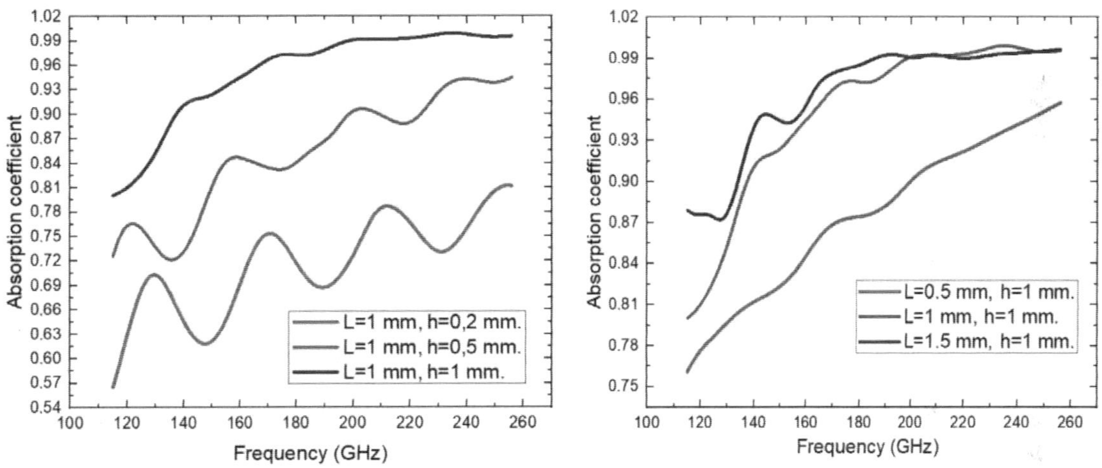

Figure 8. Frequency dependences of the absorption coefficients of the material with pyramidal fluting.

4. Measurement results and discussion

The results of an experimental study of the electromagnetic response from composite structures with technological inhomogeneities made of carbon-containing plastic using FDM 3D printing technology are presented below.

Figure 9 shows the frequency dependences of the electromagnetic response from samples with pore inclusion and with pyramidal corrugation on the surface of a solid sample in comparison with a solid flat sample. It can be seen that the experimental data qualitatively quite accurately coincide with the simulation. Small deviations from the calculated data can be attributed to the fact that the original geometry of the experimental samples has undergone some changes in the process of 3D printing.

The results obtained confirm that at frequencies up to 250 GHz, the inclusion of air pores in the polymer volume reduces the transmission coefficient, practically does not affect reflection, and increases absorption. The addition of a pyramidal corrugated layer leads to a significant reduction in the transmission and reflection coefficients, as well as to a significant increase in absorption. The studied sample is able to absorb more than 99% of the radiation in the frequency range from 200 to 635 GHz.

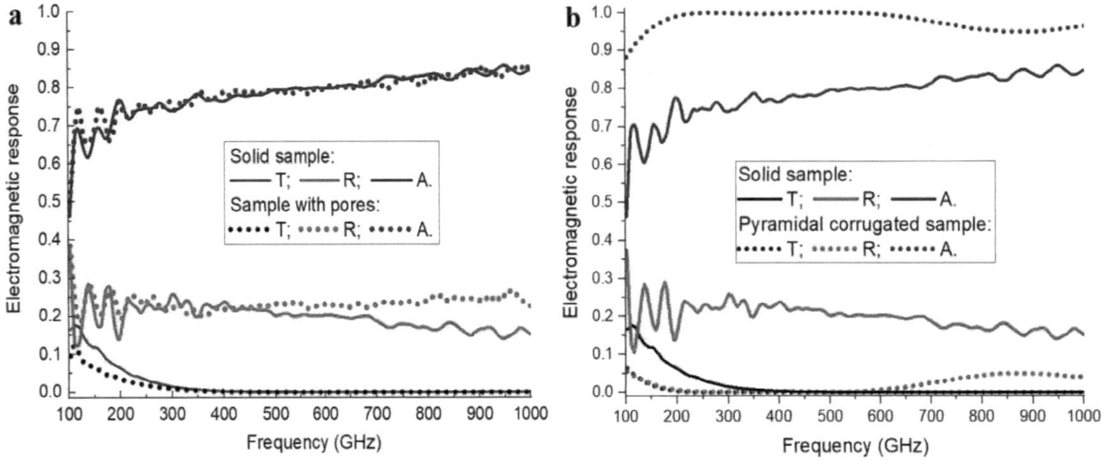

Figure 9. Frequency dependences of the electromagnetic response on samples with pore inclusion (a) and with pyramidal corrugation on the surface (b)

5. Conclusions

A model and experimental study of the electromagnetic response from composite structures made of carbon-containing plastic with technological inhomogeneities is carried out. The addition of a pyramidal corrugated layer leads to a significant reduction in the transmission and reflection coefficients, as well as to a significant increase in absorption. The studied sample is able to absorb more than 99% of the radiation in the frequency range from 200 to 635 GHz. Thus, the studied materials can be used to create thin, light and effective shielding materials of the EHF and HHF ranges.

Acknowledgment

This research was supported by Ministry of Science and Higher Education of the Russian Federation, project No FSWM-2020-0038.

References

[1] Son H-K, Chong Y-J 2020 *Int. Conf. on ICT Convergence* **9289432** 1553–1556
[2] Rumeng T, Ying S, Tong W, Wentao Z 2020 *IEEE International Symp. on EMCSI 2020* **9191518** 659–662
[3] Lei Z, Yuyang W, Tingting W, Lili C 2021 *Lect. Notes in El.l En.* **677** 191–198
[4] Mathur P, Raman S 2020 *J. of Electronic Mat..* **49(5)** 2975–98
[5] Przesmycki R, Wnuk M 2018 *Int. J. of Safety and Security Engineering* **8(2)** 223–233
[6] Lai Y-F, Wang H-Y, Peng R-Y 2021 *Military Medical Research* **8(1)** 12
[7] Youji Kotsuka 2019 *Electromagnetic Wave Absorbers: Detailed Theories and Applications* 336
[8] Green M, Chen X 2019 *Journal of Materiomics* **5(4)** 503–541
[9] Kuleshov G, Badin A 2019 *IOP Mat. Sci. and Eng.* **525(1)** 012030
[10] Gupta S, Tai N-H 2019 *Carbon* **152** 159–187
[11] Badin A, Kuleshov G, Bodazhkov D, Simonova K 2019 *Int. Conf. EDM* 8823228 143–148

Stabilization of the backward wave oscillator frequency for operation with high quality resonators

A Shansho, G E Dunaevsky, I O Dorofeev, A V Badin and E V Emelyanov

National Research Tomsk State University, Tomsk, Russia

E-mail: ahmadsho1@gmail.com

Abstract. This article describes the performance of a backward wave oscillator, stabilized with phase locked loop. The backward wave oscillator is locked to the frequency of a direct digital synthesizer using an automatic phase locked loop. The direct digital synthesizer is playing the role of tuning part in the suggested frequency synthesizer. Mathematical evaluations of the stability, phase noise and locking speed of the suggested frequency synthesizer are obtained. The optimization technique of frequency synthesizers parameters is presented. This allows to get the maximum response speed and the lowest phase noise and level of spurious with fine tuning (less than 1 Hz). It was shown that the order and type of loop filter in the frequency synthesizer will affect simultaneously the response speed, phase noise and spurious level. The results of this paper shows that the frequency synthesizer can be used in small samples properties measurement using open resonators.

1. Introduction

Currently, the use of open quasi-optical resonators allows very accurate measurements of the electro physical parameters of substances [1–3]. However, for these measurements it is necessary to observe the resonance curve and its changes with a high frequency resolution [4]. In the sub-millimeter frequency range (0.3 – 3 THz), existing generators of electromagnetic oscillations have either high stability, but low power, as in the case of quantum generators, or acceptable power, but frequency instability, as in a generator based on a backward wave oscillator (BWO). Therefore, we strive to develop a frequency synthesizer with high stability and acceptable power for use in small samples measurement.

Despite the significant development of solid-state electronics, the number of authors implementing it when creating THz frequency generators for measuring the electrophysical parameters of small samples is small. This is due to the fact that the use of solid-state electronics methods in terahertz radiation sources facing many difficulties. as an example, the application of multiple frequency multiplication requires the use of advanced technologies, however, the output signal power of such devices is still very low. BWO-based sources have a higher output power, and a frequency stabilization circuit using a phase-locked loop (PLL) and direct digital synthesizer (DDS) as reference provides a frequency resolution comparable to a microwave synthesizer with stable output.

Frequency synthesizers can be divided into two categories: traditional electronic synthesizers and photonic synthesizers. In [5], an amplifier-multiplier single-chip circuit with a frequency of 300 GHz was developed. It contains a power amplifier to drive the 300 GHz frequency doubler and two balanced doublers. The synthesizer output power ranged from 8 to 2.5 dBm, with a peak value of 2.5 dBm at 284 GHz with a 0 dBm input at 71 GHz. The disadvantage of this synthesizer design was

Content from this work may be used under the terms of the Creative Commons Attribution 3.0 licence. Any further distribution of this work must maintain attribution to the author(s) and the title of the work, journal citation and DOI.
Published under licence by IOP Publishing Ltd

the limited bandwidth and high cost. In [6], a multiplier amplifier circuit for the frequency range 315–340 GHz based on monolithic microwave integrated circuit and Schottky diodes with high output power was presented. The synthesizer output power was varied from 6 to 8 dBm. The disadvantage of this synthesizer design was the limited bandwidth of no more than 25 GHz. Photonic millimeter wave synthesizers such as optical frequency comb synthesizers have the advantage of greater synthesized bandwidth and higher stability. In [7], a stable tunable millimeter-wave frequency synthesizer (100–300 GHz) based on soliton was presented, but the output power was low, and the spectrum contains many "parasitic" harmonics.

The aim of this work is to develop an optimized phase locked BWO to be used in small samples characterization by open resonators. To do this the following problems have to be solved:

- describing of mathematical model of frequency synthesizers;
- simulation of the synthesizers in the time and frequency domains;
- optimization of speed response, phase noise, tuning step and spurs.

2. Structure of phase locked BWO frequency synthesizers

The optimized submillimeter wave frequency synthesizer (DDS/PLL/BWO frequency synthesizer) block diagram is the first step in the design (Figure 1).

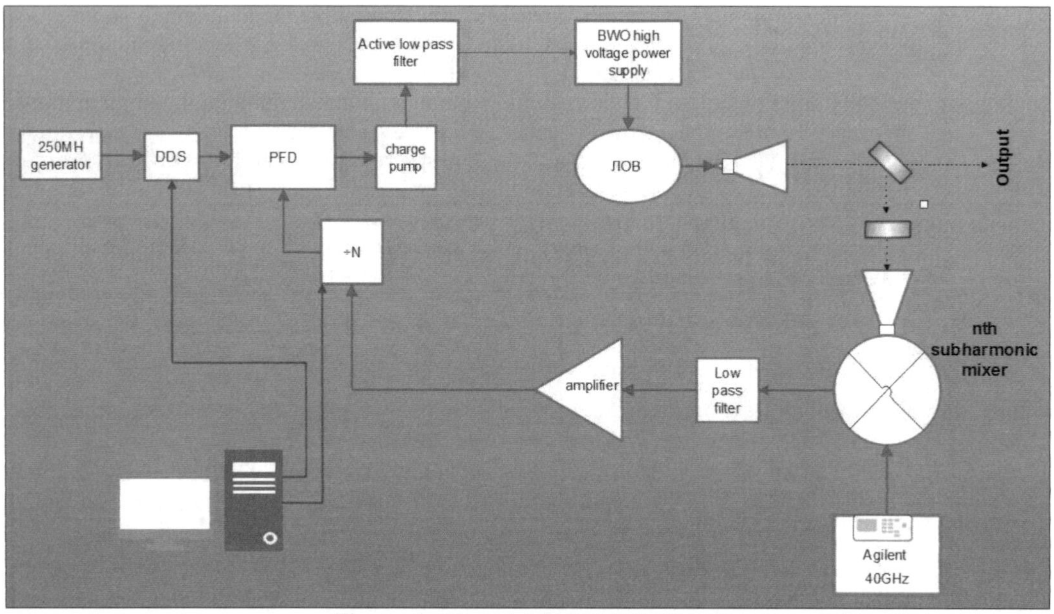

Figure 1. Phase locked BWO flow chart.

The synthesizer circuit consists of a DDS as a reference, BWO, PLL, subharmonic mixer, horn antennas for transmission and reception, filters.

The BWO stabilization scheme works as follows: the BWO continuously generates electromagnetic radiation, which is then split into two parts by a beam splitter. One part will be used as the synthesizer output, and the other part will be used for PLL feedback. This signal is mixed with the *n*th harmonic of a 40 GHz from generator (Agilent E8363B). The intermediate frequency at the output of the mixer will be filtered by a low-pass filter, amplified, and then is fed to phase-frequency detector. A phase-frequency detector compares this signal with the output of a DDS and generates a signal, which in turn is filtered to obtain a direct current (DC) voltage proportional to the phase and frequency difference between the DDS and loop feedback signals. This DC voltage is fed as a signal to the BWO power supply to control the high voltage of the power supply at BWO cathode which controls the BWO output frequency. When the BWO frequency changes, the DC voltage from the output of the phase-frequency detector will change and the frequency returns to its original value. Adjusting the BWO frequency can be done on a computer by adjusting the DDS frequency.

3. Phase locked BWO parameters analysis

The proposed synthesizer uses a fixed division ratio and two tunable stable frequency synthesizers. This allows to highly optimizing the phase noise and spurs performance, and widening the bandwidth of the loop.

3.1 Spurs performance

To maintain a high frequency resolution, we use DDS as tunable reference frequency. Using DDS will also provide low noise and the switching speed required by our synthesizer. The output frequency of the synthesizer can be calculated as follows:

$$f_{phlBWO} = f_{DDSin}\frac{W}{2^k}N + n * f_{Agil} = f_{DDS} * N + n * f_{Agil} \quad (1)$$

where f_{phlBWO}: the output frequency of the phase locked BWO.

W: DDS tuning word.

k: the length of the tuning word in bits.

N: the loop feedback division value.

f_{DDSin}: the reference frequency of the DDS.

n: the harmonic number used in the harmonic mixer.

f_{Agil}: the local oscillator frequency of the mixer.

The DDS output frequency band needed to achieve the band $B = [f_{phlBWO_{min}} \quad f_{phlBWO_{max}}]$ at the BWO output must be:

$$\left[\frac{f_{phlBWO_{min}} - n * f_{agil_{max}}}{N} \qquad \frac{f_{phlBWO_{max}} - n * f_{agil_{min}}}{N}\right] \quad (2)$$

The DDS contains digital to analog converter (DAC), thus it will produce spurs. The spur level (in dBc) in the band of interest witch is the loop bandwidth is given by:

$$P_{DDS_{spur}} = -1.76 - 6.02k - 20\log\left(\frac{f_{DDSin}}{f_{DDSout}}\right) \quad (3)$$

This spurs will be added to the spurs produced by the feedback divider, therefore the total spurs level at the phase locked BWO can be expressed by:

$$P_{DDS_{spur}} = -1.76 - 6.02k - 20\log\left(\frac{f_{DDSin}}{f_{DDS}}\right) + 20\log(N) \quad (4)$$

So to improve the spurious free dynamic rang (SFDR) of the synthesizer the feedback division ratio must be reduced and the direct digital synthesizer DAC length must be increased.

3.2 Phase noise

The phase noise power density of phase locked BWO (Figure 1) is additive, so it can be expressed as:

$$S_{\phi BWO} = \left(n^2 S_{\phi agil}(\omega) + S_{\phi mix}(\omega) + N^2(S_{\phi DDS}(\omega) + S_{\phi PFD}(\omega) + S_{\phi LF}(\omega))\right)\left|\frac{H(j\omega)}{N}\right|^2 + S'_{\phi BWO}(\omega)\left|1 - \frac{H(j\omega)}{N}\right|^2 \quad (5)$$

where $H(j\omega)$: the closed loop transfer function.

$S_{\phi agil}(\omega)$: the phase noise of Agilent generator

$S_{\phi DDS}(\omega)$ the phase noise of the DDS

$S_{\phi \mathrm{PFD}}$: the phase noise of phase frequency detector
$S_{\phi \mathrm{LF}}(\omega)$: the phase noise of active Loop filter
$S_{\phi \mathrm{mix}}(\omega)$: the phase noise of the mixer
$S'_{\phi \mathrm{BWO}}(\omega)$: the phase noise of the unlocked BWO

The modern DDS generator has phase noise fare below than the Agilent generator and, which means that $S_{\phi \mathrm{agil}}(\omega) \ll S_{\phi \mathrm{DDS}}(\omega)$. So $S_{\phi \mathrm{DDS}}(\omega)$ can be neglected. The phase noise added by mixer also can be neglected if it is passive mixer. Equation 5 can be written as:

$$\mathrm{S}_{\phi \mathrm{BWO}} = \left(\mathrm{n}^2 \mathrm{S}_{\phi \mathrm{agil}}(\omega) + \mathrm{N}^2(\mathrm{S}_{\phi \mathrm{PFD}}(\omega) + \mathrm{S}_{\phi \mathrm{LF}}(\omega))\right) \left|\frac{\mathrm{H}(\mathrm{j}\omega)}{\mathrm{N}}\right|^2 + \mathrm{S}'_{\phi \mathrm{BWO}}(\omega) \qquad (6)$$

but $\left|\frac{\mathrm{H}(\mathrm{j}\omega)}{\mathrm{N}}\right|$ have 1 value in the loop bandwidth and 0 outside it. This indicates that the phase noise of the n_{th} harmonic of Agilent generator has a major contribution to the phase locked BWO phase noise inside the loop bandwidth. While outside the loop bandwidth, the phase noise of the phase locked BWO is mainly determined by the unlocked BWO.

By using harmonic mixer in down and up conversion modes we can get further high output frequency of the loop while keeping the input frequency of the feedback divider low, which keeps the division factor low. Keeping division ratio in the loop low will proportionally reduce near phase noise. It can be said that the use of a harmonic mixer in loop feedback contributes significantly to improving the phase noise of the BWO locked phase because it reduces the dividing ratio N, provided that the phase noise performance of the local oscillator in the mixer is good.

3.3 Frequency error

The frequency of the phase locked BWO can be expressed as:

$$f_{phlBWO} = f_{DDS} * N + n * f_{Agil} \qquad (7)$$

If we express the frequency stability of the DDS with Δf_{DDS} and the frequency stability of the Agilent generator with Δf_{agil}, equation 7 can be rewritten as:

$$f_{phlBWO} = (f_{DDS} \pm \Delta f_{DDS}) * N + n * (f_{Agil} \pm \Delta f_{agil}) \qquad (8)$$

This is equivalent to:

$$f_{phlBWO} = f_{DDS} * N + n * f_{Agil} \pm (\Delta f_{DDS} * N + n * \Delta f_{agil}) \qquad (9)$$

Thus

$$\Delta f_{phlBWO} = \Delta f_{DDS} * N + n * \Delta f_{agil} \qquad (10)$$

Since the stability of the modern DDS is very high, Equation 10 indicates that the stability of the phase locked loop is proportional to the stability of the Agilent generator and the harmonic number used in the mixer.

4. Numerical calculation and simulation

The BWO, which needs to be stabilized, have the frequency rang 105–240 GHz. We designed the synthesizer to have a frequency of 600 MHz at the input of the loop feedback. If we use a 6th harmonic mixer, the output frequency of the Agilent generator should be in rang 17.5–40 GHz. To reduce the division ratio at the feedback, a 25MHz reference frequency was used. This means that the division ratio equal to 24. This 25MHz is generated by AD9913 DDS with 250 MHZ reference frequency. Since the DDS output is 25MHz the loop filter bandwidth should be 2.5MHz.

Since a spectrum analyzer at high frequency not common, it's necessary to calculate the phase noise and spurs numerically. We have all parameters needed to calculate the performance of the synthesizer. The spurs level of the synthesizer can be calculated from equation 4:

$$P_{DDS_{spur}} = -1.76 - 6.02 * 10 - 20\log\left(\frac{250}{25}\right) + 20\log(24) = -54.35\ dBc$$

A simulation for the synthesizer output of 160GHz was carried out to compare the phase noise of the synthesizer with and without harmonic mixer (Figure. 2).

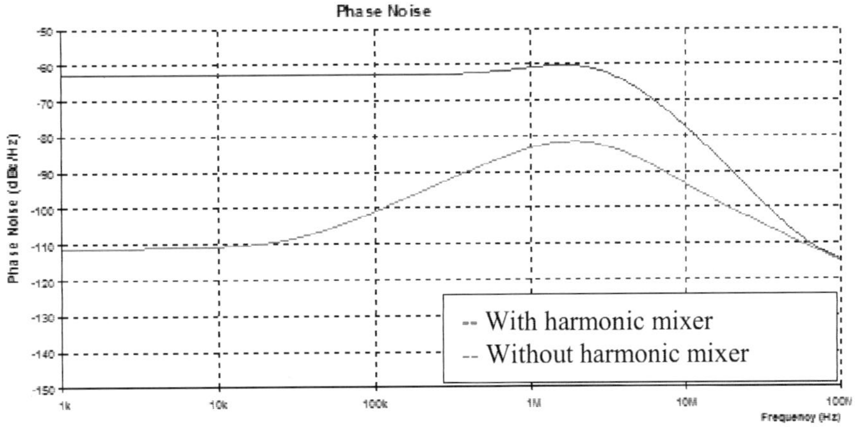

Figure 2. Phase noise comparison at 160GHz.

The simulation showed also that the lock time to 160 GHz of the synthesizer is 3.2 us (Figure 3), and it's more than we need in small samples properties measurement.

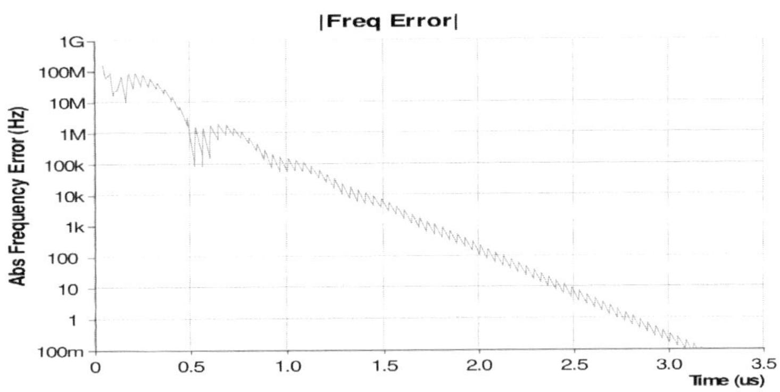

Figure 3. Lock time to 160GHz.

A performance characterization is given in Table 1. The proposed block diagram achieves a good enough spurious free dynamic range and phase noise to use in high quality resonators, and fine tuning.

Table 1. Performance characteristics of phase locked BWO.

Parameter	Value
Frequency band (GHz)	105-240
Lock time (us)	3.2
Frequency stability (PPM)	2
Phase noise (@100kHz) dBc/Hz	-111
SFDR (dBc)	-54
Minimal frequency step (Hz)	1

5. Conclusion

We could have a good performance phase locked BWO. It provides frequency control in small steps (1 Hz), which is more than sufficient for measuring electro physical parameters of small samples in high quality resonators. We can say that the proposed frequency complex can cover a frequency range of up to 1.5 terahertz, but with a different subharmonic mixer operating in the upper frequency range.

References

[1] Badin A *et al* 2015 *Russian Physics Journal* **58**(8-3) 121–123

[2] Badin A *et al.*. 2018 *International Conference on Infrared, Millimeter, and Terahertz Waves IEEE* 1-2 doi: 10.1109/IRMMW-THz.2018.8510310

[3] Karpisz T *et al.* 2019 *IEEE MTT-S International Microwave Symposium (IMS)* 1503-1506 doi:10.1109/MWSYM.2019.8700981

[4] Goodnick S, Korkin A, Nemanich R 2018 *Semiconductor Nanotechnology: advances in information and energy processing and storage* (Cham: Springer International Publishing AG) Chapter 7 pp169–197

[5] Zhou P *et al.* 2019 *IEEE Custom Integrated Circuits Conference (CICC)* 1-4 doi: 10.1109/CICC.2019.8780333

[6] Deng X *et al.* 2020 *IEEE MTT-S International Wireless Symposium (IWS)* 1-3 doi: 10.1109/IWS49314.2020.9360180

[7] Zang J *et al.* 2019 *IEEE Avionics and Vehicle Fiber-Optics and Photonics Conference (AVFOP)* 1-2 doi: 10.1109/AVFOP.2019.8908197

Journal of Physics: Conference Series 2140 (2021) 012012 doi:10.1088/1742-6596/2140/1/012012

Broadband selection of EHF radiation of metamaterial surface obtained by photolithography

K V Bilinskiy, K V Dorozhkin, V D Moskalenko, G E Kuleshov, A V Badin, E A Trofimov and A I Berdyugin

National Research Tomsk State University, Tomsk, Russia

E-mail: thzlab@mail.ru

Abstract. In paper results of research of metamaterial surface based on split-ring resonators obtained by photolithography are presented. Numerical simulation of electromagnetic response of the created structure are shown. Experimental researches of the transmission coefficient using quasi-optical methods of continuous and time-domain spectroscopy in the frequency range from 34 to 200 GHz were carried out. Area of broadband screening in the EHF range has been found.

1. Introduction

Since the end of the last century, the materials with anomalous refractive index (left-handed media) [1] in various frequency ranges - UHF [2], EHF [3], sub-terahertz and terahertz ranges [4], optical range [5] were actively investigated and used. Metamaterials are composite structures whose properties are determined by macroscopic elements of subwavelength periodically located in space. Their properties are defined by parameters such as the size of the structures, shape, constant and variable lattice period of the elements forming the structure, permittivity and conductivity of materials used in the structures. Therefore, metamaterials may have unique electromagnetic characteristics (such as negative and zero refractive index) that are unavailable to natural [6] and artificial [7, 8] materials.

Recently, much attention has been paid to frequency-selective dielectric metasurfaces. They are spatial filter consisting of flat periodically arranged scattering arrays. These arrays are able to fully control the wavefront in the reflection or transmission mode by introducing sharp phase changes in certain frequency ranges [9, 10]. They have found many potentials uses for creating lenses [11, 12], mirrors [13], radiation polarization converters [14], invisibility cloaks [15], effective absorbers [16], detectors [17], modulators [18], filters [19] and sensors [20].

However, conventional metamaterial surfaces have a narrow-band frequency response, which significantly limits their practical application in most real devices. The solution to this problem for different frequency ranges is to use multi-level design and creation of multilayer metamaterials. Since in the presence of several metallic structural layers separated by a dielectric, it is possible to increase the absorption/reflection/transmission bandwidth due to the impedance corresponding to the free space [21]. However, these classes of materials are still poorly explored.

2. Model of the metamaterial surface

A sample of composite radio material based on array of identical elementary split-ring cells obtained by double-sided photolithography on a textolite is presented on Figure 1.

Content from this work may be used under the terms of the Creative Commons Attribution 3.0 licence. Any further distribution of this work must maintain attribution to the author(s) and the title of the work, journal citation and DOI.
Published under licence by IOP Publishing Ltd

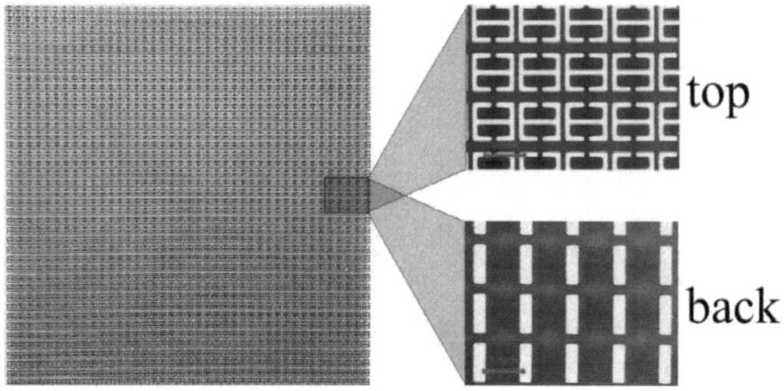

Figure 1. Micrograph of the metamaterial surface structure from two sides.

An elementary cell (Figure 2*a*) of $a \times b$ size consists of two metallized layers placed on a substrate of FR4 dielectric [22] with a thickness of $h = 0.3$ mm with a dielectric constant of $\varepsilon' = 4.3$, and the tangent of dielectric loss angle $tg\ \delta = 0.025$. The upper metallized layer (Figure 2*a*) consists of two electrically interconnected circuits with a small gap. The dimensions of the contour system are $L_a \times L_b$, the width of the conductors is w, the length of the gap in the contours is g. The thickness of the metallization is 18 microns. The bottom layer (Figure 2*b*) is a piece of conductor $a_1 \times b_1$ in size, placed orthogonally to the common wall of the contours.

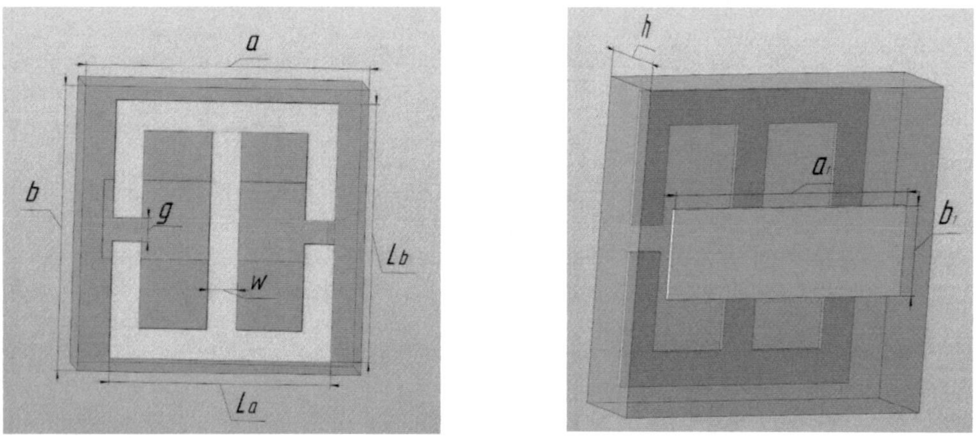

a) b)

Figure 2. Unit cell structure: upper metallization layer (a), lower metallization layer (b).

The values of topology parameters for three samples of composite materials were given in table 1.

Table 1. Metamaterial topology parameters.

a, mm	b, mm	h, mm	L_a, mm	L_b, mm	w, mm	g, mm	a_1, mm	b_1, mm
1.188	1.226	0.3	1.075	0.958	0.126	0.1	0.329	0.972

3. Modeling of electromagnetic response parameters

In the CST Studio engineering program, the calculation of the electromagnetic response was carried out. The simulation was performed in the range of 30 – 200 GHz for a structure consisting of 15×15 elementary cells (Figure 3).

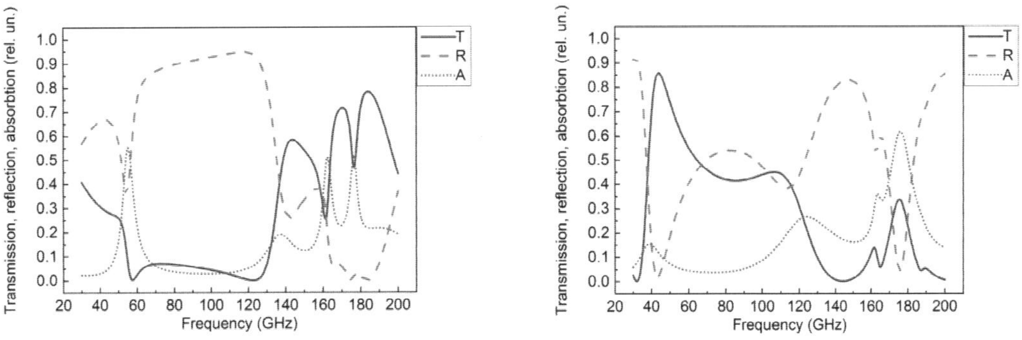

a) b)

Figure 3. The results of modeling the coefficients of transmission (*T*), reflection (*R*), absorption (*A*) for the structure of a composite material from a 15×15 unit cell: a) $\vec{E}$ is codirected with the conductor on the lower layer; b) $\vec{E}$ is codirected with the common wall of the contours.

As can be seen from the results, for a metamaterial sample a minimum transmission coefficient was observed in the range of 54-130 GHz (less than 10%), if $\vec{E}$ is codirected with the conductor on the lower layer. At the same time, the reflection coefficient at these frequencies reaches a maximum (90% and higher). A minimum of the transmission coefficient in the region of 145 GHz and absorption maxima at frequencies of 163 and 176 GHz were observed with orthogonal polarization ($\vec{E}$ is codirected with the common contour wall of the contours). The real structures of the metamaterial were manufactured according to the obtained parameters as a result of electromagnetic modeling of the structure of 15×15 elementary cells.

A three-layer composite material can be represented as a homogeneous medium with effective electromagnetic characteristics, namely effective complex permeability ($(\mu(f)_{eff} = \mu' + i\mu'')$) and permittivity ($(\varepsilon(f)_{eff} = \varepsilon' + i\varepsilon'')$). The wave resistance of a given medium is determined as follows: $Z(f) = \sqrt{\frac{\mu(f)_{eff}}{\varepsilon(f)_{eff}}}$. Thus, by changing the geometric parameters of the topology of the upper and lower metallized layers (L_a, L_b, a, b, g, w, a_1, b_1) it is possible to select a certain value of the wave resistance $Z(f)$, which will either differ significantly from the wave resistance of the free space Z_0. Then the medium will have reflective properties in a wide frequency range with a minimum transmission coefficient (losses in the medium itself will be minimal, Figure 3*a*, frequency range 54–130 GHz).

The distribution of electric and magnetic fields ($\vec{E}$ is codirected with the conductor on the lower layer) at a frequency of 70 GHz are shown on Figures 4*a* and 4*b* respectively. If the values of $Z(f)$ are close to Z_0, then a decrease in the transmission coefficient in a narrow frequency range is possible due to a sharp increase in absorption properties (Figure 3*b*, in the frequency range 176 GHz). The absorption properties are determined to a large extent by dielectric losses caused by the interaction of narrow-band split-ring resonator structures with incident radiation. There are also conduction losses in the conductor segment on the lower metallization layer. The values of $\mu(f)_{eff}$ and $\varepsilon(f)_{eff}$ in this case ($\vec{E}$ is codirected with the common contour wall on the upper layer) set $Z(f)$ approximately equal to Z_0. They are accompanied by maximum of the electric field at the top of the metallization at the points of contour breaks (Figure 4*c*). For the magnetic field, the maximums of the electric field are observed on the lower layer in the conductor segment (Figure 4*d*).

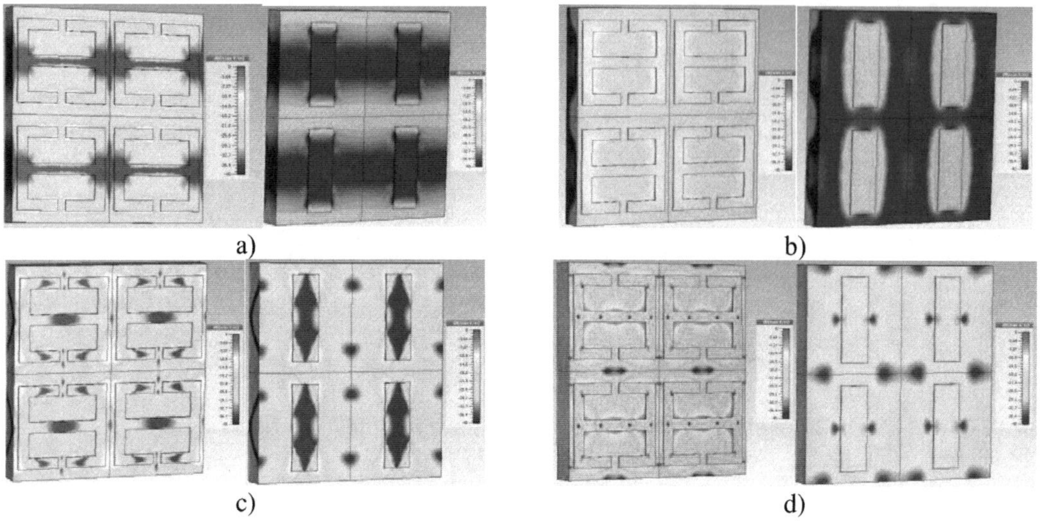

Figure 4. The results of modeling the electromagnetic field: $\vec{E}$ is codirected with the conductor on the lower layer (frequency 70 GHz, distribution of electric (a) and magnetic (b) fields; $\vec{E}$ is codirected with the common wall of the contours on the upper layer (frequency 176 GHz, maxima of electric (c) and magnetic (d) fields).

The maximum values of the electric and magnetic fields ($\vec{E}$ is codirected with the common contour wall on the upper layer) lead to significant dielectric losses on the upper layer and losses on the lower layer, respectively, in a narrow frequency range with minimal reflection since $Z(f)\sim Z_0$.

4. Measurements and results

Measurements of transmission power in the frequency range 34 – 177 GHz on a terahertz spectrometer STD-21 were carried out. Four backward wave oscillators (BWO) were used as a source, the transmitted radiation was detected by an acousto-optic converter (Golay cell). To confirm the reliability of the results, measurements of transmission power were additionally carried out on a time-domain spectrometer T-Spec 1000. The results obtained using the two techniques are consistent (Figure 5).

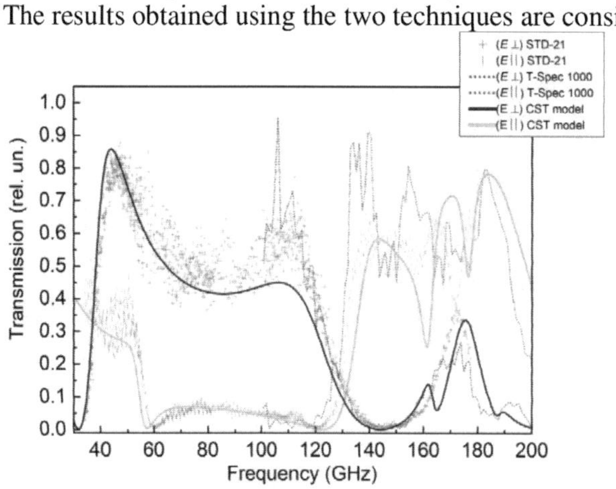

Figure 5. The results of measurements of the transmission coefficient at two different polarizations $\vec{E}$.

The metamaterial sample ($\vec{E}$ is codirected with the conductor conductor on the lower layer) is characterized by the presence of a minimum transmission coefficient (less than 5–10%) in a wide frequency range of 54–130 GHz. With orthogonal polarization of the incident radiation ($\vec{E}$ is codirected with the common wall of the contours), a minimum transmission coefficient is observed in the region of 145 GHz (about 5%). Based on the obtained transmission coefficient spectra, it can be concluded that the results of electromagnetic modeling and real experiment coincide qualitatively and quantitatively.

5. Conclusion

A selective feature of the interaction of electromagnetic radiation of the EHF band with a metamaterial of the split-ring resonator type was demonstrated. According to the calculation, the maximum absorption of the presented configuration is achieved at a frequency of 55 Hz. At the same time, in the range of 54 – 130 GHz, the power of the transmitted radiation (less than 5-10%) suggests that a blocking filter was obtained. The resulting structure of the composite radio material makes it possible to create a material with high reflectivity in a wide frequency range, and this material will also have narrow-band absorption maximum. The use of two such materials with different topologies makes it possible to distinguish a narrow band of a useful EHF signal, since each of the materials separately is a band-blocking filter with a sufficiently high squareness ratio.

Acknowledgments

This research was supported by Ministry of Science and Higher Education of the Russian Federation, project No FSWM-2020-0038.

References

[1] Veselago V G 2003 *Electromagnetic Materials* 115–122
[2] Landy N, Mock J and Padilla W 2008 *Physical Review Letters* **100** 207402
[3] Gokkavas M, Guven K and Bulu I 2006 *Opt. Express* **14** 8685–8693
[4] Bingham C, Tao H and Zhang X, 2008 *Opt. Express* **16(10)** 7181–7188
[5] Li T, Liu H, Wang F, Dong Z, Zhu S and Zhang X 2006 *Opt. Express* **14(23)** 11155–11163
[6] Eleftheriades G V, Balmain K G 2005 *Negative-Refraction Metamaterials: Fundamental Principles and Applications* (Toronto: Novel Microwave Technologies Thrust at the University's Nortel Institute for Telecommunications) p 440
[7] Dunaevskii G E *et al* 2016 *Proceeding of the* 41st International Conference on Infrared, Millimeter, and Terahertz waves (IRMMW-THz) 1–2 doi: 10.1109/IRMMW-THz.2016.7758771
[8] Dunaevskii G E *et al* 2014 *Proceeding of the 39th International Conference on Infrared, Millimeter, and Terahertz waves (IRMMW-THz)*,1–2
[9] Yang Y, Xu Y, Zhang B, Duan J, Yan L, Xu H, Liu Y and Shi Y 2019 *Optics Communications* **438** 39–45
[10] Chen Y, Cheng J and Liang C 2020 *Advances in Condensed Matter Physics* **3902835** 1–6
[11] Liu C, Wu J 2021 *J. Phys D* **54(395102)** 1–8
[12] Wu G-B, Zeng Y-S, Chan K, Qu S-W and Chan C 2019 *IEEE Trans. Antennas. Propag.* **67(7)** 4429–4437
[13] Lee Y, Moon K and Kwon S-H 2019 *Plasmonics* **14(5)** 1227–1232
[14] Lin R, Lu F, He X, Jiang Z, Liu C, Wang S and Kong Y 2021 *Opt. Express* **29(19)** 30357–30370
[15] Jing X et al 2020 *Opt. Express* **28(24)** 35528–35539
[16] Vahidi A, Rajabalipanah H, Abdolali A and Cheldavi A 2018 *Appl. Phys. A* **124(337)** 1–10
[17] Bigioli A *et al* 2020 *Appl. Phys. Lett.* **116(161101)** 1–5
[18] Wang Q, Gao B, Raglione M, Wang H, Li B, Toor F, Arnold M and Ding H 2019 *Laser and Photonics Rev.* **13(11)** 1900071
[19] Asl A, Rostami A and Amiri I 2020 *Optical and Quantum Electronics* **52(3)** 1–13

[20] Saadeldin A, Hameed M, Elkaramany E and Obayya S 2019 *IEEE Sensors Journal* **19(18)** 7993–7999

[21] Liu Z, Qi L, Shah SMA, Sun D and Li B 2019 *Mater* **16(6)** 1–11

[22] KB-6165F datasheet 2020 Available at: https://www.rezonit.ru/upload/spetsifikatsii/KB-6165F.pdf

Enhancing performance in a LOS MIMO communication using a passive repeater

Mahmoud Eissa and D Sukhanov

Tomsk State University, Tomsk, Russia

E-mail: Mahmoud.eissa@stud.tsu.ru

Abstract. This paper presents a technique for obtaining a well-conditioned channel matrix in a line of sight multiple input multiple output (MIMO) environment. The technique is based on the implementation of a back-to-back antenna system as a passive repeater to enhance performance in MIMO systems. The flexible configuration with no need for a phase controller allows to spread the proposed repeater in MIMO communications to ensure spatial multiplexing and enhance capacity. A condition number and matrix rank are proposed as metrics to demonstrate the validity of the proposed method.

1. Introduction

With the ability to improve channel capacity and reliable transmission, Multiple-Input Multiple-Output (MIMO) techniques are still deployed in the current and future wireless networks. MIMO can provide diversity gain that improves link reliability [1] or can improve spectral efficiency using spatial multiplexing [2]. The capacity of the MIMO systems arises from scattering, and that the greater the amount of scattering, the greater the amount of capacity [1]. However, in the case where the transmit and receive antenna arrays are within line-of-sight (LOS) of each other and there is no scattering in the channel, the effective of the MIMO systems can become strongly influenced [3]. Different approaches have been introduced to enhance the performance of MIMO communications in LOS environments. In [3], the optimum antenna separation at the transmit and receive ULAs in a LOS MIMO system is derived to maximize the spatial degree of freedom and the channel capacity. In [4], a new structure consists of a back reflector and a side reflector to enhance the capacity of short-range MIMO communications is proposed. The proposed structure increases all eigen values and equalizes their distribution. Authors in [5] proposed the use of switched parasitic antennas (SPA), with employing the water-pouring algorithm to modify the covariance matrix in order to improve the channel capacity. In this article, a condition number is proposed to select the optimal pattern configuration for MIMO SPAs. In [6], an intelligent reflecting surface (IRS), consisting of reconfigurable metamaterials, is used to enrich the propagation environment by adding multipaths with distinctively different spatial angles. Authors demonstrated that the proposed structure helps to achieve multiplexing gain and improve matrix rank. Passive repeaters are also used to improve propagation environments in MIMO systems. The principle behind using passive repeaters is to create multipath propagation to increase the effectiveness of the MIMO system or to deliver signals to blind spots that may exist [7–10].

The back-to-back antenna system is very known in antenna technology and is used as a passive repeater in point-to-point communications [11]. In this paper, we propose to use the back-to-back antenna system as a passive repeater in LOS environments in MIMO communications to increase

Content from this work may be used under the terms of the Creative Commons Attribution 3.0 licence. Any further distribution of this work must maintain attribution to the author(s) and the title of the work, journal citation and DOI.
Published under licence by IOP Publishing Ltd

capacity and enhance MIMO performance. The proposed passive repeater consists of two parts connected through a microwave connection line. Each part comprises an antenna and a parabolic reflector (Figure1). The first part with narrow beam is directed to the base station with LOS communication, while the other part is directed toward users. The repeater is completely passive and can be widely used in MIMO communications to provide multipath propagation and improve MIMO performance. This improvement can be verified by obtaining a well-conditioned channel matrix.

The validation of the proposed approach is made by analytical analysis and numerical computation using two metrics: rank of the channel matrix and condition number. Simulation results show that implementing the proposed passive repeater helps to get high rank of the channel matrix and lower value for condition number.

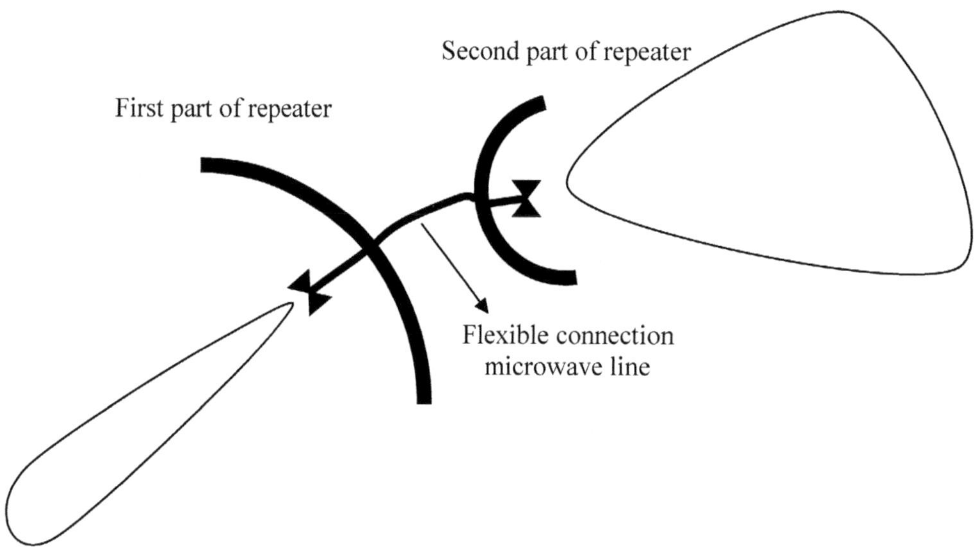

Figure 1. Passive repeater as back-to-back antenna system.

2. Rank matrix and condition number

Multipath channels are capable of supporting the transmission of multiple independent data streams and that increases capacity without increasing the bandwidth. The capacity of MIMO system can be expressed as [1]:

$$C = \sum_{i=1}^{r} log_2(1 + \frac{\rho}{N_t}\lambda_i) \quad (1)$$

Where r is the rank of the channel matrix. ρ is the signal to noise ratio at the receiver. N_t is the number of antennas in the transmitter and λ_i is the ith eigenvalue. Equation (1) shows that it is possible to transmit up to r data streams over a MIMO channel, and that indicates the importance of having a large channel rank. Channel rank is a quantitative way to characterize the scattering richness of a MIMO channel and is an indicator of how many data streams can be spatially multiplexed on the channel. From (1), it is also possible to notice that the capacity of MIMO systems is maximized when all eigenvalues are equal, i.e. the channel matrix is orthogonal. This also demonstrates the importance of the condition number ($k(H) = \sqrt{\frac{\lambda_{max}}{\lambda_{min}}} \geq 1$) as a metric of the quality of the channel matrix.

3. Application of the proposed passive repeater in LOS environments

The high gain provided by the parabolic reflectors in addition to the directivity pattern allow to cover users with acceptable strength. The flexible connector between the two parts of the repeater allows to direct each part mechanically with no need for a phase controller. For these features, the spreading of this repeater in the propagation environment helps to provide multipath propagation and increase the aperture size of the base station. As a result, the number of spatial channels is increased and the orthogonality of the channel matrix is enhanced. Therefore, we suggest using this repeater in a LOS environment where there is no scattering to improve MIMO performance. A simulation is performed to validate the proposed method.

4. Simulation and Discussion

To investigate the impact of adding the proposed repeater to the propagation environment on the performance of MIMO systems, the condition number and matrix rank are examined and compared. The considered simulation environment is shown in Figure 2.

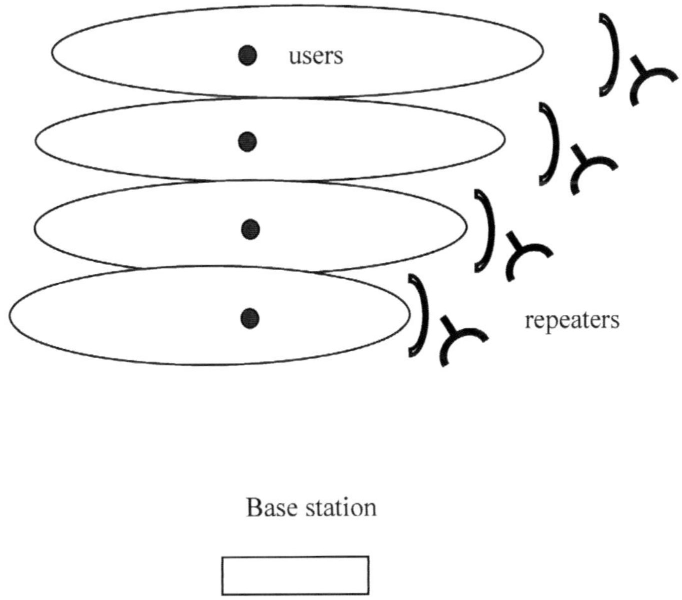

Figure 2. The simulation environment.

The idea is to use directive repeaters to cover the target area to ensure acceptable signal strength delivered to the users. The distance between repeaters should be sufficient for the base station to select desired repeaters separately using field signature corresponding to user's location, which can be calculated as:

$$A_n = \frac{U_n^*(x_0,y_0)}{|U_n(x_0,y_0)|} \quad (2)$$

Where (x_0, y_0) is the user's location coordinates, $U_n(x_0, y_0)$ is the field distribution at point (x_0, y_0) resulting from all repeaters in the case when only the nth antenna of the base station transmits a signal.

For the numerical simulation, the simulated area is chosen to be 1200 m^2 with $N_x = 256$ and $N_y = 256$ cells in x and y directions respectively. Number of repeaters is M = 4. Number of antennas in the base station is N = 16. For the users' arrangement shown in Figure 2, base station cannot distinguish users, so one channel can be established as shown in Figure 3. Proposed repeaters can be used to create spatial channels as shown in Figure 4*a*. the selection of repeaters according to user's location coordinates is shown in Figure 4*b*. It can be seen that the base station selects the 2nd repeater, because it covers user's location with strongest signal.

To demonstrate the improvement on matrix rank and condition number, Table 1 includes a comparison between the values of the parameters before and after adding repeaters. It can be seen that without repeaters the condition number goes to infinity and the matrix is rank deficient. However, the channel matrix (resulted from the direct path and that through the repeaters) became more orthogonal with full rank and improved condition number after adding repeaters. In Table 1, eigenvalues are normalized to the largest one.

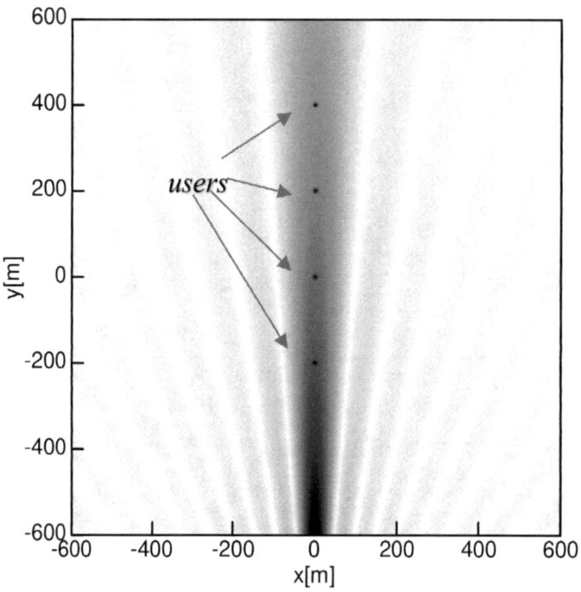

Figure 3. LOS communication with no use of repeaters.

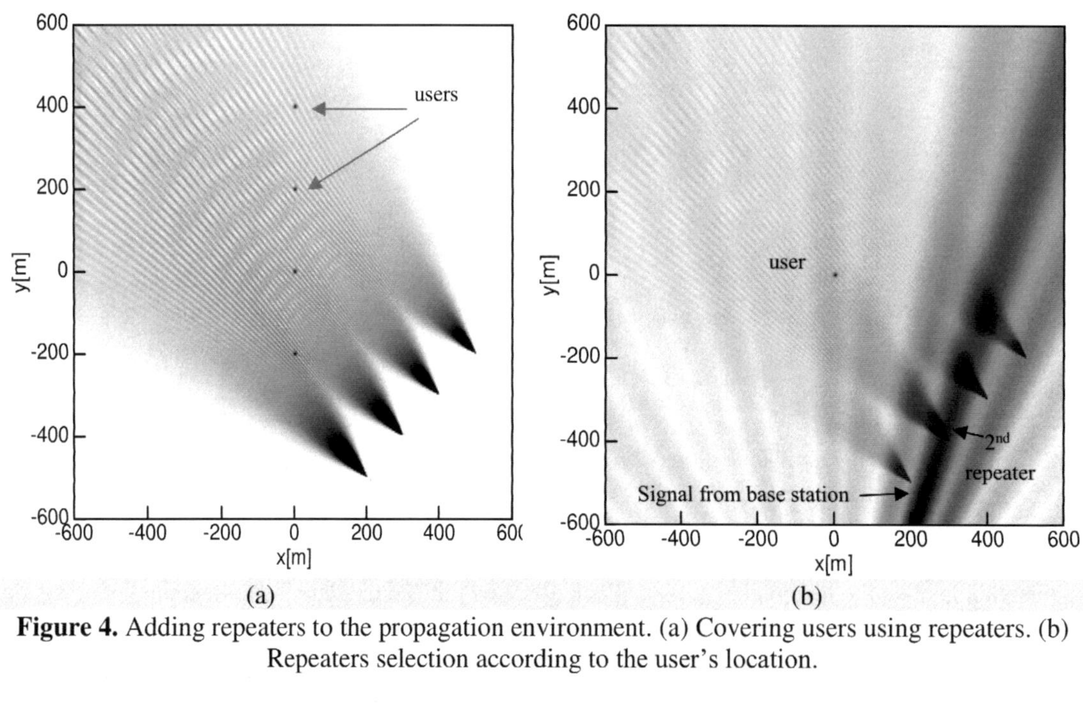

Figure 4. Adding repeaters to the propagation environment. (a) Covering users using repeaters. (b) Repeaters selection according to the user's location.

Table 1. Comparison of channel matrix parameters before and after adding repeaters to the communication environment.

	rank	λ_0	λ_1	λ_2	λ_3	K(H)
Without repeaters	The matrix is rank deficient	1	$2.621 * 10^{-4}$	≈ 0	≈ 0	$\rightarrow \infty$
With repeaters	Full rank	1	0.242	0.017	0.037	7.6

5. Conclusion

In this paper, we demonstrated the ability to improve the quality of the channel matrix using the back-to-back antenna system as a passive repeater. We have shown that the use of the proposed repeater helps to enrich the propagation environment by adding multipaths. Results show that the proposed method improves the channel matrix by getting higher channel rank and lower condition number, which means better channel matrix quality and higher capacity.

Acknowledgments

The work was supported by government task of Ministry of science and Education of Russia, project N 0721-2020-0038.

References

[1] Hampton J R 2014 *Introduction to MIMO Communications* (New York: Cambridge University Press)

[2] Larsson E G, Edfors O, Tufvesson F and Marzetta T L 2014 *IEEE Communications Magazine* **52** (2) 186–195

[3] Castaneda Garcia M H, Iwanow M and Stirling-Gallacher R A 2018 *IEEE 88th Veh. Technol. Conf.* 1–5

[4] Hirayama H, Matsui G, Kikuma N and Sakakibara K 2011 *IEICE Trans. Commun.* **E94-B** (5) 1280–1283

[5] Pal P K and Sherratt R S 2018 *ETRI Journal* **40** (2) 197– 206

[6] Ozdogan O, Bjornson E and Larsson E G 2020 *Proc. IEEE Int. Conf. Acoust. Speech Signal Process ICASSP* 9160–9164

[7] Wang L, Qu S, Li G, Chen Q, Yuan Q and Sawaya K 2011 *IEEE Antennas and Wireless Propagation Letters* **10** 752–755

[8] Honma, N, Takahashi Y and Tsunekawa, Y 2014 *Proc. IEEE Antennas and Propagation Society International Symposium (APSURSI)* 1129–1130

[9] Wang L, Qu S, Jianfeng L, Qiang C, Qiaowei Y and Sawaya K 2011 *IEEE Antennas and Wireless Propagation Letters* **10** 752–755

[10] Honma N, Takahashi Y and Tsunekawa Y 2014 *Proc. of IEEE Asia-Pacific Microwave Conference (APMC)* 504–506

[11] Norton M 1962 *IRE Transactions on Communications Systems* **10**(3) 304–311

Grouping of particles in a wideband ultrasonic field

D Y Sukhanov, S N Rosliakov

National Research Tomsk State University, Tomsk, Russia

E-mail: sdy@mail.tsu.ru

Abstract. In this paper, we propose to expand the capabilities of wideband levitation and show the possibility of forming a structure of a complex shape based on focusing a wideband field in a given area. Focusing the field of planar radiating arrays makes it possible to form a region of stable levitation in a plane parallel to the arrays. The counter radiation of the two arrays creates a standing wave, at the nodes of which the particles are grouped. The use of a wideband signal makes it possible to create many stable nodes of standing waves in specified areas, and to realize the required shape of the levitating object. Simultaneous monitoring of multiple particles in a wideband ultrasonic field may become a new direction in the development of methods of acoustic trapping and control of particles, as well as technologies of acoustic tweezers.

1. Introduction

Acoustic levitation technologies can find application in medicine, mixing of chemicals, prototyping, and sorting of particles. At the present time, scientists from different countries offer their versions of the implementation of particle levitation by an acoustic field [1–6]. The proposed solutions are based on the interaction of two or four oppositely directed narrow-band ultrasonic arrays located along the sides of the cube. In [7–9], a setup was considered in which emitters are located on segments of a sphere. This configuration allows the particles to be levitated, controlled in a three-dimensional plane, and sorted. However, such a configuration and, in general, the use of a narrow-band field leads to the fact that the region of particle levitation has a periodic structure and is related to the frequency of the signal used. In [10], an implementation of an installation based on the solution described above is proposed, which trappings, transfers and positions electrical components on a printed circuit board, and then they are soldered with a laser. Also, earlier we carried out experiments on narrow-band levitation of small-sized particles and particles larger than the wavelength [11]. The disadvantage of such systems is the use of a monochromatic signal, which leads to the fact that the particles levitate in a periodic grid. If the required distribution is formed, an auxiliary array is used to "knock out particles" or remove them mechanically.

We have proposed a method for levitation of particles in air under the action of wideband ultrasonic waves emitted by opposite focused arrays [12]. The linear frequency modulated (LFM or chirp) signal has been shown to be suitable for acoustic levitation. The possibility of localizing particles in a given area using two and four arrays of wideband ultrasonic emitters has been shown experimentally. The use of four opposing array along the sides of the cube made it possible to control the levitating particles in three-dimensional space. The movement of particles in the plane of levitation is carried out by manipulating the time delay of the radiation of the signal of the opposing arrays.

Content from this work may be used under the terms of the Creative Commons Attribution 3.0 licence. Any further distribution of this work must maintain attribution to the author(s) and the title of the work, journal citation and DOI.
Published under licence by IOP Publishing Ltd

This paper proposes the development of the method of levitation in a wideband field with the possibility of simultaneous controlled grouping of particles in several areas.

2. Formulation of the problem

It is proposed to expand the capabilities of wideband levitation and to show the fundamental possibility of forming a complex region of particle trapping. Consider two planar counter-directed arrays Figure 1. It is assumed that on each array with dimensions of 200 mm by 200 mm there are emitters in a filled array with a step of 5 mm. Each emitter is controlled separately and is capable of emitting an arbitrary wideband signal in the frequency range from 20 to 40 kHz. The distance between the arrays is 200 mm.

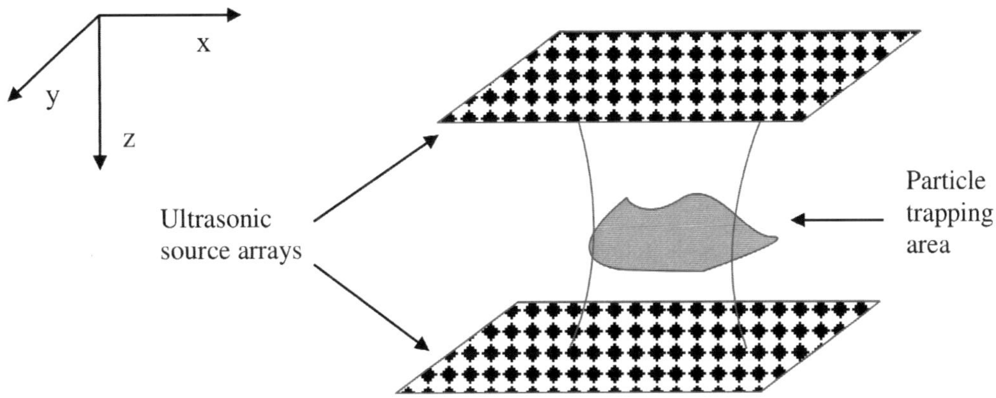

Figure 1. Placement of source arrays for levitation of particles in wideband ultrasonic field.

Let's set the focus area as a scalar two-dimensional function $M(x, y)$. This function is equal to one at the points where it is required to concentrate the particles, and is equal to zero at the points where there should be no particles. It is necessary to create such signals in the emitters so that the ultrasonic fields are focused at the nonzero points of the function $M(x, y)$. In this case, the signals of converging waves should be in the form of signals with LFM. Then, as it was shown earlier [12], the grouping of particles in a given area will be provided. Consider an LFM signal of the form:

$$S(t) = A\sin\left(\omega_1 + \frac{[\omega_2 - \omega_1]}{2T} \cdot \left(t - \left\lfloor \frac{t}{T} \right\rfloor \cdot T\right)\right), \tag{1}$$

where $\lfloor ... \rfloor$ – rounding down; T is the repetition period of the chirp signal; ω_1 – the lower cyclic frequency of the chirp signal; ω_2 – the upper cyclic frequency of the chirp signal; A – signal amplitude.

It is necessary to calculate the signals in the sources of one array, which are necessary for focusing the field according to the function $M(x, y)$, so that at nonzero points of this function in the plane $z = 0$ (in the middle between the radiating arrays), signals are realized (1). As in [12], we consider the case when signals of opposite signs are fed to opposite arrays in order to provide a node of standing waves in the plane in the middle between the arrays. As you know, particles will be grouped into nodes of standing waves. Let the coordinates of the radiators of the upper array be given by the vector: $\mathbf{R}_n = (X_n, Y_n, Z_n)$, and the lower array $\tilde{\mathbf{R}}_n = (X_n, Y_n, -Z_n)$. Then the signal in the n-th emitters of the array will be specified as the sum of the backpropagation fields from the focusing area:

$$W_n(t) = \iint_{XY} M(x,y) S\left(t + \frac{|\mathbf{R}_n - \mathbf{r}|}{c}\right) dxdy, \tag{2}$$

where c – speed of sound; $W_n(t)$ – signal of n-th element of the top array; $\mathbf{r} = (x, y, 0)$ – point on the focusing area.

The source field for two arrays is calculated according to the solution in retarded potentials. We assume that the sources in the arrays are non-directional. The field in the focusing area between the upper and lower arrays can be written as:

$$U(x,y,z) \sim \sum_n \frac{W_n\left(t - \frac{1}{c}|\mathbf{R}_n - \mathbf{r}'|\right)}{|\mathbf{R}_n - \mathbf{r}'|} - \frac{W_n\left(t - \frac{1}{c}|\tilde{\mathbf{R}}_n - \mathbf{r}'|\right)}{|\tilde{\mathbf{R}}_n - \mathbf{r}'|}, \tag{3}$$

where $\mathbf{r}' = (x, y, z)$ – point of the field observation.

3. Numerical Simulations

Consider, as a special case, focusing a field at two different points $r_1 = (25\text{ mm}, 0, 0)$ and $r_2 = (-25\text{ mm}, 0, 0)$. Consider the frequency range of a chirp signal: f_1= 20 kHz, f_2= 40 kHz with period T = 1 ms. According to formula (2), the signals in the emitters were calculated, and then from formula (3) the field of acoustic waves in the considered volume was calculated. Figure 2 shows nodes and antinodes of standing waves at a time of 0.5 ms. The nodes of the standing waves are distinctly expressed. Darker color on Figure 2 corresponds to higher magnitude. As shown by experimental studies [12], levitating particles should be grouped into a central stable node of standing waves.

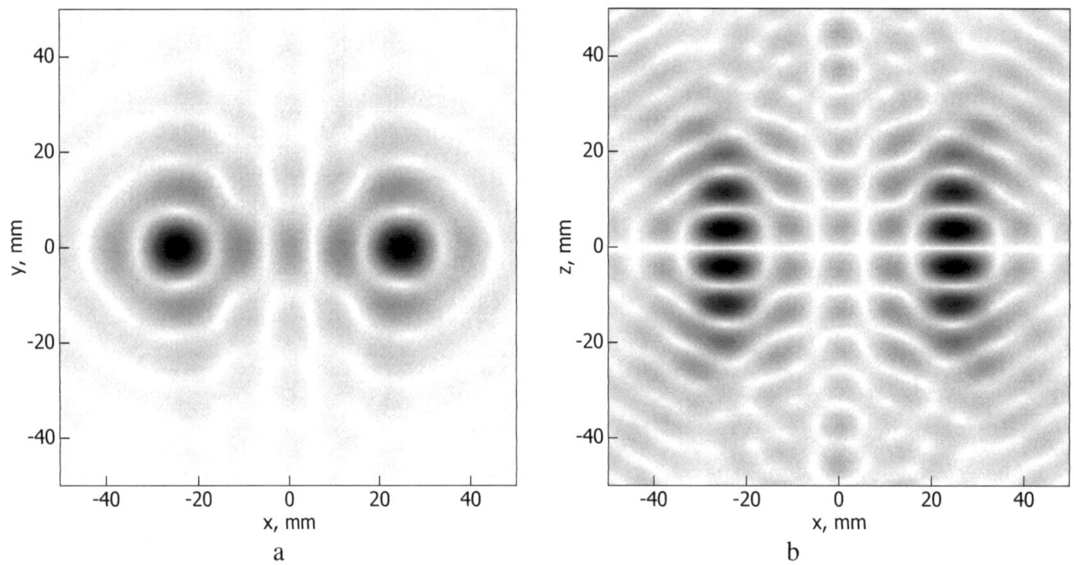

Figure 2. Acoustic pressure magnitude of wave field formed by planar counter-directed arrays (a – section by plane z=0; b – section by plane y=0).

With the passage of time, the nodes of the standing waves shift towards the focusing points, which should ensure the gradual movement of particles to the given nodes of the standing waves in the z = 0 plane. Figure 3 shows the visualization of the displacement of the standing waves trapping nodes in time for 0.6 ms. It can be seen that the farther from the focusing points, the greater the displacement speed of the nodes of the standing waves. However, at these points the field of acoustic waves is not focused and has a small amplitude; therefore, the probability of particle trapping in these regions is small. Near the focusing points, the speed of the nodes is low and the field amplitudes are high enough. Hence, we can conclude that the particles will be reliably trapped near the focus points and gradually move to the focus points.

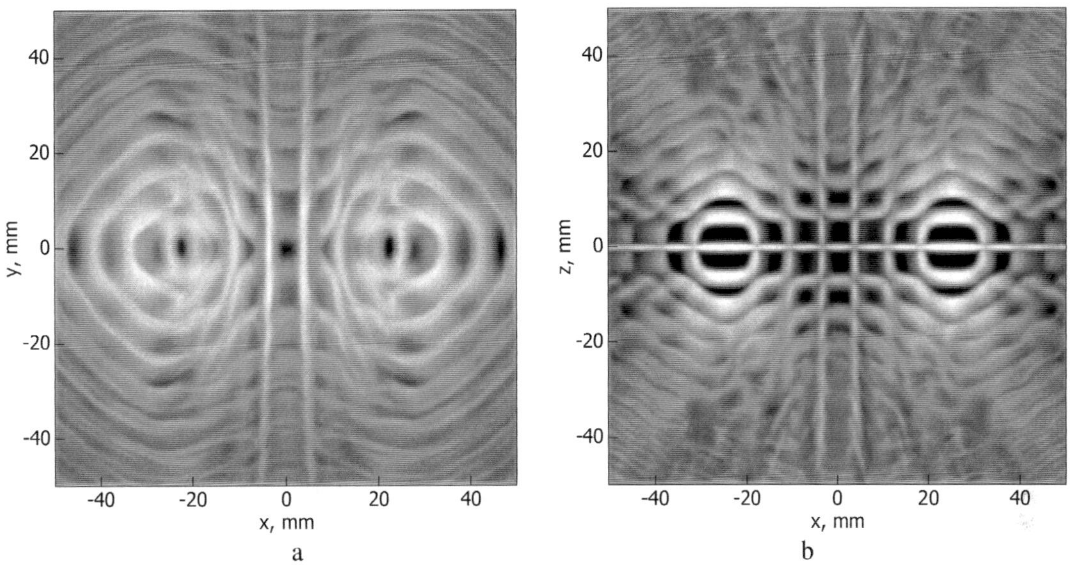

Figure 3. Visualization of the displacement of the nodes of standing waves with time - a smooth increase in the amplitude corresponds to an increase in the considered moment in time (a – section by plane z=0; b – section by plane y=0).

On the vector diagram of the displacement of the nodes of the standing waves (Figure 4), one can see that the particles trapped by the wideband focused ultrasonic field will have to drift to the given focal point.

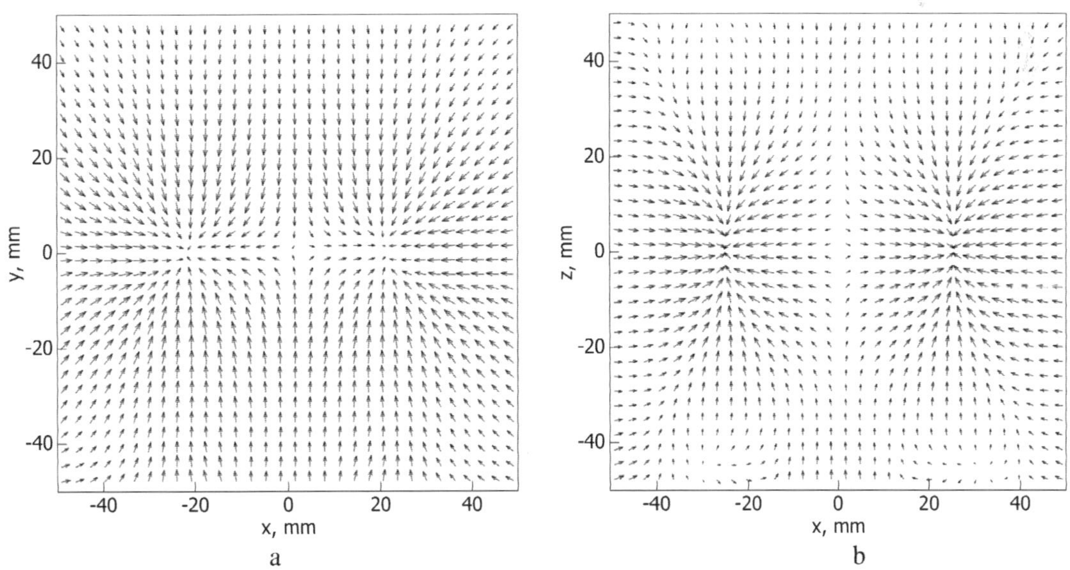

Figure 4. Vector visualization of the direction of particle displacement when entering a focused field of wideband waves (a – section by plane z=0; b – section by plane y=0).

By numerically simulating wideband acoustic fields, we have demonstrated the possibility of creating such a field distribution that will ensure the trapping and displacement of levitating particles to specified focus points. In general, there can be more than two focus points, but we have considered only two points for visual demonstration.

4. Conclusion

A method for controlling a wideband ultrasonic field is proposed for the formation of such trajectories of motion of the nodes of standing waves in order to ensure the drift of levitating particles to a given set of focal points. Numerical simulation of the fields of wideband arrays of emitters for two focusing points is carried out. It is shown that the displacement vectors of the nodes of the standing waves provide the movement of the levitating particles to the focal points.

Acknowledgments

The study was carried out with the financial support of the Russian Foundation for Basic Research within the framework of a scientific project No 20-32-90066\20 от 17.08.2020

References

[1] Chen X, Lam K H, Chen R, Chen Z, Qian X, Zhang J, Yu P, Zhou Q 2019 *Appl. Phys. Lett.* **114** 054103 doi: 10.1063/1.5066615
[2] Ochiai Y, Hoshi T, Rekimoto J 2014 *PLOS ONE.* **9**(7) e102525 doi: 10.1371/journal.pone.0097590.g001
[3] Marzo A, Seah S A, Drinkwater B W, Sahoo D R, Long B, Subramanian S 2015 *Nature Communications* **6** 8661 doi: 10.1038/ncomms9661
[4] Kozuka T, Yasui K, Tuziuti T, Towata A, Iida Y 2007 *Japanese Journal of Applied Physics.* **46**(7B) 4948–4950; doi: 10.1143/JJAP.46.4948
[5] Fushimia T, Hill T L, Marzo A, Drinkwater B W 2018 *Appl. Phys. Lett.* **113** 034102 doi: doi:10.1063/1.5034116
[6] Andrade M A B, Skotis G D, Ritchie S, Cumming D R S, Riehle M O, Bernassau A L 2016 *IEEE Transactions on Ultrasonics Ferroelectrics and Frequency Control* **PP(99)** 1593–1600 doi: 10.1109/TUFFC.2016.2608759
[7] Marzo A, Ghobrial A, Cox L, Caleap M, Croxford A, Drinkwater B W 2017 *Appl. Phys. Lett.* **110** 01410 doi: 10.1063/1.4972407
[8] Marzo A, Barnes A, Drinkwater B W 2017 *Review of Scientific Instruments* **88** 085105 doi: 10.1063/1.4989995
[9] Liu P, Ming D, Tan C S, Lin B 2019 *Applied Acoustics* **155** 216–221 doi:10.1016/j.apacoust.2019.06.008
[10] Putkis O 2017 Patent: Contactless manipulation apparatus, assembly method and 3d printing WO 2017/068435
[11] Sukhanov D Y, Roslyakov S N, Emel'yanov F S 2020 *Acoustical Physics* **66(2)** 137–144 doi: 10.1134/S1063771020020104
[12] Sukhanov D, Rosliakov S 2021 *Applied Acoustics* **178** 108004 doi: 10.1016/j.apacoust.2021.108004

Two-dimensional THz reflectometry of a periodic structure obtained by additive technology

A V Badin, A I Berdyugin, V D Moskalenko, K V Simonova and R P Gursky

National Research Tomsk State University, Tomsk, Russia

E-mail: thzlab@mail.ru

Abstract. This paper considers the development and application of a system of reflectometry for the analysis of the homogeneity of structures manufactured by additive technologies. A system of reflectometry based on a backward wave oscillator, a two-dimensional object positioning system and an optoacoustic detector (Goley cell) is described. The results of reflectometry of the hexagonal periodic structure of cells based on acrylonitrile butadiene styrene at a wavelength of 343 microns are presented.

1. Introduction

Over the past decade, significant progress in the development and creation of structures using additive technologies that effectively interact with electromagnetic radiation of the extreme high-frequency and terahertz (THz) ranges [1,2] has been made. First of all, this to manufacture homogeneous composite materials and structures that can effectively interact with electromagnetic radiation [3] uniformly throughout the entire plane, as well as having such advantages as lightness, tensile strength, elasticity to deformation, thermal stability and the presence of functional features (for example, electrical conductivity) [4–6]. The 3D printing method makes it easy to reproduce construction with high accuracy while observing all stages of the manufacturing process. Modern additive technologies make possible manufacturing of various electronic devices with a complex topology: photodetector arrays, electrically small antennas [7], hemispherical solar cells [8], toroidal inductors [9], load cells [10]. When creating structures, one of the main criteria for their quality is the homogeneity of their structure. To control the homogeneity of structures obtained by 3D printing, an optical method (cameras in combination with computer image analysis) [11] and industrial X-ray tomography [12] are traditionally used. However, the described methods have the following disadvantages: limited information about the tested object (optical) and unsafe-ionizing diagnostic methods (X-ray).

Currently, there are various tools for non-destructive testing and evaluation of the uniformity of the material [13]. Among them, several methods are distinguished based on the registration of the response of electromagnetic radiation transmitted [14] and reflected from the material (reflectometry [15]). Spatial resolution and depth of penetration into the sample, depending more on the wavelength of the radiation used, are important parameters in the diagnosis of materials homogeneity. The THz radiation of the frequency range is used to assess the quality of periodic structures manufactured using additive technologies because: the radiation is non-ionizing [16], has a deep penetrating ability, and also allows you to determine the properties of the constituent substance by frequency response. In the time-domain spectroscopy methods common today, the THz radiation reflected from the material carries a greater amount of information about its internal structure and

Content from this work may be used under the terms of the Creative Commons Attribution 3.0 licence. Any further distribution of this work must maintain attribution to the author(s) and the title of the work, journal citation and DOI.
Published under licence by IOP Publishing Ltd

spectral features [17, 18]. However, diffraction processes and multiple scattering on the internal structure make it difficult to process the data object, which does not allow making clear conclusions about the structure of the object [19]. This emphasizes the need to use sources of continuous monochromatic radiation as a diagnostic setup for samples of materials manufactured using additive technology.

2. Quasi-optical system

To measure the two-dimensional distribution of the reflection coefficient, a quasi-optical system based on a source of continuous THz radiation - a back wave oscillator (BWO), a detector (Golay cell), a mechanical amplitude modulator, a set of film attenuators, a wire grid beam divider was made. To exclude the re-reflection of radiation towards the detector an cone absorber was installed in front of the parabolic mirror (Figure 1).

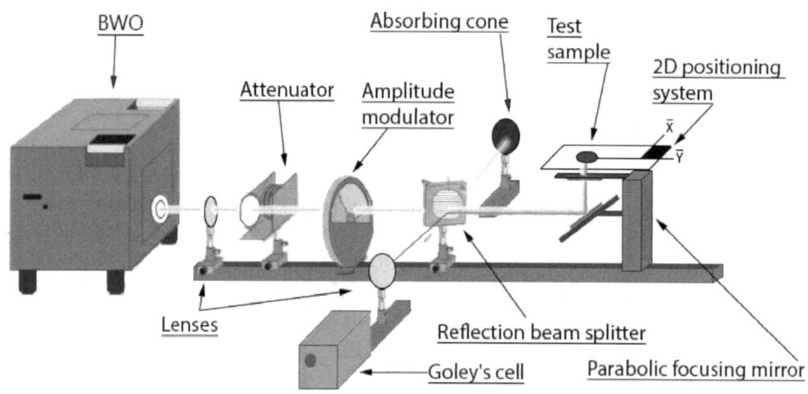

Figure 1. Scheme of quasi-optical THz reflectometer.

Using a personal computer (PC) via a digital-to-analog converter (DAC) of the STD-21 spectrometer and a 12-bit I/O module E-154 by L-card communication with the elements of the THz reflectometer is carried out. It allows to control the positioning system and register an amplitude-modeled detector signal with an accuracy of ± 2.4 mV. The source of monochromatic BWO-THz radiation is configured by controlling the high-voltage of the BWO cathode from a controlled voltage source (PS 10V in Figure 2) by supplying a DC voltage from the DAC in the range of ± 10V.

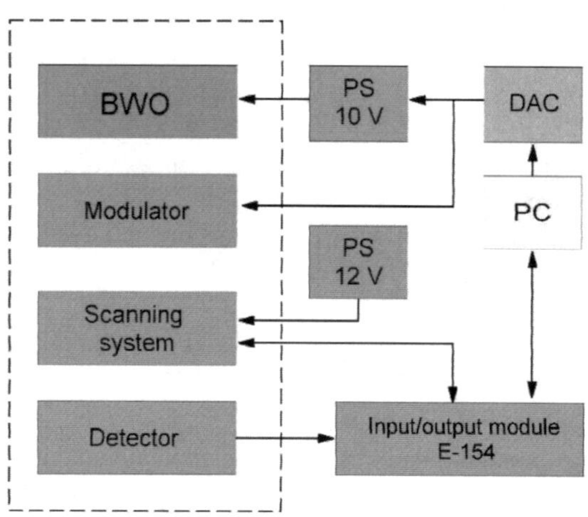

Figure 2. Functional diagram of the installation of a THz frequency band reflectometer.

Based on the software for recording the intensity of electromagnetic radiation [20, 21], a program in the LabVIEW programming environment for controlling the THz reflectometer was developed. By recording the amplitude of the reference signal reflected from the copper plate $P_{reference}$ and the reflected radiation signal at each point of a given area of the surface of the test sample $P_{sample}(x, y)$, a two-dimensional distribution of the reflection coefficient $R(x, y)$ was calculated according to the expression:

$$R(x,y) = P_{sample}(x,y) / P_{\text{reference}} \qquad (1)$$

3. Results and discussions

For this work, a three-dimensional model (Figure 3) of a periodic structure with hexagonal through holes (cells), made in the form of a cylinder with diameters of 48 mm, was created as a test object.

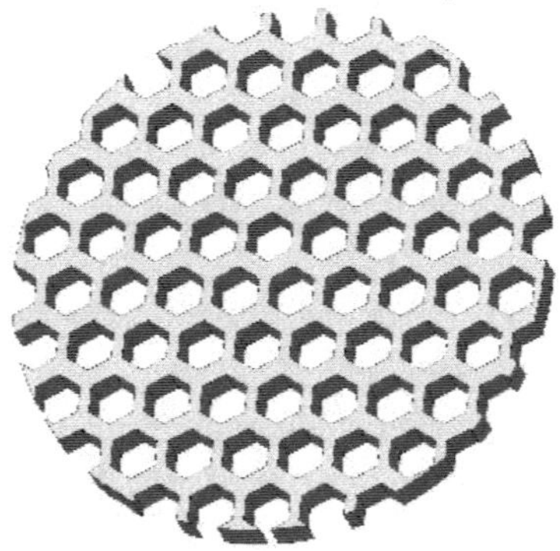

Figure 3. Three-dimensional model of a periodic structure based on hexagonal cells.

The cell repetition period was 4 mm, the thickness of the structure was 3.2 mm, the width of the face was 3 mm, the thickness of the adjacent wall was 1 mm, the mass was 1.924 g. By fused deposition modeling of acrylonitrile butadiene styrene modified with nanofibers at a temperature and nozzle diameter of 230 ° C and 300 μm, respectively, the manufacture of experimental samples was carried out. The temperature of the 3D printer bed was 110 °C.

Figure 4 shows the results of measuring the two-dimensional spatial distribution of the reflection coefficient using a THz reflectometer.

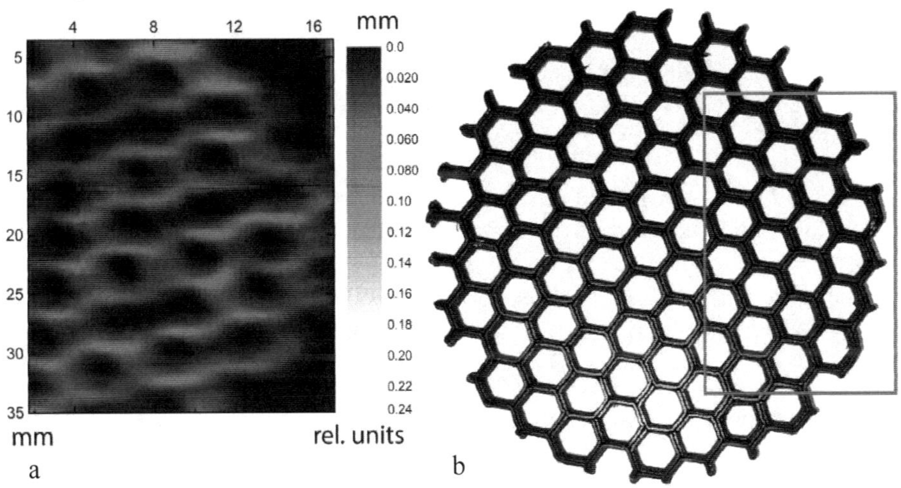

Figure 4. Two-dimensional distribution of the reflection coefficient of a periodic structure based on hexagonal cells at a frequency of 874 GHz (a) and an optical image of sample (b).

Analysis of the two-dimensional distribution of the reflection coefficient shows that a periodic hexagonal structure based on an electrically conductive polymer with a cell face size of 3 mm is well detailed when using an electromagnetic radiation source with a wavelength of 343 μm. The maximum value of the reflection coefficient was 0.09 rel. units at the cell boundary. Horizontal fuzzy cell boundaries in the sample may be associated with a low resolution of the positioning system along the *X*-axis (1.6 px/mm) compared to *Y* (7.6 px/mm) in combination with the effect of diffraction and multiple scattering on the internal structure of the object. It makes visually difficult to interpret the results and evaluate the geometry of the structure of the test object.

To illustrate the structure of the sample, two-dimensional graphs were constructed (Figures 5*a,b*) of the reflection coefficient in a given profile (red marker), which can be observed on visualization (Figures 5*c,d*). This graph shows how the reflection coefficient changes along the hexagonal holes. According to the graph, it is possible to obtain the distance between the midpoints of the walls of the material, which corresponds to the distance between the maxima of the reflection coefficient. Due to the different number of points per unit length, it became necessary to approximate the data, the results of which are shown in Figures 5*b,d*. The approximation consisted in equalizing the number of pixels per unit length by calibrating the positioning system.

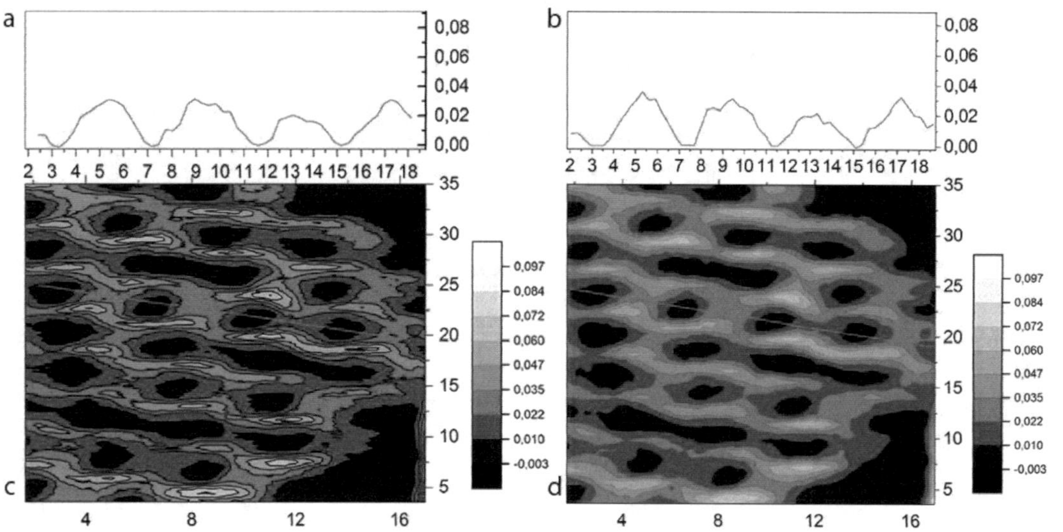

Figure 5. Radiation intensity distribution and a graph of the reflection coefficient profile of a periodic structure based on hexagonal cells at a frequency of 874 GHz: a, c) without using approximation; b, d) using approximation.

The average distance between the middle of the walls without using approximation was 4.1 mm, and using approximation - 4 mm. The approximation makes it possible to eliminate bursts of the reflection coefficient caused by diffraction and internal scattering. The influence of the described effect can be observed on the graphs (Figure 5*a, b*) in the vicinity of 9 mm.

4. Conclusion

THz reflectometer using a source of continuous monochromatic electromagnetic radiation based on a back wave oscillator is presented. Quasi-optical schemes of the reflectometer were developed. To register the two-dimensional distribution of the reflected radiation intensity in the THz range software was written. Samples of the periodic structure based on hexagonal cells using 3D printing technology were obtained. The geometric shape and repetition period of the hexagonal cells of the test sample were evaluated based on the distribution of reflected THz radiation received by the reflectometer.

The system of quasi-optical THz reflectometry has shown its promise for visualization and diagnostics of the internal structures of objects obtained by additive technologies.

Acknowledgments
The reported study was funded by grant of the president of the Russian Federation for state support of young Russian scientists MK-1709.2020.8

References
[1] Tubío C R, Nóvoa J A, Martín J, Guitián F, Salgueiro J R, Gil A 2019 *Ceramics International* **45**(5) 6223–6227 doi:10.1016/j.ceramint.2018.12.100
[2] D'Auria M, Otter W J, Hazell J, Gillatt B T, Long-Collins C, Ridler N M, Lucyszyn S 2015 *IEEE Transactions on Components, Packaging and Manufacturing Technology* **5**(9) 1339–1349 doi:10.1109/TCPMT.2015.2462130
[3] Liang M, Xin H 2015 *Proceeding of the Asia-Pacific Microw. Conf. (APMC)* **2** 2–3 doi:10.1109/APMC.2015.7413206
[4] Kim H, Oh E, Hahn HT, Lee K-H 2015 *Compos A Appl Sci Manuf* **71** 72–83 doi:10.1016/j.compositesa.2014.12.014
[5] Saba N, Jawaid M 2018 *J Ind Eng Chem* **67** 1–11 doi:10.1016/j.jiec.2018.06.018
[6] Mittal G, Rhee KY, Mišković-Stanković V, Hui D 2017 *Compos B Eng* **138** 122–139 doi:10.1016/j.compositesb.2017.11.028
[7] Kim O S 2013 *IEEE Antennas and Propagation Society International Symposium (APSURSI)* 776–777 doi:10.1109/APS.2013.6711047
[8] Sim K, Chen S, Li Z, Rao Z, Liu J, Lu Y, Yu C 2019 *Nature Electronics* **2**(10) 471–479 doi:10.1038/s41928-019-0304-4
[9] Yan Y, Moss J, Ngo K D, Mei Y, Lu G Q 2017 *IEEE Transactions on Industry Applications* **53**(6) 5709–5714 doi:10.1109/TIA.2017.2729504
[10] Bogue R 2016 *Sens. Rev.* **36**(4) 333–338 doi:10.1108/SR-07-2016-0114
[11] Nuchitprasitchai S, Roggemann M C, Pearce J M 2017 *Journal of Manufacturing and Materials Processing* **1**(1) 2 doi:10.3390/jmmp1010002
[12] Khosravani M R, Reinicke T 2020 *Journal of Nondestructive Evaluation* **39**(4) 1–17 doi:10.1007/s10921-020-00721-1
[13] Dwivedi S K, Vishwakarma M, Soni A 2018 *Materials Today: Proceedings* **5**(2) 3690–3698 doi:10.1016/j.matpr.2017.11.620
[14] Berdyugin A I, Badin A V 2021 *Journal of Physics: Conference Series* **1862** 012030 doi:10.1088/1742-6596/1862/1/012030
[15] Chen C C, Lee D J, Pollock T, Whitaker J F 2010 *Optics express* **18**(4) 3477–3486 doi:10.1364/OE.18.003477
[16] Vohra N, Chavez T, Troncoso J R, Rajaram N, Wu J, Coan P N, Jackson T A, Bailey K, El-Shenawee M 2021 *Journal of Medical Imaging* **8**(2) 023504 doi:10.1117/1.JMI.8.2.023504
[17] Alekseev S V, Andreev Y M, Losev V F, Lubenko D M 2020 *Russian Physics Journal* **62**(11) 2151–2155 doi:10.1007/s11182-020-01960-5
[18] Hangyo M 2015 *Japanese Journal of Applied Physics* **54**(12) 120101 doi: 10.7567/JJAP.54.120101
[19] Bespalov V, Gorodetsky A, Grachyov Y, Kozlov S A, Novosyolov Y 2011 *Scientific and Technical Bulletin of Information Technologies, Mechanics and Optics* **1**(71) 19–23
[20] Zhakupov S N, Badin A V, Berdugin A I 2019 *Proceeding of the International Siberian Conference on Control and Communications (SIBCON)* 1–4 doi:10.1109/SIBCON.2019.8729563
[21] Badin A V, Dorozhkin K V, Suslyaev V I, Berdyugin A I, Vigovskiy V Y 2017 *Proceeding of the 23rd International Symposium on Atmospheric and Ocean Optics: Atmospheric Physics* **10466** 1046625

Investigation of the peculiarities of lattice structures forming in films magnetic fluid for various intensities of the orienting magnetic field

V Yurchenko[1], D Zyatkov[1,2] and V Cherepanov[1]

[1]National Research Tomsk State University Tomsk, Russia
[2]Research Institute of Semiconductor Devices, Tomsk, Russia

E-mail: yur_med@mail.ru

Abstract. The processes of ordering and structuring of particles in a magnetic fluid (MF) arise in three cases: interactions of magnetic particles, internal forces of liquid and an external magnetic field of constant or variable magnitude. Processes of magnetic particles ordering in a magnetic fluid by interaction an external field are considered, and threshold of occurrence of a periodic lattice from particles with different size is established.

1. Introduction

To date, the bibliography on magnetic fluids (MF) has about seven thousand articles in journals and periodicals and about three thousand patents for various devices. By the early 1980s, the one-particle Langevin approximation could not explain clustering in MFs and their new manifestations.Research on multi-particle clusters, chains, lattices is most interesting, and a large number of publications.However, often the results of calculations and experiments do not coincide.There is an intensive search for new solutions to this problem. One of these directions is the study of the behavior of MW depending on the characteristics of the external magnetic field and the various parameters of the particles used. The objective of the study is to determine the conditions under which clusters are distributed periodically with a given period.

2. Interaction aspects

Three main aspects determine the characteristics of the MF, namely the external magnetic field, Brownian motion, and interparticle interactions. Magnetic particles interact with each other through a carrier liquid (hydrodynamic interactions), through microscopic magnetic fields (magnetic dipole interactions), through protective shells (steric interactions) and due to van der Waals forces.

There are very interesting to study magnetic viscosity under the influence of a magnetic field of various magnitudes and configuration of particles carried out at Perm University and Ural Institute.
Very interesting are the works of the Institute of Continuous Media Mechanics of the Ural Branch of Russian Academy of Science and Perm State University on the study of the magnetic viscosity or the lag of the magnetization of the MF from the magnetic field of various magnitudes and configurations of particles [1–5]. The experiments confirmed the presence in the magnetic fluid of a large number of aggregates with a characteristic size of 50–70 nm, these sizes are approximately three to four times larger than the average diameter of individual particles. The contribution of aggregates to the magnetic susceptibility of ferrocolloids increases exponentially with a decrease in temperature and becomes

Content from this work may be used under the terms of the Creative Commons Attribution 3.0 licence. Any further distribution of this work must maintain attribution to the author(s) and the title of the work, journal citation and DOI.
Published under licence by IOP Publishing Ltd

predominant at temperatures below 300–320 K. This conclusion is in good agreement with the data obtained from diffusion experiments and the data obtained in experiments with crossed magnetic fields. The amount of aggregated particles is highly dependent on the magnetic fluid production technology and can vary widely limits. In this sense, the results may not always be transferred to other types of magnetic fluid.

3. Analysis results

The processes of ordering in magnetic fluid arise in three cases: due to self-organization of particles under the action of internal forces of the fluid [6], due to an external uniform field [7, 8] and due to the external regular field [9].

The MF consists of many oriented particles of different size and shape, and is a complex inhomogeneous magnetic system. Shape (configuration) of magnetic powder particles is determined by ratio of maximum and minimum transverse size of particles in different directions [9]. Depending on this ratio, the following typical forms of magnetic powder particles are distinguished: spherical (from 1.0 to 1.2); rounded (from 1.2 to 2.0), angular (from 2.0 to 5.0 with curved surfaces, sharp corners and edges); rod (from 5.0 to 25.0); needle (over 25.0).

Due to the wide particles size distribution, the formation of short random bonds (from weak to strong) of individual isolated particles near large particles into extended finite spatial figures occurs, as a result of the magnitude of their magnetization [2]. It is necessary to investigate the movement of clusters from random irregular shapes to regular fractal or linear periodic lattices.

In reality (Figure 1), the operation of processe and the type of structures formed are influenced by the amount (concentration) of all colloidal particles of different types, configurations of magnetic nanoparticles [3], the ratio of the sizes of magnetic particles, as well as the magnitude and orientation of the magnetic field. Computer experiments by the molecular dynamics method showed [3] in a wide range of changes in the magnitude of the magnetic moment and particle concentration, the formation of clusters becomes weak if the anisotropy of the particle shape increases.

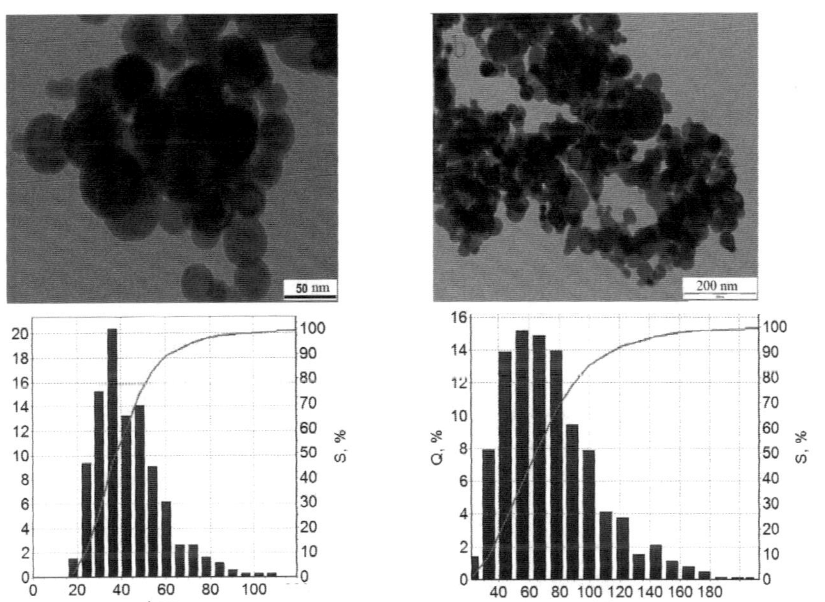

Figure 1. The process of accumulation of various formations around large particles.

Structures are formed at reaching some critical value [6] of external magnetic field anddecay when it is turned off. However, the agglomeration of the particles can be made irreversible if the MF is polymerized. In general, the structure of an MF largely depends on the qualitative and quantitative

transformation of the cluster into a system, primarily on the critical size of the cluster.The following factors were taken into account when analyzing structuring processes:
1.In the chain model, the interaction of chains with the environment and the total concentration of magnetic particles in the solution is an essential parameter;
2. The chain model takes into account the presence of clusters in a magnetic fluid with a complex spatial structure in particular fractal clusters;
3. Chains contain hundreds and thousands of particles and always grow in MF at different speeds.

4. Experimental data

In the works on the project [10–17], magnetic fluids (MF) and composite fluid media (CFM) with the obtained solid dispersed fillers (SDF) were synthesized and the conditions of their stability in an external magnetic field were investigated.

- MF based on PFMS-4 with particles from NiTi with concentrations of 5% and 10%;
- MF based on PFMS-4 with AMAG alloy particles with concentrations of 5% and 10%;
- MF based on PFMS-4 with particles from Fe3O4 with concentrations of 5% and 10%;
- MF based on PFMS-4 with particles of Fe3O4 burned at a temperature of 300 ^{0}C with concentrations of 5% and 10%;
- MF based on PFMS-4 with particles of Fe3O4 burned at a temperature of 500 ^{0}C with concentrations of 5% and 10%;
- MF based on PFMS-4 with particles from alloy 5BDSR with concentrations of 2.5%, 5% and 10%;
- MF based on motor oil with Fe3O4 particles fired at 300 degrees Celsius with concentrations of 5% and 10%;
- MF based on motor oil with 5BDSR alloy particles with concentrations of 2.5% and 5%;
- CFM with SDF TiO2 with concentrations of 5% and 10%.

Investigations of the stability of MF and CFMin an external magnetic field have been carried out. The stability of magnetic fluids in an external magnetic field created by Helmholtz rings was investigated by the method of dynamic light scattering. Definitely the minimum magnetic field (20 mT) corresponding to the stability of the magnetic fluid in the magnetic field. Figure 2 shows the change in the direction of clustering when the direction of the magnetic field lines in the film plane changes, as well as the disorientation of the clusters during the rapid rotation of the magnetic field.

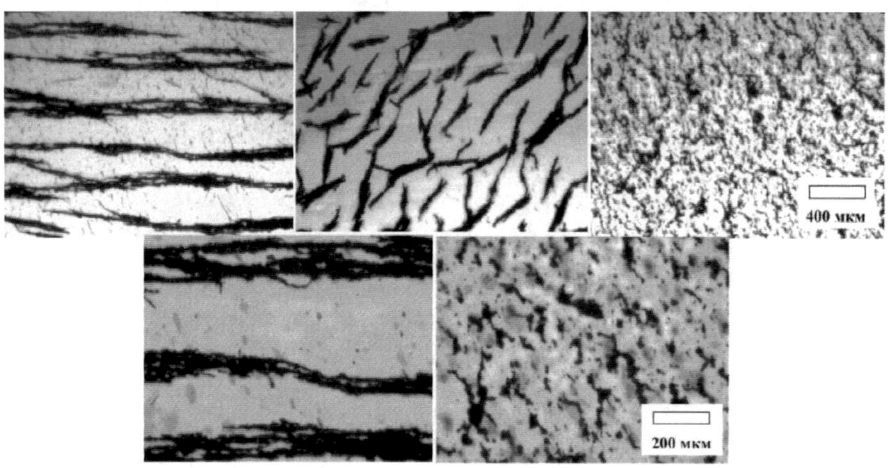

Figure 2. Microscopic image of clusters 5BDSR in oil when a magnetic field of 3000 gs is applied and without it (right images).

Figure 2 shows that irregular, but oriented along the field lines, filamentous clusters with a length from several hundred microns to several 1-2 mm are formed in a homogeneous magnetic field, the distance between clusters is from 30–50 to 200–400 microns with an increase in concentration, the distance between them decreases [19, 20].

The magnetic properties of MFs are very ambiguous. This may be due to different (different for different samples) particle concentrations in the samples, the small number of samples created, and the significant structural differences between them. However, the results already obtained suggest new samples as of undeniable interest for further study.

5. Discussion

Based on the model of the periodic structure of cylindrical magnetic domains [18] in work [19] was obtained that to create a given spatial distribution of magnetic clusters, it is necessary to apply a magnetic field with a corresponding distribution in space. This dependence shown uses the most likely average particle size. However, as previously performed analysis showed, large-sized particles make the greatest contribution to organization agglomerate. Introduction of magnetic conductor into the field does not lead to regularization of the cluster structure, since an inhomogeneous field is observed only near the magnetic conductor, while an oriented structure is formed in the plane at a significant distance (up to 1–2 cm). However, as the results of rigorous mathematical modeling of lattices [18] show, when interacting with an external field, spatially TEM waves are excited in them. This is due to the influence of the skin effect at frequencies up to 30 THz and the resulting violation of the homogeneity of the magnetization distribution in the magnetic conductors of the lattice. Under resonance conditions, the depth of penetration of the field into the metal decreases sharply within the width of the resonance curve. The frequency of a single magnetic particle depends on its geometric shape, size and can vary significantly from particle to particle (each particle has its own frequency).The dipole interaction of the magnetic moments of the individual particles transforms the magnetic fluid into a system of a large number of connected resonators. Width of resonance curve is determined by total resonance curves of magnetic particles and their magnetic losses. Based on this, there is a need for adequate probability models for the analysis of effective electromagnetic parameters.

When considering the regularities of regular ordering in the macrovolume, it is necessary to consider the well-studied results obtained for meta materials. The ratio of the diameter D of the conductors to the period of the lattice T and the diameter D of the conductors to their length L significantly affects the anisotropy of the magnetic lattice and, accordingly, the internal magnetic field, which is determined by the anisotropy of the shape of the conductor and the packing density of the lattice [18, 21].

6. Conclusion

1. Analysis of literature data and research results on the project on the formation of lattice structures confirmed a significant effect in homogeneous and inhomogeneous magnetic fields on the direction of cluster growth.
2. Numerous theoretical and experimental studies have shown that, depending on the direction and magnitude of the critical magnetic field, the processes of linear structuring are due to the configuration and large particle sizes.
3. There are no data on the possibility of controlling the lattice periods, except for quasiperiodic needle clusters [8].

Acknowledgments
This study was supported by Russian Science Foundation (RSF) (project No. 18-19-00268П)

References

[1] Pshenichnikov A, Gile V 1997 *Colloid Journal* **597**(3) 382–389 (in Russian)
[2] Ivanov A, Kantorovich S 2003 *Colloid journal* **65**(2) 189–200 (in Russian)
[3] Gudkova A, Pyazina E 2017 *Solid state physics* **59**(11) 2159–2162(in Russian)
[4] Pshenichnikov A, Lakhtina E 2006 *Colloid journal* **68**(3) 327–337 (in Russian)
[5] Pshenichnikov A F. Smolina Yu A. 2010 *Bulletin of Perm University* **1** 17–21 (in Russian)

[6] Luca E, Cotae C, Calugaru G 1978 *Rev. Roum. Phys* **23** 1173–1178
[7] Dikansky Yu, Gladkikh D, Kunikin S *et.al 2012 Journal of Technical Physics* **82**(6) 135–139 (in Russian)
[8] DikanskyYu I, Nechaeva O A 2003 *Colloid journal* **65**(3) 338–342 (in Russian)
[9] Biryukov P, Vavilov V, Voronkov V *et al* 2003 *Non-destructive testing system. Types (methods) and technology of non-destructive testing. Terms and definitions: Reference manual* (Moscow: State Unitary Enterprise "Scientific and Technical Center for Safety in Industry of Gosgortekhnadzor Russia)
[10] Zyatkov D O, Yurchenko A V, Yurchenko V *et al* 2018 *IOP Conf. Ser.: Mater. Sci. Eng* **363** 012023 doi:10.1088/1757-899X/363/1/012023
[11] Svetlichnyi V A, Shabalina A V, Lapin I N *et al* 2018 *Applied Surface Science* **462** 226–236 doi:10.1016/j.apsusc.2018.08.116
[12] Zyatkov D O, Kochnev Z S, Knyazkova A I *et al 2019 Russ Phys J* **62** 400–405 doi:10.1007/s11182-019-01726-8
[13] Svetlichnyi V A, Balashov V B, Lapin I N *et al* 2019 *Russ Phys J* **62** 411–415 doi:10.1007/s11182-019-01728-6
[14] Zyatkov D, Balashov V, Yurchenko V *et al* 2019 *Progress In Electromagnetics Research M* **80** 103–110 doi:10.2528/PIERM18092303
[15] Zyatkov D, Balashov V, Borisov A *et al 2018 Progress in Electromagnetics Research Symposium* 843–847 doi:10.23919/PIERS.2018.8597816
[16] Zyatkov D, Kochnev Z, Knyazkova A *et al* 2019 *PhotonIcs & Electromagnetics Research Symposium - Spring (PIERS-Spring)* doi:10.1109/PIERS-Spring46901.2019.9017500
[17] Ezhov D M, Kochnev Z S, Fakhrutdinova E D *et al* 2020 *PROCEEDINGS OF SPIE* **11582** doi:10.1117/12.2580496
[18] Yurchenko A V *et al* 2018 *Russian Journal of Nondestructive Testing* **54**(7) 528–533
[19] Mekhtiyev A D *et al* 2018 *News of the National Academy of Sciences of the Republic of Kazakhstan, Series of Geology and Technical Sciences* **5**(431) 66–75
[20] Serikov T G *et al* 2017 *Proceedings of the 11th International Forum on Strategic Technology, IFOST-2016* 510–515
[21] Baryakhtar V, Gorobets Yu 1988 *Cylindrical magnetic domains and their lattices* (Kiev:NaukovaDumka)

Improvement of the all-weather profiler

V Y Zhukov[2], I N Rostokin[1] and G G Shchukin[1, 2]

[1]Vladimir State University named after A.G. and N.G. Stoletovs Murom Institute, Murom, Russia
[2]Military Space Academy named after A.F. Mozhayskogo, St.-Petersburg, Russia

E-mail: rostockin.ilya@yandex.ru

Abstract. The article considers the issues of improving the quality of vertical wind profile measurements. In order to make these measurements under any meteorological conditions, data from profilometers operating in different wave ranges, from optical to centimetre, are combined. At the same time, the resolving power of the latter is considerably inferior to that of the former. We propose a way to improve the quality of measurements in the radio band by applying a new method of information processing, using estimates of the width of the spectrum and the peculiarities of their spatial distribution.

1. Introduction

Remote measurement of the vertical wind profile is required for many practical tasks. For this purpose, special technical means are created - profilometers. In fact, these are coherent radars, which measure the Doppler frequency shift of signals coming from different altitudes and calculate the vertical wind profile. Carrier frequency of these devices is selected depending on the range of heights of interest to the information consumer [1].

The exception is the lower layer of the atmosphere up to heights of 15 km, where due to the high variability of the conditions of propagation of electromagnetic waves it is impossible to work on a single wavelength. Thus, in airports, where the heights up to 500 m are of interest, two profilers are installed - optical and millimeter ranges, working in the absence or presence of hydrometeors in the atmosphere, respectively. If measurements at higher altitudes are required (for example, at spaceport) two profilers are not enough - at least three of them are required [2].

It is desirable that all of them are located on the same platform and are united by a single information processing system.

An example of such an all-weather profiler has already been implemented (Figure 1) and presented at a number of conferences, in particular, in 2015 in Tomsk [2].

Content from this work may be used under the terms of the Creative Commons Attribution 3.0 licence. Any further distribution of this work must maintain attribution to the author(s) and the title of the work, journal citation and DOI.
Published under licence by IOP Publishing Ltd

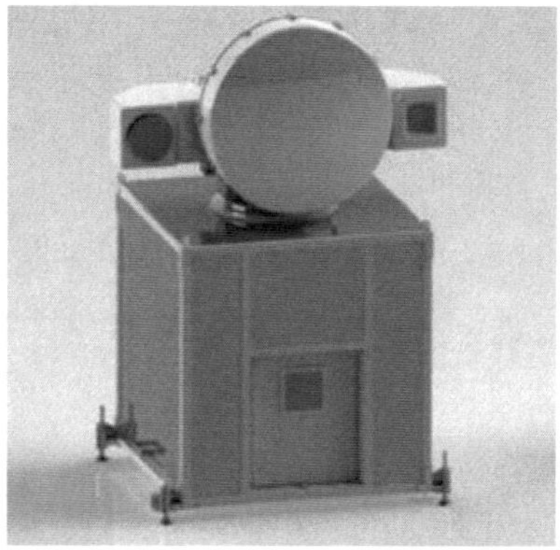

Figure 1. General view of the lidar-radar meteorological complex, combining the lidar of the infrared range, two radars of X and Ka ranges.

However, effective processing of data obtained by different channels is possible only at low altitudes. Further radiofrequency devices, unlike optical ones, cease to meet the requirements for accuracy and resolution. Therefore, the search for new methods of radar signal processing, capable of eliminating the emerging contradiction is relevant.

2. Question posed

The problems with radio profilers begin when the linear size of their resolution element becomes larger than the minimum possible size of the wind field heterogeneity. For wind shear, this limit is about 30 meters (100 feet). The problems are the distortion of the information obtained, which becomes greater the greater the ratio of the resolution element size to the specified limit.

Under these conditions, the signal received by the profiler can be represented as two components formed by the two volumes of space above and below the wind shear layer. The form of the spectra according to the norms accepted in radio-meteorology should be considered Gaussian [3].

Under the given conditions, the average frequency of the total signal is defined by the formula [4]:

$$\omega_{\text{cp}} = \frac{A_1}{A_1+A_2}\omega_1 + \frac{A_1}{A_1+A_2}\omega_2, \tag{1}$$

where A_1 and A_2 are the powers of the components of the total signal; ω_1 and ω_2 are the average frequencies of their spectrum.

The reflection from the layer hosting the heterogeneity in the vast majority of cases can be neglected due to either the smallness of its power relative to the power of the components mentioned, or because its spectrum is almost uniformly distributed between the spectra of the components, which minimizes its impact on the value ω_{cp}.

The formula (1) shows that the difference in the meteorological reflectivities of each of the layers has a significant influence on the obtained result. An example of how the information about the radial velocity distribution of elementary reflectors is distorted in the case of observing a thirty-meter layer with a change in wind speed from 10 to 20 m/s, located at a height of 1200 m, is shown in Figure. 2.

It follows from this graph that as the ratio of the resolution element size of the radio-profile meter to the width of the wind shear layer increases, system errors increase:

- in estimating the magnitude of the wind shear;
- in estimating the height of its location.

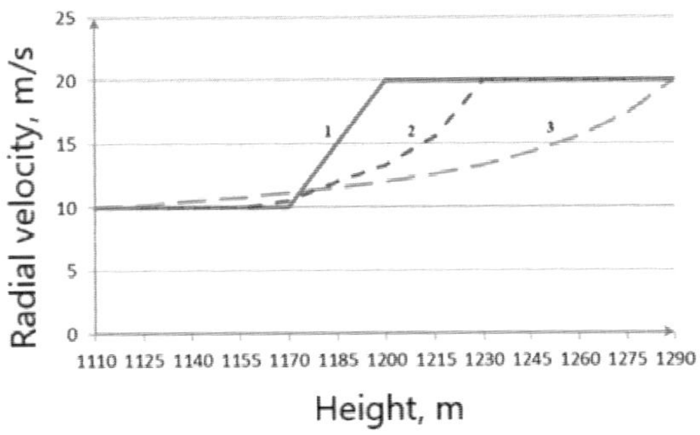

Figure 2. High-altitude wind profile real (1) and obtained by the profilometers at linear size of the resolution element two (2) and four (3) times the size of the heterogeneity layer, at a power ratio of 4 to 1.

Combating them by narrowing the radiation pattern is practically impossible - too large an antenna is required to achieve its acceptable width. It remains to use another way, based on the fact that information about the velocity distribution of elementary reflectors in the case we are considering does not disappear without a trace. It is "pumped" into the spectral region. In the already described case of two air masses simultaneously observed with a wind shear layer between them, the spectrum of the received signal is the sum of two Gaussian spectra with different values of average frequency, width and power (Figure 3).

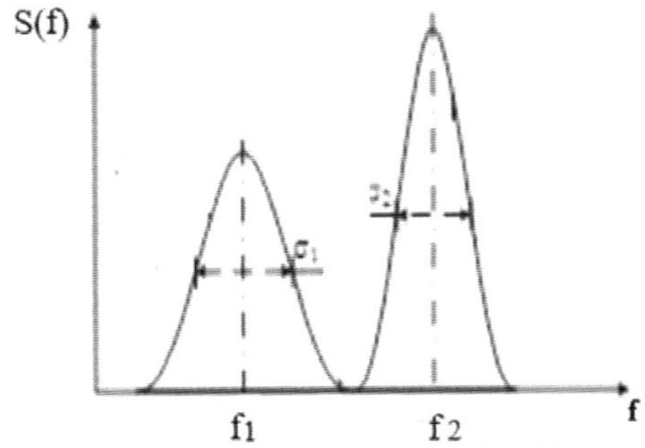

Figure 3. Example of the energy spectrum of the signal received in the case of two masses of air with different wind speed parameters hitting the resolved volume.

However, in practice, the spectrum of the signal calculated from a limited sample has a much less presentable appearance, and extracting the data of interest from it is a difficult and, most importantly, poorly formalizable task. Therefore, it is better to solve the problem by using a more analyzable parameter, namely the width of the spectrum of the received signal. It is on the estimations of the latter that the new method for the detection of dangerous wind shifts, described below, is based.

3. Method description

The width of the spectrum shown in Figure 3, is determined by the formula [5]:

$$\sigma = \sqrt{\frac{A_1}{A_1+A_2}\sigma_1^2 + \frac{A_2}{A_1+A_2}\sigma_2^2 + \frac{A_1 A_2}{(A_1+A_2)^2}(\omega_1 - \omega_2)^2}, \qquad (2)$$

where σ_1 and σ_2 are the spectral widths of the components of the total signal.

If the difference in the average frequencies of these components reaches a sufficiently large value that exceeds the widths of their spectra, the third term in formula (2) begins to play a decisive role in determining the value of σ.

Since the average frequency of the spectrum is related by the known relation to the average velocity of reflecting particles:

$$\omega_i = \frac{2\pi V_i Cos(\alpha-\varphi_i)}{\lambda}, \quad (3)$$

where V_i – is the mean velocity of particles in the i-th layer, α is the azimuth of the profiler antenna, φ_i – is the direction of air velocity in the i-th layer, λ – is the wavelength of the carrier vibration, the value of the third term of formula (2) is not constant and depends on the angle of antenna rotation in the horizontal plane.

The maximum absolute value is reached when the antenna axis is parallel to the velocity difference vector of particles in the two observed layers, while the minimum value is reached when these two vectors are perpendicular. Besides, the increased values of the spectrum width exist only in the place where the antenna beam crosses the boundary between the two air masses, i.e. in the limited range area, as it is shown in Figure 4.

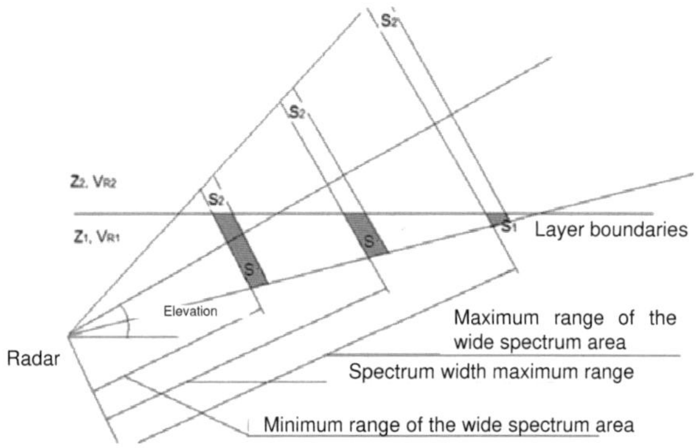

Figure 4. Schematic of the formation of the broad spectrum zone when the antenna beam crosses the boundary of atmospheric layers, characterized by the reflectivity Z_1, Z_2 and the radial velocity of particles V_{R1}, V_{R2}.

It follows that the zones of large spectral widths form on the maps of conic sections of the parameter in question concentric arcs symmetric about the center or, if the wind shift in magnitude is added to the shift in direction, spiral segments. This is confirmed by numerous cases of in-situ observations, examples of which are shown in Figure 5.

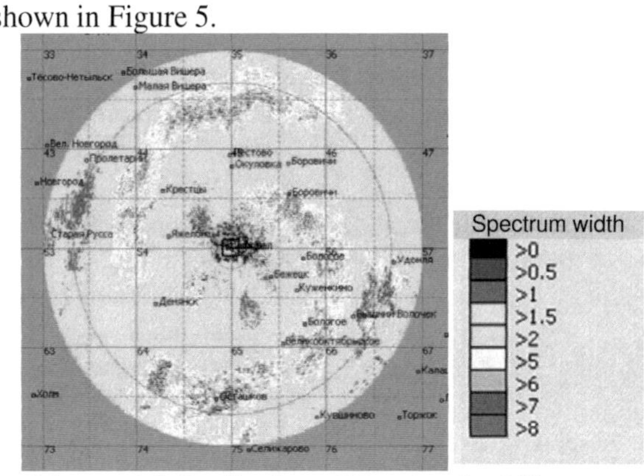

Figure 5. Example of the spatial distribution of the signal width on a cone section map when observing wind shear.

The parameters of the region formed in this way (maximum parameter value, band width, spiral "curl" velocity, distances to the region edge) allow us to obtain data on the magnitude of the observed shear and its height [5]. For example, the wind speed difference at the edges of the shear layer is determined by the formula:

$$\Delta V = \sqrt{4\sigma_{V\,max}^2 - 2\sigma_{V1}^2 - 2\sigma_{V2}^2}, \quad (4)$$

where $\sigma_{V\,max}$ – is the maximum value of the velocity spectrum width detected in the wind shear layer, σ_{Vi} – is the average value of the velocity spectrum width in the i-th layer (below and above the shear), the velocity spectrum is recalculated from the Doppler frequency shift spectrum according to formula (3).

The helix «curl» velocity $d\alpha/dR$ is related to the magnitude of the wind shift in the direction Ω by the relation:

$$\frac{d\alpha}{dR} = \frac{\Omega}{4}\left(\beta + \frac{\pi}{4}\Theta\right), \quad (5)$$

where β - is the location angle of the profilometers antenna, Θ - is the width of its directional diagram, R - is the range of the point where measurements are made.

The information processing algorithms based on these formulas were tested on the data of the weather radar DMRL-C installed in Valday, Novgorod region, and were compared with the data of the airborne station «Bologoye» [6]. As a result, the high efficiency of the new method in detecting wind shifts even of insignificant (from 1 m/s per 100 ft) value was proved.

4. Conclusion

In the vast majority of cases, the task of profilers is to detect dangerous wind shifts. It is solved by reconstructing the vertical wind profile by estimating the average Doppler reflection frequency shift for each of the analyzed heights. It is this method of processing the incoming information that makes the radio profiler considerably inferior to the lidar one due to a much wider antenna directional pattern compared to the latter.

The method considered in this article, based on estimates of the width of the received signal spectrum, eliminates this drawback. When it is used, the width of the antenna beam is not essential. The accuracy of the results in this case is determined to a greater extent by the sample size and the procedures for estimating the spectral characteristics of the random process under study. As a result, the quality of the radio profilometers, of course, does not reach the level of optical technology, but remains within the limits of the requirements at sufficiently high altitudes. At the same time, it should be noted that the range of the lidar is severely limited because of the energy performance. For radio range devices, this limitation is much weaker, allowing them to operate at higher altitudes.

In those rare cases where the wind profile itself is of interest, the method proposed by the authors is also applicable as a supplement to the basic method based on average Doppler frequency shift estimates. The latter gives acceptable results even with a too wide directivity diagram in conditions of smooth change of wind parameters with altitude. Errors appear when observing sharp inhomogeneities. This is where the new method is able to highlight this heterogeneity and make corrections to the basic data.

Acknowledgments

The research was carried out at the expense of a grant from the Russian Science Foundation № 21-19-00378, https://rscf.ru/en/project/21-19-00378/

References

[1] Gorelik A, Kolomiets S, Shchukin G 2011 *Notes of Russian Hydrometeorological University* **1** 80–89

[2] Boreisho A, Zhukov V, Ilyin M *et al.* 2015 *Proceedings of Higher Education Institutions. Physics* **58** (10) 100–104

[3] Doviak R, Zrnic D 2006: *Doppler radar and weather observations* (New York: Dover Publications Inc.)

[4] Denisenkov D, Zhukov V, Karavaev D *et al* 2016 *Proceedings of Higher Education Institutions. Physics* **12**(2) 15–19

[5] Devyatkin A, Denisenkov D, Zhukov V *et al* 2018 *Russian Meteorology and Hydrology* **43**(1) 56–61

[6] Denisenkov D, Zhdanova M, Zhukov V *et al* 2016 *Scientific Notes of Russian Hydrometeorological University* **45** 113–118

Numerical modeling of anisotropic properties of a solid by particle dynamics method

D Ya Sukhanov, A E Kuzovova

National Research Tomsk State University, Tomsk, Russia

E-mail: sdy@mail.tsu.ru

Abstract. A method is proposed for the numerical simulation of acoustic processes in solids based on the particle dynamics approach for describing the anisotropic properties of a solid. It is proposed to consider a solid body in the form of an array of particles located in a cubic body-centered crystal lattice. To set the anisotropic properties of a solid, it is proposed to use its own proportionality force to shift coefficient for each direction. Based on the results of numerical simulation, the dependence of the longitudinal wave velocity on the direction is shown.

1. Introduction

One of the widespread methods of numerical modeling of acoustic processes in solids is the method of finite differences in the time domain [1, 2]. This method is based on discretization of differential wave equation in space and time. The particle dynamics method is widely used to describe the behavior of particles [3]. In the method of particle dynamics, the medium under consideration is represented as a set of interacting particles (material points or solids), for which the classical equations of dynamics are written. The interaction of particles is described by means of interaction potentials, the main property of which is repulsion when approaching and attraction when moving away.

The advantage of the particle dynamics method in comparison with methods based on the representation of the medium as a continuous one is that it requires significantly fewer a priori assumptions about the properties of the material. Considering the interaction of particles only through the simplest interaction potential (for example, of the Lennard-Jones type) [4, 5] and insignificant dissipation, it is possible to simulate such complex effects as plasticity, cracking, fracture, temperature change in material properties, phase transitions [6, 7]. To describe these effects in the framework of a continuous medium, a separate theory is needed for each of the effects, while in molecular dynamics modeling; these effects are obtained automatically, as a result of integrating the equations of motion.

Anisotropy during the propagation of acoustic waves in matter is characteristic of solid crystalline bodies. It is proposed to simulate the anisotropic properties of a solid in a particle model by specifying different dependences of the interaction forces on shear for each direction. In this work, only the longitudinal forces of interaction between particles are considered, however, the model does not exclude the possibility of taking into account the shear forces.

Content from this work may be used under the terms of the Creative Commons Attribution 3.0 licence. Any further distribution of this work must maintain attribution to the author(s) and the title of the work, journal citation and DOI.
Published under licence by IOP Publishing Ltd

2. Method description

In this paper, it is proposed to use the particle dynamics method to model the anisotropic properties of a solid, representing a solid in the form of a set of interacting particles located in a cubic body-centered crystal lattice. In such a crystal lattice, each particle has 14 nearest neighbors: 6 at a distance d and 8 at a distance $d\sqrt{3}/2$ (Figure1).

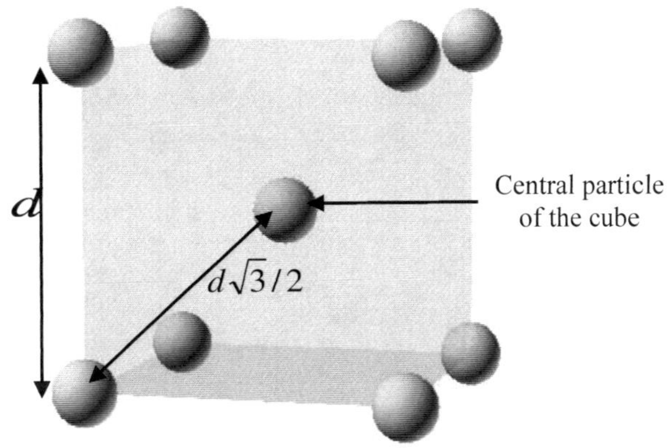

Figure 1. Body-centered cubic crystal lattice cell.

Each particle in the crystal lattice has its own mass $m = \rho d^3/2$ depending on the density of the substance under consideration ρ, the step of placing particles in the crystal lattice d (d^3 – the volume of the cube, divided by 2 due to the presence of particles in the center of each cube, which leads to a doubling of the total number of particles).

The elastic properties of a substance are determined by the dependence of the force of interaction between particles on the distance. We consider linear dependence of force on shift in approximation of small deformations. For each particle, its acceleration is calculated according to Newton's law as the ratio of the applied force to the particle mass. The speed and coordinate of the particle are calculated by numerical integration of it's acceleration over time. A change in the coordinate of a particle entails a change in the force applied to it due to interaction with neighboring particles. This interaction provides the possibility of propagation of acoustic waves in the system under consideration.

Particle mass and interaction force determine the type of material. The force of interaction depends on the distance between particles, but the main influence is exerted by the dependence of the force of interaction near the equilibrium point.

We describe the force interaction between particles by the function $F(x)$ where x - is the distance between particles. On distance d or $d\sqrt{3}/2$ particles from each other are in a state of stable equilibrium and are not affected by forces from their neighbors.

The dependence of the force of attraction between particles on the distance is proposed to be characterized by the following formula:

$$F(x) = k_n(x - R) \tag{1}$$

where R – is the equilibrium distance between particles ($R = d$ or $R = d\sqrt{3}/2$), x – actual distance between particles, $k_n = c_n^2 m / R^2$ – proportionality coefficient between force and distance in linear approximation in N/m, n – is the number of neighbouring particles symmetric pair, c_n – the speed of sound in the medium for specific direction. The value of the coefficient k follows from the reduction of the one-dimensional model of interacting particles to the wave equation. In formula (1), the value of the coefficient k is specified for each pair of interacting particles. We consider that two opposing particles had the same force coefficient k_n to ensure pulse preservation law. If k_n is different in different directions than it should allow to simulate anisotropic waves propagation in the media. On the Figure 2 is shown the placement of 14-n neighboring particles with coefficients k_n that corresponds to each of them.

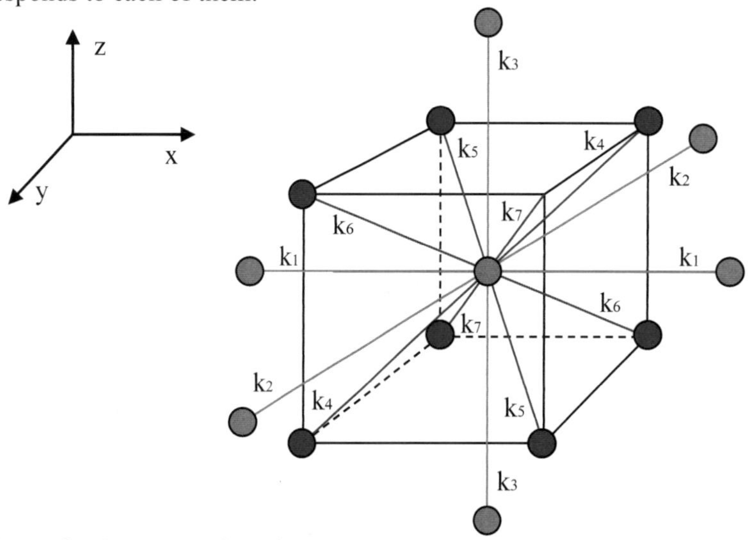

Figure 2. Placement of particles and the corresponding coefficients of interaction with the central particle.

3. Numerical modeling

In the model under consideration, each particle has 14 neighboring particles. Since the interaction coefficient k_n is the same for opposing particles, there are only 7 independent values of the k_n coefficients. The anisotropy of a rigid body is taken into account by setting different proportionality coefficients for different directions.

In order to test the possibility of modeling wave processes in an anisotropic medium. The modeling of the structure was carried out, which is a body-centered lattice, consisting of 161040 particles. Distance between particles (grid step) R =50 mkm time step dt = 2.5 ns. A spherical wave of short-pulse source was generated from the middle of the simulated volume. For computation of k_1 we considered c =6000 m/s, $\rho = 2500$ kg/m^3.

An isotropic source in the form of a symmetric shift of particles from the center was considered as an introduced source. The shift of particles decreased with distance according to the Gaussian function with a half-width of 2 mm. At the initial moment of time, a shift is introduced, and then the process of numerical simulation of particle dynamics is started. Introducing an initial offset is equivalent to applying a short pulse.

If we take into account that 6 neighboring particles are located at a distance d, and 8 particles are at a distance $d\sqrt{3}/2$, then the equality of all coefficients will not ensure complete isotropy of the medium. For isotropy, it is necessary to provide a ratio between the coefficients k_{1-3} and k_{4-7} as 2/3. Numerical simulation was carried out for the case when all coefficients are the same (Figure 3*a*) for simulation time 4 mks. It can bee seen that the wave is propagated with different speed in different directions that indicates on anisotropy.

At the same time, we modeled the case when there is no interaction between particles along the coordinate axes $k_{1-3} = 0$. The simulation results for 7.5 mks are shown in Figure 3*b*. In this case, the amplitude of the longitudinal wave depends on the direction, which also indicates anisotropy.

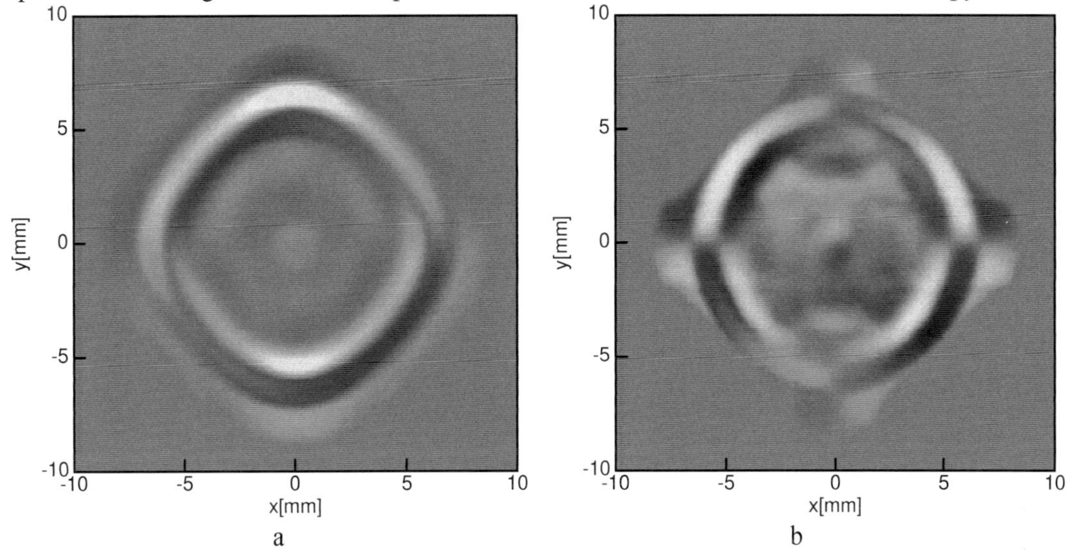

Figure 3. Distribution of particles velocity after short pulse isotropic source introduced. The red color represents X component of particle velocity; the green color – Y component; the green color – Z component; (a – the case of the media with $k_1 = k_2 = k_3 = k_4 = k_5 = k_6 = k_7$; b – the case of the media with $k_1 = k_2 = k_3 = 0$, $k_4 = k_5 = k_6 = k_7 \neq 0$)

Figure 4 shows the spatial distribution of particle velocity after 3750 ns of simulated time for two different types of anisotropic media. It can be seen that the field of an initially isotropic source was transformed into a wave propagating at different speeds in different directions. This indicates the anisotropic nature of the simulated medium.

With an increase in the interaction coefficient k_n, the speed of propagation of waves in the corresponding direction increases. We mainly consider longitudinal waves in a particle system. But transverse waves also arise due to interactions k_{4-7}, the speed of transverse waves is significantly less than the speed of longitudinal ones.

It should be noted that the model of particle dynamics is not limited to the body-centered crystal lattice we have considered with longitudinal interaction forces. It is possible to introduce transverse forces of interaction; consideration of more neighboring particles; change in the crystal lattice. An increase in the number of model parameters allows considering a wider class of acoustic effects. The advantage of the particle dynamics method is the possibility of embedding it in the simulation of the macroscopic motion of solids, the simplicity of describing the interaction of several bodies, various states of aggregation within the framework of one model.

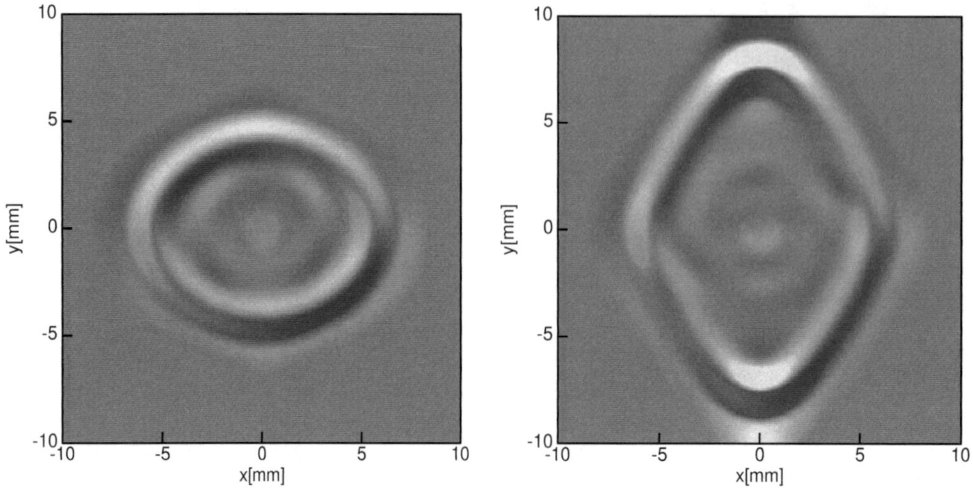

Figure 4. Distribution of particles velocity after short pulse isotropic source introduced. The red color represents X component of particle velocity; the green color – Y component; the green color – Z component; (a – the case of the media with $k_1 = k_3 = k_4 = k_5 = k_6 = k_7$ and $k_2 = k_1 / 2$; b – the case of the media with $k_1 = k_3 = k_4 = k_5 = k_6 = k_7$ and $k_2 = 2k_1$;)

4. Conclusion

A method is proposed for the numerical simulation of the propagation of acoustic waves in anisotropic media based on the dynamics of interacting particles. A body-centered crystal lattice is considered. A linear dependence of the force of interaction between neighboring particles on the shift from the equilibrium position is assumed. Each particle has 14 neighbors. Due to the need to fulfill the law of conservation of momentum, it is necessary that the interaction for opposing particles be described in the same way, so there are 7 coefficients left to describe the anisotropy of the medium. Numerical modeling has shown the possibility of modeling waves in the structure under consideration.

Acknowledgments

The reported study was funded by RFBR, project number 20-32-90074.

References

[1] Barkhatov V A 2007 *Russian Journal of Nondestructive Testing* **43**(9) 605–618
[2] Avdeev D A *et al* 2016 *Vestnik PNU* 92–96
[3] Krivtsov A M 2007 *Deformation and fracture of solids with microstructure* (Moscow: Fizmatlit) p 307
[4] Sharma S S *et al* 2019 *Journal of Applied Physics* **125** (21) 1–8
[5] Natsik V D *et al* 2018 *Low Temperature Physics* **44**(7) 688–695
[6] Starikov S *et al* 2020 *Comp. Materials Sci* ***179*** 1–14

Journal of Physics: Conference Series **2140** (2021) 012019 doi:10.1088/1742-6596/2140/1/012019

Spatiotemporal structure of the electromagnetic field in a forest environment at nanosecond location

B Ch Dorzhiev, O N Ochirov

IPMS SB RAS, Ulan-Ude, Russia

E-mail: bchdorj@yandex.ru

Abstract. The results of measurements of the spatio-temporal distribution of direct and reflected signal levels inside a pine forest are presented. Circular diagrams are built on the basis of which it is possible to assess the levels of direct and reflected signals at a bistatic nanosecond location. The obtained diagrams a give spatial picture of the total energy of the electromagnetic field in the forest environment.

1. Introduction

Along with studies of the mechanisms of radio wave propagation in forest environments, such as scattering, diffraction and attenuation of waves, the determination of the spatio-temporal structure of the electromagnetic field in the forest environment itself is of undoubted interest from a practical point of view. A number of works are devoted to experimental studies of the reflection of radio waves from forests, including scattering both on elements of forest vegetation and from tree trunks as a function of the azimuthal angle. To calculate the mean field in a discrete medium, Tverskoy's theory of multiple scattering [1] is used, which has some limitations. Briefly, they can be formulated as follows. With a low forest density and small transverse dimensions of trees in comparison with the average distance between them, one can neglect the transverse dimensions of trees and assume that the scattering is due to point scatterers. Then, the field at the receiving point is defined as the superposition of the fields of the straight line and the field components scattered on the trees. A sufficiently detailed analysis of the problem data is given in [2], where solutions for the cases of single and multiple scattering are considered. In this case, the assumption is introduced that the multiple mutual scattering in terms of the contribution level is less than the component of the incident wave. Thus, two questions remain open - the time of arrival of the point of reception of the direct and reflected signals and the ratio of their levels. In this paper the results of experimental studies to determine the spatial distribution of the field in a forest environment in the case of a bistatic nanosecond location are presents.

2. Test site and measurement technique

The area of sparse pine wood with the side of 20 m wide and 40 m long was selected to carry out the measurements. On Figure 1 are shown a test site of the forest. To determine the main average parameters of forest, this site was divided into squares (10×10 m). The location of trees in each square was determined and their diameters were measured (Figure 2). The average height of the forest is about 12m, while the crown of the trees occupies 1/3 of the height of the forest. In the case of a horizontal forest location, the most significant parameter is the average distance between trees is 4.5 m. The average tree diameter is ~ 30.4 cm.

Content from this work may be used under the terms of the Creative Commons Attribution 3.0 licence. Any further distribution of this work must maintain attribution to the author(s) and the title of the work, journal citation and DOI.
Published under licence by IOP Publishing Ltd

Figure 1. General view of the test site of the forest.

Thus, for this forest area, the above condition for applying the theory multiple scattering is satisfied. In this case, the transverse dimensions of the scatterers will affect the characteristics of the scattered signals, which will then be clearly presented in the circular diagrams of the scattered signal.

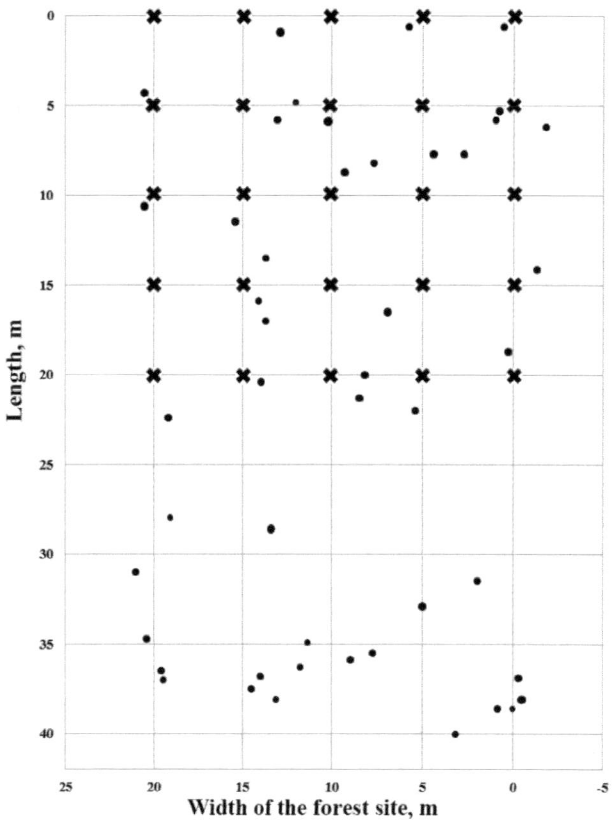

Figure 2. Forest site plan and receiver installation locations (marked with crosses).

The experimental technique was as follows. The radar transmitter is installed at a distance of 30m from the forest edge at a height of 2m, and the receiver is located at certain points inside the forest (Figure 2). The receiver and transmitter units have parabolic antennas with a 5° beamwidth at half power. Carrier frequency of radar is 10 GHz, pulse duration is 10 ns. To determine the diagram of the secondary radiation at each point, a signal was recorded every 30° with the rotation of the receiving antenna in azimuth from 0° to 360°. The initial direction of the receiving antenna is oriented towards the radar and perpendicular to the forest edge. The height of the pole-mounted antenna was 2.15m. Thus, for a given geometry of the problem, scattering is mainly due to tree trunks.

3. Results

The intensity of the resulting field at the receiving point is determined by elementary reflectors, which are located inside one pulse resolved volume of space. To determine the spatial picture of the field, the receiving points were distributed evenly within the test area. A total of 25 measurements were carried out on vertical polarization and the same amount on horizontal polarization. When registering a signal on the oscillogram, both the level of the direct signal (when the receiving antenna is oriented to the transmitting one) and the re-reflected signals (when the antenna rotates) are noted. An example of a useful signal oscillogram on horizontal polarization is shown in Figure 3. On the abscissa axis is marked the signal delay time, on the ordinate axis is marked the input signal level.

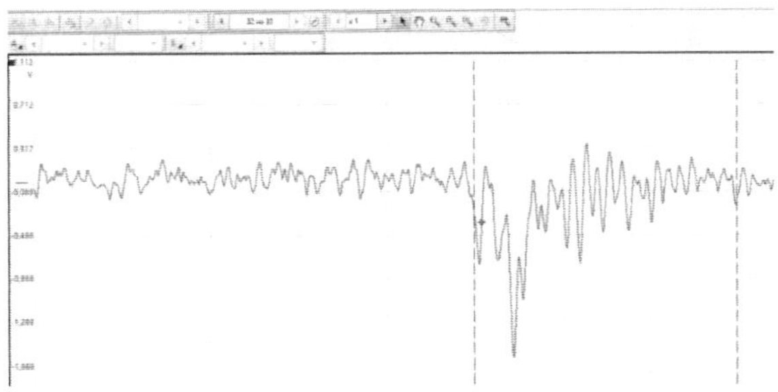

Figure 3. Signal oscillogram at one of the receiver installation points at one of the azimuth angles.

Twelve signal records corresponding to azimuth angles was carry out at each point of the receiver installation. The results of processing the signal records for all 12 azimuth angles for one observation point are shown in Figure 4, where the signal values are plotted along the line of direction to the source. Different color gradations of the signal level on the diagram mean levels with different signal time delays, which can be observed on the oscillogram in Figure 4. At the same time it should be noted, that first four significant levels were selected for processing. The main signal received without time delay is shown in gray. As can be seen from the diagram, the level of the direct signal significantly exceeds the level of the scattered signal in all cases.

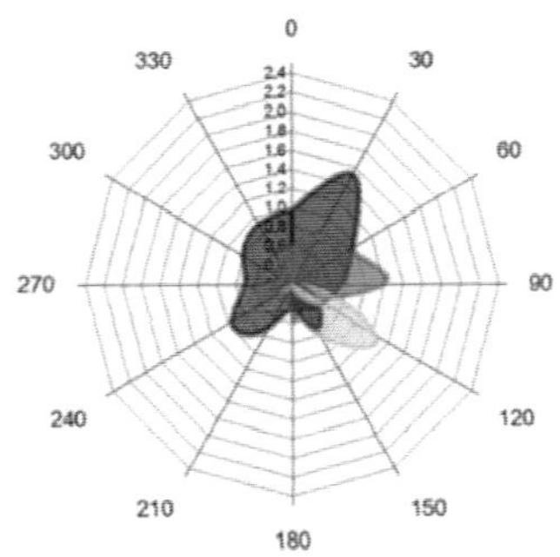

Figure 4. Circular diagrams of direct and scattered signals.

The resulting circular diagrams for all points of the receiver installation were combined with the plan of the forest site (Figure 5).

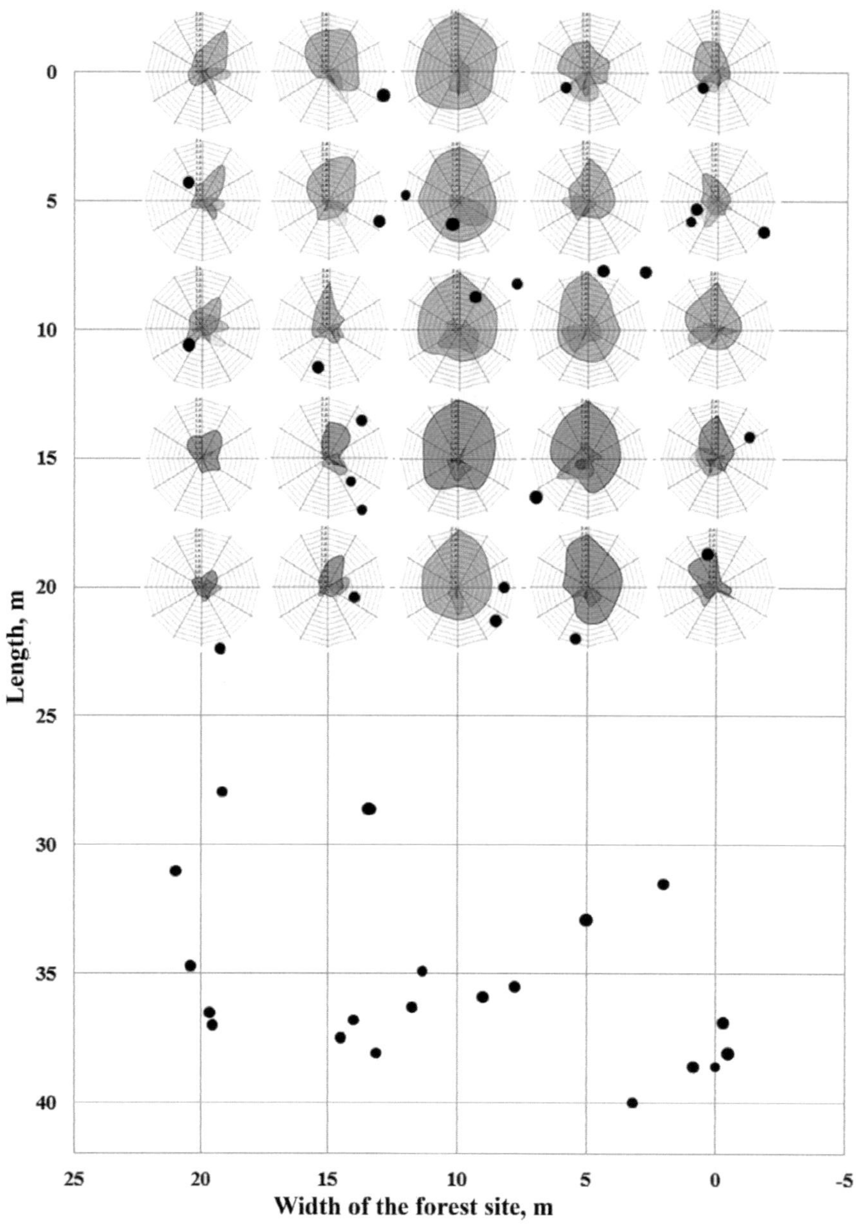

Figure 5. Spatial distribution of the electromagnetic field.

As can be seen from the complete picture of the scatterplots, the direction of the maximum of the "secondary" scatterplots falls on more distant tree trunks. The levels of the “secondary” signals are much less than the level of the first signal, and they are mainly oriented to the opposite half-space region and are scattered signals from tree trunks. The Figure 5 also shows a plan of a forest area with an indication of the location of trees, the size of the points corresponds to the real diameter of the trees.

In general, the diagrams obtained paint a spatial picture of the total backscattering energy. The results of measurements at linear polarizations, presented in the form of circular diagrams, show that in the centimeter wavelength range the multiple scattering model can be used to calculate the mean field at both polarizations. However, when calculating the mean field using the multiple scattering model, it is necessary to take into account that the reflected / scattered signals do not arrive at the receiving point simultaneously. The effect of localization of electromagnetic radiation or the demonstration of close to localization wave states during the interaction of radio emission with vegetation elements was not found.

5. Summary

Based on the obtained spatio-temporal distribution of the electromagnetic field in the forest environment, the following conclusions can be drawn:

• the maximum levels in the backscatter circular diagrams are oriented towards the radiation source. Thus, despite the presence of randomly located reflectors, the main direction of wave propagation lies within the antenna radiation pattern. In this case, the presence of a narrow directional pattern of the antenna causes an almost "channel" wave propagation;

• the signal level at the first tens of meters remains practically constant, which indicates a low attenuation when propagating in a forest environment in the case of a short-pulse signal;

• comparison of the spatial distribution of the field with the plan of the forest site shows that the maximum levels of reflected signals are caused by tree trunks located within the antenna pattern and standing along the line of sight. The "secondary" signal levels are due to more distant trees.

Acknowledgments

The work was carried out with financial support of the research project No. 0270-2021-0004

References

[1] Ishimaru A 1981 *The propagation and scattering radiowaves in random media* (Moscow: Mir) p 280

[2] Popov V I 2015 *Propagation of radiowaves in forest*s (Moscow: Hot line–Telecom) p 392

Journal of Physics: Conference Series **2140** (2021) 012020 doi:10.1088/1742-6596/2140/1/012020

Wave structure at radar irradiation of homogeneous absorbing media

V P Butukhanov, Ju L Lomukhin and E B Atutov

Institute of Physical Materials Science SB RAS, Ulan-Ude, Russia

E-mail: vbut1951@gmail.com

Abstract. The work is devoted to the study of the structure of waves excited in bordering media under radar irradiation of both smooth and rough interfaces. It is found that counter propagating waves are excited in bordering absorbing media, which determine backward reflection at the interface. On the other hand, the reflection of the counter propagating wave excites waves with a negative angle of refraction. It was found in this work that when the interface is irradiated with a plane wave during polarization, when the electric field strength vector lies in the plane of incidence, the backward reflection and the refracted wave are increases, and the specular reflection decreases. Electrodynamics models of the back reflection coefficients are developed for both the case of smooth and for the case of uneven interfaces between the media.

1. Introduction

It is well known and reliably established that when a plane wave is incident on a plane interface between two clear media, refracted and reflected waves are excited in them. Together with the incident mode, this system of modes satisfies the wave equation and boundary conditions. In this case, the refraction and reflection of these waves obeys well-known laws [1, 2]. However, even in such a simplest model of bordering media, the possibility of excitation of the fourth mode was found, the direction of which obeys Snell's law - this is a mode with a negative angle of refraction [3].

Discussions about the physical mechanisms of the origin of unusual refraction and the propagation of the last two waves either subside or resume [4–7]. Without going into the analysis of the proposed mechanisms of unusual refraction and propagation, let us point out that there are no transparent material media in nature, i.e. all material environments in one way or another have losses. This means that the permittivity and permeability in Maxwell's equations must always be complex. Therefore, in particular, the assumption of the existence of media with negative permeability's causes some doubts.

2. Multimode structure of waves on the border media

In this work, we initially assume that the media are absorbing, isotropic, and homogeneous. In [8], it is indicated that more than four of the above-mentioned modes can be excited in adjacent media. In this paper, we will show that this is indeed the case, write down and analyze the excited modes, and point out the fundamental importance of some of them in the models of backward radar reflection used in methods of sounding natural and artificial media.

Let us consider the possibility of constructing the structure of waves in adjoining media taking into account the excitation of counter propagating waves by the example of irradiation by a plane wave of the interface between two media, one of which is a void, the other is an isotropic absorbing medium (Figure 1). Note that the presence of counter propagating waves does not contradict Maxwell's

Content from this work may be used under the terms of the Creative Commons Attribution 3.0 licence. Any further distribution of this work must maintain attribution to the author(s) and the title of the work, journal citation and DOI.
Published under licence by IOP Publishing Ltd

equations. The physical mechanism of excitation of counter propagating backward waves is described in [9, 14].

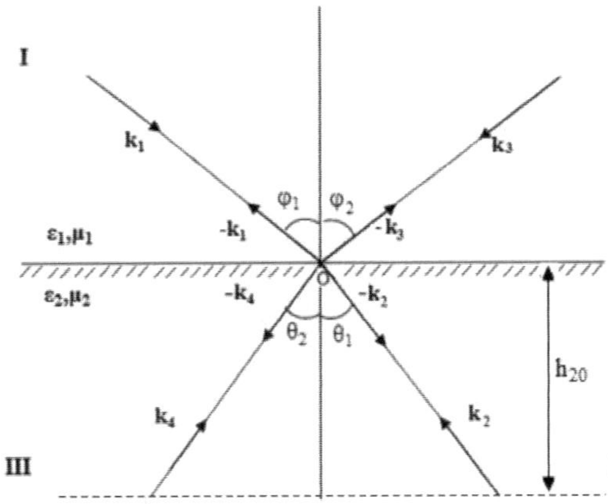

Figure 1. Wave structure in bordering media.

Let a plane wave fall on the interface from the first medium at an angle φ_1, Figure 1. Medium are characterized by parameters $\varepsilon_1 = \varepsilon_1^{'} - i\varepsilon_1^{''}$, $\varepsilon_2 = \varepsilon_2^{'} - i\varepsilon_2^{''}$, $\mu_1 = \mu_2 = 1$. In this case $\varepsilon_1^{''} << \varepsilon_1^{'}$, $\varepsilon_1^{'} = 1$. In accordance with [15], we can search for a wave structure consisting of the following modes.

1. Incident wave:

$$\mathbf{E}_1^{p,s} = \mathbf{E}_0^{p,s} \exp\left[-i\mathbf{k}_1\mathbf{r}_1\right]. \tag{1}$$

2. Refracted wave:

$$\mathbf{E}_2^{p,s} = \mathbf{E}_1^{p,s}\left(1 - V_{12}^{p,s}(\varphi_1)\right)\exp\left[-i\mathbf{k}_2\mathbf{r}_2\right] = \mathbf{E}_1^{p,s}T_{12}^{p,s}(\varphi_1)\exp\left[-i\mathbf{k}_2\mathbf{r}_2\right]. \tag{2}$$

3. Backward wave:

$$\begin{gathered}\mathbf{E}_4^{p,s} = \mathbf{E}_1^{p,s}T_{12}^{p,s}(\varphi_1)\exp\left[-i(\mathbf{k}_2\mathbf{h}_3)\right]\exp\left[-i((-\mathbf{k}_2)(-\mathbf{h}_3))\right]\times \\ T_{21}^{p,s}(\theta_1)\exp\left[-i((-\mathbf{k}_1)(-\mathbf{r}))\right] = \mathbf{E}_1^{p,s}\left[1 - V_{12}^{p,s}(\varphi_1)^2\right]\exp\left[-2i(\mathbf{k}_2\mathbf{h}_2)\right].\end{gathered} \tag{3}$$

4. Reflected wave:

$$\begin{gathered}\mathbf{E}_5^{p,s} = \mathbf{E}_1^{p,s}\left[V_{12}^{p,s}(\varphi_1) + T_{12}^{p,s}(\varphi_1)\exp\left[-2i(\mathbf{k}_2\mathbf{h}_4)\right]V_{21}^{p,s}(\theta_1)\exp\left[-2i(\mathbf{k}_2\mathbf{h}_4^{p,s})\right]T_{21}^{p,s}(\theta_2)\right]\times \\ \exp\left[-i(\mathbf{k}_3\mathbf{r}_3)\right] = \mathbf{E}_1^{p,s}V_{12}^{p,s}(\varphi_1)\left\{1 + \left[1 - V_{12}^{p,s}(\varphi_1)^2\right]\exp\left[-4i(\mathbf{k}_2\mathbf{h}_4)\right]\right\}\exp\left[-i(\mathbf{k}_3\mathbf{r}_3)\right].\end{gathered} \tag{4}$$

Here

$$k_1 = k_3 = \frac{\omega}{c}, k_2 = k_4 = \frac{\omega}{c}\sqrt{\varepsilon_2}, \varphi_1 = \varphi_2, \theta_1 = \theta_2,\ T_{12}^{p,s}(\varphi_{1,2}) = 1 - V_{12}^{p,s}(\varphi_{1,2}),$$

$$T_{21}^{p,s}(\varphi_{1,2}) = 1 - V_{21}^{p,s}(\theta_{1,2}), V_{21}^{p,s}(\theta_{1,2}) = -V_{12}^{p,s}(\varphi_{1,2}), V_{12}^{p} = \frac{n_1\cos\theta_1 - n_2\cos\varphi_1}{n_1\cos\theta_1 + n_2\cos\varphi_1},$$

$$V_{12}^{s} = \frac{n_1\cos\varphi_1 - n_2\cos\theta_1}{n_1\cos\varphi_1 + n_2\cos\theta_1}.$$

The p and s symbols indicate the polarization of the waves, respectively, when the vector **E** is parallel to the plane of incidence and when the vector **E** is perpendicular to the plane of incidence.

From (3) it follows that the backward wave is the result of refraction at the interface of the counter propagating wave with the vector $\mathbf{k}_2$, (Figure 1). From (3) it follows that the refracted wave with the vector $-\mathbf{k}_4$ is the result of reflection on the interface of the same counter propagating wave. The second

Journal of Physics: Conference Series **2140** (2021) 012020 doi:10.1088/1742-6596/2140/1/012020

term in curly brackets in expression (4) is the refraction of the counter propagating wave to the refracted wave with the vector $-\mathbf{k}_4$.

Equations (3) – (4) contain unknown parameters $h_3^{p,s}$, $h_4^{p,s}$. The physical meaning of these parameters is that the depth to which the waves pass into the second medium as long as they are coherent in polarization of the incident wave at the interface. This phenomenon takes place for the following reason.

It is known that refracted waves are excited in the second medium due to the influence of electromagnetic radiation as a result of the interference of waves from secondary sources and incident radiation, as well as fluctuation electromagnetic waves due to the interaction of charged particles with each other [10]. Along with these fields, any medium has its own thermal isotropic unpolarized radiation.

Thus, as the depth of immersion in the second medium increases, the intensity of the refracted wave decreases due to the loss of energy for work on the displacement of charges, i.e. on heat and radiation by moving charges, including in the direction of the opposite refracted wave. At a certain depth, the intensity of the refracted wave becomes equal to the intensity of its own thermal radiation, and in this case the refracted wave completely transforms into a fluctuation state. In the future, taking into account this effect, we determine the parameters $h_3^{p,s}$, $h_4^{p,s}$ corresponding to equations (3), (4).

3. Backward reflection

Let the interface be a flat surface. According to (3), the intensity of the backward wave (the intensity of radiation from the second medium) is written in the form:

$$J_1^{p,s}(\varphi_1) = J_0^{p,s}\left[1 - V_{12}^{p,s}(\varphi_1)^2\right]^2 \exp\left(-4\operatorname{Im}(k_2)\right)h_{30}^{p,s}. \tag{6}$$

The intensity absorbed in the second medium can be represented as

$$J_2^{p,s}(\varphi_2) = J_0^{p,s}\left[1 - V_{12}^{p,s}(\varphi_1)^2\right]. \tag{7}$$

Then, according to Kirchhoff's law, we have

$$J_1^{p,s}(\varphi_1) = \Theta(\omega,T)J_2^{p,s}(\varphi_1). \tag{8}$$

Here $\Theta(\omega,T)$ is the Planck function.

Taking into account (6) and (8), we write the equality

$$J_0^{p,s}\left[1 - V_{12}^{p,s}(\varphi_1)^2\right]^2 \exp\left(-4\operatorname{Im}(k_2)\right)h_{30}^{p,s} = J_0^{p,s}\Theta_0(\omega,T)\left[1 - V_{12}^{p,s}(\varphi_1)^2\right]. \tag{9}$$

From here

$$h_{30}^{p,s} \equiv h_{30}^{p,s}(\varphi_1) = -\frac{1}{4\operatorname{Im}}\ln\left|\frac{\Theta_0(\omega,T)}{\left[1 - V_{12}^{p,s}(\varphi_1)^2\right]}\right|. \tag{10}$$

Substituting (10) into (6) we write backward reflection intensity at the smooth boundary in the form:

$$J_1^{p,s}(\varphi_1) = J_0^{p,s}A^{p,s}\Theta_0(\omega,T)\left[1 - V_{12}^{p,s}(\varphi_1)^2\right]. \tag{11}$$

The coefficient $A^{p,s}$ is determined from the conditions that for $\varphi_1 = 0$ there must be

$$J_1^{p,s}(0) = J_0^{p,s}\left|V_{12}^{p,s}(0)\right|^2,\ J_2^{p,s}(0) = J_0^{p,s}\left|V_{12}^{p,s}(0)\right|^2$$

wherefrom

$$A^{p,s} = \left|V_{12}^{p,s}(0)\right|^2 \Big/ \Theta_0(\omega,T)\left[1 - V_{12}^{p,s}(0)^2\right].$$

Thus, the intensity of the backward reflection for a flat interface between the media and a plane incident wave is

$$J^{p,s} = J_0^{p,s} \frac{\left|V_{12}^{p,s}(0)\right|^2}{\left[1 - V_{12}^{p,s}(0)^2\right]} \left[1 - V_{12}^{p,s}(\varphi_1)^2\right]. \tag{12}$$

If the interface is rough (Figure 2)

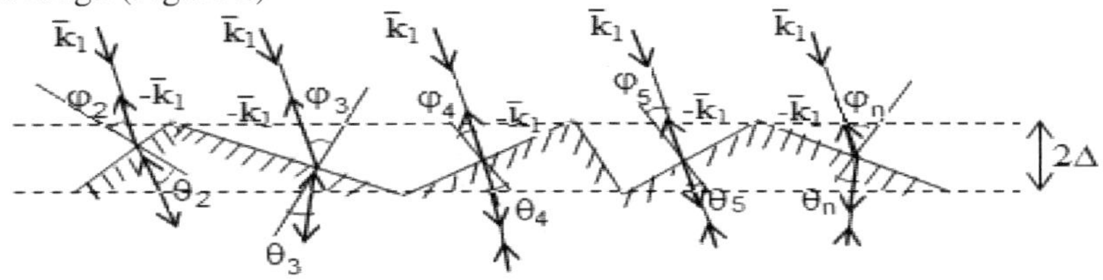

Figure 2. Backward reflection of a plane wave from a rough interface.

In the case of a rough interface, the Fresnel coefficients can be represented in the form [11]

$$V_{12}^{p,s}(\varphi_1) \to V_{12}^{p,s}(k_1\Delta, \varphi_1) = V_{12}^{p,s}(\varphi_1)\exp\left(-2(k_1\Delta)^2 \cos^2\varphi_1\right), \tag{13}$$

where $\exp\left(-2(k_1\Delta)^2 \cos^2\varphi_1\right)$ is the Rayleigh factor, Δ is the standard deviation of rough. Since in this case the backward reflection at a rough interface is divided into backward reflection with individual randomly oriented sites, then we introduce the function

$$\Phi(k_1\Delta, \varphi_1, \omega, T) = F(k_1\Delta, \varphi_1) + \Theta(\omega, T). \tag{14}$$

Here is the function that takes into account the randomly distributed radiation of the medium in the form of backward waves, i.e. it is somewhat similar in meaning to the Planck function describing discrete radiation quantum objects medium. Based on Kirchhoff's law, we write an equation similar to (8).

$$J_1^{p,s}(k_1\Delta, \varphi_1) = \Phi(k_1\Delta, \varphi_1, \omega, T) J_2^{p,s}(k_1\Delta, \varphi_1), \tag{15}$$

where

$$J_1^{p,s}(k_1\Delta, \varphi_1) = J_0^{p,s}\left[1 - V_{12}^{p,s}(k_1\Delta, \varphi_1)^2\right]^2 \exp\left(-4\operatorname{Im}(k_2) h_3^{p,s}(k_1\Delta, \varphi_1)\right),$$

$$J_2^{p,s}(k_1\Delta, \varphi_2) = J_0^{p,s}\left[1 - V_{12}^{p,s}(k_1\Delta, \varphi_1)^2\right].$$

Let us $F(k_1\Delta, \varphi_1)$ choose the function

$$F(k_1\Delta, \varphi_1) = A \int_0^{\pi/2} G(\varphi_1 - \varphi_n)\left[1 - \exp\left(-2(k_1\Delta)^2 \cos(\varphi_n)\right)\right] d\varphi_n. \tag{16}$$

Here φ_n is a random variable (Figure 3), distributed in the interval $0 \le \varphi_n \le \pi/2$.

We put in (16) $G(\varphi_1 - \varphi_n) = \delta(\varphi_n - \varphi_1)$, $\delta(\varphi_n - \varphi_1)$ is the delta function. Taking into account (16), equation (15) will take the form:

$$J_0^{p,s}\left[1 - V_{12}^{p,s}(k_1\Delta, \varphi_1)^2\right]^2 \exp\left(-4\operatorname{Im}(k_2) h_3^{p,s}\right) =$$

$$J_0^{p,s}\left[A^{p,s}\left[1 - \exp\left(-2(k_1\Delta)^2 \cos^2\varphi_1\right)\right] + \Theta(\omega, T)\right] \times \left[1 - V_{12}^{p,s}(k_1\Delta, \varphi_1)\right].$$

From here

$$h_3^{p,s} = -\frac{1}{4\,\mathrm{Im}(k_2)}\ln\left|\frac{A\left[1-\exp\left(-2(k_1\Delta)^2\cos^2\varphi_1\right)\right]+\Theta(\omega,T)}{\left[1-V_{12}^{p,s}(k_1\Delta,\varphi_1)^2\right]}\right|. \qquad (17)$$

Then the intensity of the back reflection will be written as

$$J_1^{p,s}(k_1\Delta,\varphi_1) = J_0^{p,s}\left[1-V_{12}^{p,s}(k_1\Delta,\varphi_1)^2\right]\left\{A^{p,s}\left[1-\exp\left(-2(k_1\Delta)^2\cos^2\varphi_1\right)\right]+\Theta(\omega,T)\right\}.$$

For $\varphi_1 = 0$, there should be $J_1^{p,s}(k_1\Delta,0) = J_0\left|V_{12}^{p,s}(k_1\Delta,0)\right|^2$. From here, we determine the parameter $A^{p,s}$:

$$A^{p,s} = \left[\frac{\left|V_{12}^{p,s}(k_1\Delta,0)\right|^2}{\left[1-V_{12}^{p,s}(k_1\Delta,0)^2\right]} - \Theta(\omega,T)\right]\frac{1}{1-\exp\left(-2(k_1\Delta)^2\right)}.$$

Thus, the intensity of the backward reflection with a rough boundary has the form:

$$\left[\left[\frac{\left|V_{12}^{p,s}(k_1\Delta,0)\right|^2}{\left[1-V_{12}^{p,s}(k_1\Delta,0)^2\right]} - \Theta(\omega,T)\right]\times\frac{1-\exp\left(-2(k_1\Delta)^2\cos^2\varphi_1\right)}{1-\exp\left(-2(k_1\Delta)^2\right)} + \Theta(\omega,T)\right]. \qquad (18)$$

If $k_1\Delta = 0$, then (18) coincides with (12).

4. Brewster's phenomenon in backward reflection

The well-known Brewster phenomenon is that when the interface is irradiated in a specular reflected field, the polarization component parallel to the plane of incidence decreases [1]. In accordance with macroscopic electrodynamics, this effect is caused by the existence of a minimum of the Fresnel reflection coefficient at an angle of incidence $\varphi_i = \varphi_{Br} = arctg\left|\frac{n_2}{n_1}\right|$, i.e. the condition is satisfied $\left|V_{12}^{p}(\varphi_i = \varphi_{Br})\right| < \left|V_{12}^{p}(\varphi_i)\right|$. In the case of clear media $\left|V_{12}^{p}(\varphi_i = \varphi_{Br})\right| = 0$.

In the multimode reflection and refraction we are considering, the coefficient $V_{12}^{p}(\varphi_1)$ enters into all mode equations, except for the incident wave, where φ_1 is the angle of incidence. Therefore, we will consider the behavior of the modes at $\varphi_1 = \varphi_{Br}$.

From (4) it can be seen that specular reflection consists of two parts: reflection from the surface and reflection from the volume of the medium. Let us assume that these contributions are independent. Then the parameter $h_4^{p,s}(\varphi_1)$ is determined from the equation following from Kirchhoff's law.

$$\left|V_{12}^{p,s}(\varphi_1)\right|^2\left[\left[1-V_{12}^{p,s}(\varphi_1)^2\right]\right]^2\exp\left(-8\,\mathrm{Im}(k_2)h_4^{p,s}(\varphi_1)\right) = \Theta(\omega,T)\left[1-V_{12}^{p,s}(\varphi_1)^2\right].$$

From here $h_4^{p,s} = -\frac{1}{8\,\mathrm{Im}(k_2)}\ln\left|\frac{\Theta(\omega,T)\left[1-V_{12}^{p,s}(\varphi_1)^2\right]}{\left|V_{12}^{p,s}(\varphi_1)\right|^2\left[\left[1-V_{12}^{p,s}(\varphi_1)^2\right]\right]}\right|.$

The intensity of the specular reflection is

$$J_{omp}^{p,s}(\varphi_1) = J_0^{p,s}\left\{\left|V_{12}^{p,s}(\varphi_1)\right|^2 + \Theta(\omega,T)\left[\left[1 - V_{12}^{p,s}\right]\right]\right\}. \tag{19}$$

If the interface is rough, then in (19) should be replaced, $V_{12}^{p,s}(\varphi_1)$ (13).

It follows from (19) that the secularly reflected field at $\varphi_1 = \varphi_{Br}$ has a minimum (as expected). This reflection also decreases with an increase in the parameter $k_1\Delta$, i.e. with an increase in roughness. However, there is no complete attenuation of the reflection. This is explained by the fact that for rough surfaces the imaginary part of the reflection coefficient changes very slowly at the Brewster angle, without reaching zero. For example, in conducting media with the appearance of small roughness on a flat boundary, the Brewster angle shifts towards smaller angles of incidence [12], which leads to the restoration of the reflection coefficient of an equivalent relatively smooth boundary.

It can be seen from (18) that the back reflection, on the contrary, increases with $\varphi_1 = \varphi_{Br}$. An increase in the intensity of vertically polarized backscattering was noticed in many works, however, the closest in the mechanism of excitation of back reflection is work [13], where the enhancement of backscattering in the volume of an "ice-water" medium was established. The explanation of this effect, according to the authors, is the transfer of field energy from the volume of air to the specified medium.

The back reflection increases with increasing $k_1\Delta$ at $\varphi_1 > 0$. At $\varphi_1 = 0$, the back reflection decreases with increasing $k_1\Delta$. This conclusion is confirmed by the results of many theoretical and experimental studies on the angular dependence of the back reflection of various media [14, 15].

From (2) it can be seen that at $\varphi_1 = \varphi_{Br}$ the amplitude of the refracted wave increases, it also increases with growth $k_1\Delta$.

Thus, a broader manifestation of the Brewster effect occurs when the interface is irradiated by a plane wave. At the Brewster angle, the backward and refracted waves are increased; the specularly reflected wave is decreased.

5. Calculated and experimental results

Let us present and compare the calculated and experimental results carried out on an important practical example of studying the angular dependence of the back reflection coefficients in radar sounding of natural and artificial environments.

As you know, a characteristic feature of the radar is the presence of a narrow radiation pattern, in which a narrow beam (beam) of radiation is formed. This radiation falls on a limited area of the surface. Therefore, the reflected signal is formed from the local area of the second medium (Figure 3)

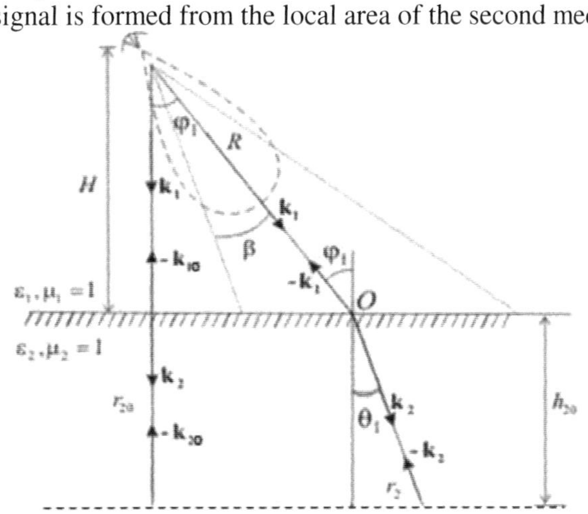

Figure 3. Scheme of radar sounding of bordering media.

Reflection from part of the surface is clipped by the radiation pattern. To calculate the reflected field in the case of a radar one, first, we represent the field of a point source in the form of an expansion in plane waves (4):

$$\mathbf{E}^{p,s}\left(k_1\Delta,\varphi_1\right)=\mathbf{E}_0^{p,s}\frac{\exp\left(-2ikr\right)}{R^2}=-\frac{ik_1}{2\pi}\mathbf{E}_0^{p,s}\iint_{\infty}\exp\left(-2i\left(\mathbf{k}_2\mathbf{r}_2\right)\right)W_{12}^{p,s}\left(k_1\Delta,\varphi_1\right)\frac{dk_{2x}dk_{2y}}{k_{2z}}. \quad (20)$$

$$W_{12}^{p,s}\left(k_1\Delta,\varphi_1\right)=\left[1-V_{12}^{p,s}\left(k_1\Delta,\varphi_1\right)^2\right]\exp\left(-2i\left(\mathbf{k}_2\mathbf{h}_3^{p,s}\right)\right),\ \mathbf{k}_2=\left\{k_{2x},k_{2y},k_{2z}\right\},\ \mathbf{r}_2=\left\{x,y,z\right\},$$

$\mathbf{h}_3^{p,s} \parallel \mathbf{r}_2$, $h_3^{p,s}$ defined by the expression (17).

Calculating expression (20), for example, by the method of the fastest descent and taking into account the directivity of the radiation of a point source, we obtain that the intensity of the reflected field entering the radar receiver is

$$J^{V,H}\left(k_1\Delta,\varphi_1\right)=J_0^{V,H}\left(k_1\Delta,\varphi_1\right)\left\{\left|V_{12}^{V,H}\left(k_1\Delta,0\right)\right|^2\exp-\left(\frac{\varphi_1}{\beta}\right)^2+\right.$$

$$\left.\left[\left[1-V_{12}^{V,H}\left(k_1\Delta,\varphi_1\right)^2\right]\left[\exp-\left(2\left(k_1\Delta\right)^2\cos^2\varphi_1\right)-1\right]\right]\cos\varphi_1\left(1-\exp-\left(\frac{\varphi_1}{\beta}\right)^2\right)\right\}, \quad (21)$$

where β is the width of the antenna directional pattern, Δ is the standard deviation of rough. The V and H symbols denote the components of the vertical and horizontal polarization of the reflected field. Hence the coefficients of the backward reflection of the copolarizations

$$\sigma_{VV}\left(k_1\Delta,\varphi_1\right)=\frac{J^V\left(k_1\Delta,\varphi_1\right)}{J_0^V},\quad \sigma_{HH}\left(k_1\Delta,\varphi_1\right)=\frac{J^H\left(k_1\Delta,\varphi_1\right)}{J_0^H} \quad (22)$$

Comparison of the results of the back reflection coefficients according to model (22) is in good agreement with the measured data from [16, 17]. The Figure 4 shows an estimate of the comparison of experimental data backward reflection coefficients from [17] with calculations (22), depending on the roughness of the interface between the media.

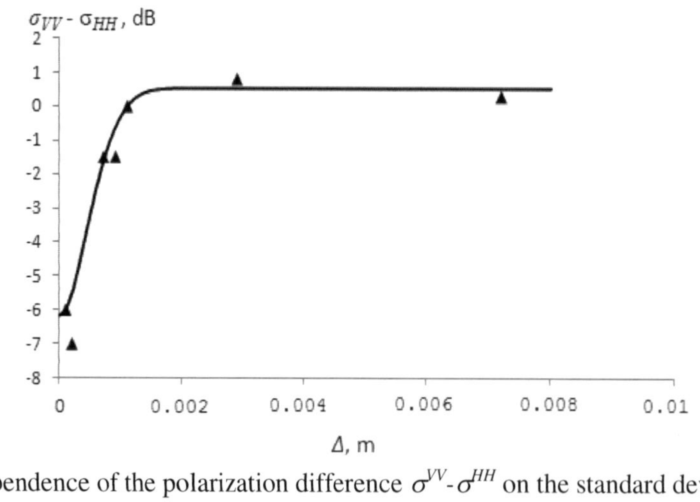

Figure 4. Dependence of the polarization difference σ^{VV}-σ^{HH} on the standard deviations Δ of the roughness heights at a radiation frequency f = 94 GHz and an angle of incidence φ_1 = 70°.

It can be seen from the figure that at the frequency of millimeter wave radiation, the polarization difference in the backward reflection coefficients strongly depends on the roughness of the interface between the media. With an increase in roughness, the polarization difference noticeably increase up to a certain limit, which is determined by the Rayleigh factor, i.e. when the wavelength and the size of the roughness are equal, the polarization difference tends to 0, and then, with an increase in the roughness, it remains constant.

6. Conclusion

Within the framework of macroscopic electrodynamics, the multimode mechanism of wave propagation in neighboring homogeneous absorbing media includes reflected and refracted waves, as well as counter propagating waves that form back reflection.

It is shown that the back reflection consists of two parts: radiation from the interface and radiation from the volume of the near-surface layer. The volume component of the back reflection is the radiation of secondary sources excited by a refracted wave in the subsurface layer with a thickness determined from equation (17).

Backward reflection analytical expressions are obtained for homogeneous, isotropic, absorbing media.

The formation of backward reflection on the surface of media strongly depends on their irregularities, and the dependence is ambiguous for different polarizations. The effect of amplification of the backward reflection for a Brewster angle is established.

References

[1] Sivukhin D V 1980 *Overall course of physics* (Moscow: Nauka)
[2] Kizel V A 1973 *Reflection of light* (Moscow: Nauka)
[3] Volpyan O D, Kuzmichev L M 2012 *Negative wave refraction* (Moscow: Publishing house Avers) p 359
[4] Veselago V G 1967 *Uspekhi fizicheskikh nauk* **92**(3) 517–526 doi: 10.3367/UFNr.0092.196707d.0517
[5] Mandelstamm L I 1947 – 1955 *Complete collection of works* (Leningrad: Publishing house of the Academy of Sciences of the USSR)
[6] Agranovich V M Gartstein Yu N 2006 *Uspekhi fizicheskikh nauk* **106**(10) 1051–1068
[7] Silin R A 2006 *Uspekhi fizicheskikh nauk* **176**(5) 562–565 doi: UFNr.0176.200605j.0562
[8] Rautian S G 2008 *Uspekhi fizicheskikh nauk* **128**(10) 1017–1024
[9] Lomukhin Yu L, Atutov E B, Butukhanov V P 2018 *IEEE Transaction on Antennas and Propagation* **66**(4) 1838–1845 doi: 10.1109/TAP.2018.2800643
[10] Rytov S M, Kravtsov Yu A, Tatarsky V I 1978 *Introduction to Statistical Radiophysics Part 2: Random fields* (Moscow: Nauka)
[11] Bass F G, Fuks I M 1972 *Scattering of waves on a statistically uneven surface* (Moscow: Nauka)
[12] Saillard M and Maystre D 1990 *Journal of the Optical Society of America A* **7**(6) 982–990
[13] Atwood D K, Gunn G E, Roussi C 2015 *IEEE Transactions on Geoscience and Remote Sensing* **53**(11) 5972 5982 doi: 10.1109/TGRS.2015.2429917
[14] Volosyuk V K, Kravchenko V F 2008 *Statistical theory of radio engineering systems for remote sensing and radar* (Moscow: Publishing house "Fizmatlit")
[15] Ulaby F T, Long DG 2014 *Microwave Radar and Radiometric Remote Sensing* (Ann Arbor: University of Michigan Press)
[16] Wei J C, Chen H, Qin X 2017 *IEEE Transactions on Antennas and Propagation* **65**(6) 3154–3161 doi:10.1109/tap.2017.2690307
[17] Adib N, Ulaby F T, Sarabandi K 1996 *IEEE Transactions on Geoscience and Remote Sensing* **34**(2) 561–572

Stability of synchronous modes of the system of two connected self-oscillators

S S Novikov, V S Kosterova

Tomsk State University, Tomsk, Russia

E-mail: nss@sibmail.com

Abstract. In this paper investigate the dynamic instability of a system of two self-oscillators with a strong resonant coupling. The analysis of influence of resonance properties of coupling on stability of synchronous modes was carried out. In the course of the computational experiment, it was possible to observe the different behavior of the system depending on the given initial conditions. The boundaries of destruction of coherent modes are determined.

1. Introduction

In systems of coupled self-oscillators with close frequencies, depending on the mutual coupling, various modes of oscillation: both ordinary coherent and incoherent [1]. The properties and features of the synchronous realization of the mode in self-oscillating systems with many diagnostic functions of the type have been sufficiently studied and can be classified. Such a classification is carried out according to design features that take into account the type of active elements, as well as circuitry methods for combining them into generating structures using transmission line segments, resonant volumes, dissipative load elements that form mutual coupling channels [2–4]. At the same time, attempts to classify should also be based on criteria arising from the concept of stability (or instability) of synchronous movements. Optimal from the point of view of stability of synchronous modes are resistive connections due to the introduction of dissipative load elements into the coupling channels of the self-oscillators. Depending on the type (sign) of the resistive coupling in the system of two self-oscillators, either in-phase or anti-phase (or close to them) synchronous oscillations are stable. In the case of purely reactive coupling, these modes are stable at the same time, which causes instability. In practice, especially in the field of ultrahigh frequencies, complete coupling circuits, in addition to external elements (sections of transmission lines, loads, etc.), include frequency-dependent elements of the output circuits of oscillators. Therefore, real coupling circuits have resonant capabilities, impact on stability [3]. Experience shows that the strong resonant properties of the coupling channels of a system of two self-oscillators can radically change the behavior of the system. An assessment of the effect of resonance properties on the stability of synchronous oscillations was carried out in [5]. Later, in [6–9], it was theoretically and experimentally shown that with strong resonant couplings, the basic synchronous regimes lose stability and a regime of dynamic chaos arises in their place. A significant feature of the model is that each of the self-oscillators in autonomous mode is an absolutely stable single-frequency system.

One of the important elements of the study of coherent multi-oscillators systems is the analysis of the dynamics of switching synchronous movements when changing control parameters. In this paper, the dynamic processes of the development of instabilities in a system of two self-oscillators with a

Content from this work may be used under the terms of the Creative Commons Attribution 3.0 licence. Any further distribution of this work must maintain attribution to the author(s) and the title of the work, journal citation and DOI.
Published under licence by IOP Publishing Ltd

strong resonant coupling are investigated by the method of numerical simulation. Using the analytical mode of stability of in-phase oscillations, the parameters of the instability of synchronous modes are estimated.

2. Resonant properties of coupling circuits, a criterion for the instability of the in-phase mode

As an example of typical resonance properties, Figure 1 shows [9] the profile of the frequency characteristics of the bond conductance $y_{12}(j\omega)$ of the simplest circuit shown in the inset. In this diagram, self-oscillators are connected to each other and to the total load $g_н$ by transmission line sections with electrical lengths θ_1 and θ_2 and with wave conductivity g_0. The resonant properties of the coupling conductance follow from the formula

$$y_{12} = \frac{-2s}{(\cos\alpha_S - 2s\cos\theta) + j\sin\alpha_S} g_0, \quad (1)$$

obtained in [2]. Where s and αS are module and argument of the scattering matrix element $s_{12} = s * \exp(-j\alpha_s)$; $\alpha_S=\theta_1+\theta_2\equiv\theta_\Sigma\sim\omega\tau$, τ is the signal propagation time in the coupling channel, $\theta=\theta_2-\theta_1$ is the asymmetry parameter, $g_н=2g_0$ is the conductivity of the matched total load. The s value lies in the range (0–0.5), the upper value corresponds to the absence of dissipative losses in the line segments.

The characteristic $Re\ y_{12}(j\omega)$ is an alternating function of frequency. The frequency range of series resonance, where $Re\ y_{12}<0$, corresponds to the first type of resistive coupling. With this type of connection, in-phase or similar vibration modes are stable. In the frequency range with resistive coupling of the second type ($Re\ y_{12}>0$), antiphase or similar modes are stable.

In the problem of local stability of synchronous oscillations in a system of two self-oscillators [1], the resonance properties are described by the parameter

$$C_{12} = d(\mathrm{Im}\, y_{12}(j\omega))/d\omega\big|_{\omega_0}. \quad (2)$$

The higher the quality factor of the resonance and the steeper the frequency dependence $Im\ y_{12}(j\omega)$, the greater the value of the parameter C_{12}. As can be seen from Figure 1, a strong resonant coupling of the first type is realized in narrow frequency intervals near $\alpha_S \approx 2\pi,\ 4\pi,\ldots$ where $Re\ y_{12}<0$. In these intervals, in the case of an almost symmetric chain ($\theta \approx 0$) and $s \to 0.5$, the parameter C_{12} is positive ($C_{12}>0$) and can theoretically reach arbitrarily large values.

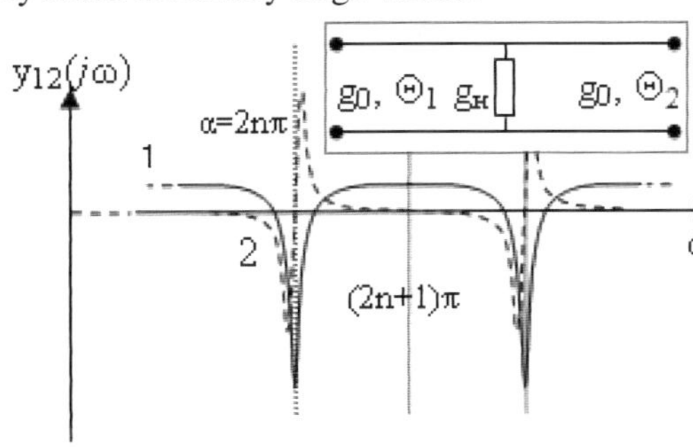

Figure 1. Resonant properties of the coupling conductivity: *1 – Rey₁₂; 2 – Imy₁₂ [9]*.

Analytical criterion was formulated in work [1]

$$C_{12}>C_{os}, \quad (3)$$

where C_{os} is the capacitance of the equivalent resonant systems of self-oscillators. When (3) is fulfilled, the in-phase regime becomes unstable and a transition occurs either to the antiphase regime (if it is stable) or to the dynamic chaos regime. In the latter case, the destruction of coherence takes place [10].

The indicated strong resonant properties of the mutual coupling of self-oscillators are also realized in circuits representing sections of the transmission line and loads: antisymmetric circuits, circuits with many loads in the wave coupling channel and also in circuits that include reactive elements [11].

Journal of Physics: Conference Series **2140** (2021) 012021 doi:10.1088/1742-6596/2140/1/012021

The scheme of the system under study, in accordance with which the self-oscillators are connected to each other and to the full load R_l through series resonant circuits, is shown in Figure 2. Let us analyze the resonant properties of the conductivity $y_{12}(j\omega)$ of the four-port network. We will consider it symmetrical for the purpose of simplification. In this case, the frequency characteristics are described by the analytical expression

$$y_{12}(j\omega) = -\frac{1}{\left(2 + \frac{z(j\omega)}{R_l}\right) z(j\omega)}, \qquad (4)$$

where $z(j\omega) = r + j\left(\omega L - \frac{1}{\omega C}\right) = r + jz(\omega)$ is the complex resistance of series $R\,L\,C$–loops.

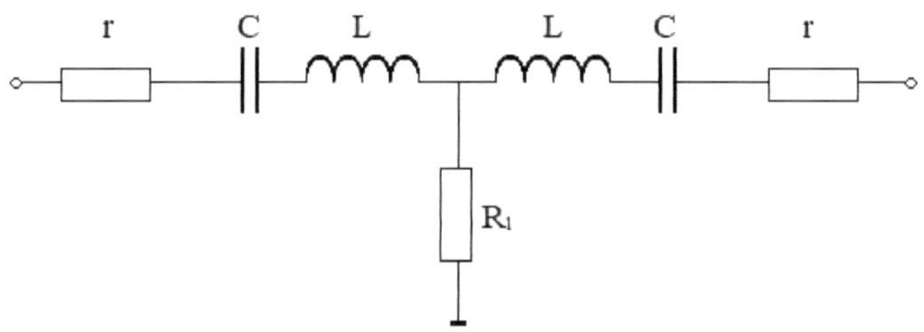

Figure 2. Resonant coupling circuit diagram.

The frequency characteristics profile is shown in Figure 3. As you can see, *Re* $y_{12}(j\omega)$ is an alternating function of frequency. The nature of the coupling in resonance region 2 corresponds to the type 1 resistive coupling: *Re* $y_{12}<0$. At the resonant frequency $\omega_0 = 1/\sqrt{LC}$, the coupling conductance is equal to the real negative value

$$y_{12}(j\omega_0) = -\frac{1}{\left(2 + \frac{r}{R_l}\right) r}.$$

The width of the resonant region is determined by the frequencies from the equation

$$\left(\omega L - \frac{1}{\omega C}\right)^2 = (2R_l + r)r \qquad (5)$$

It follows from (5) that the width of the resonance region narrows with decreasing r. It is also obvious that a decrease in the resonant frequency ω_0 by increasing the inductance L narrows the specified region, which corresponds to an increase in the resonance quality factor. With a significant detuning of the oscillator frequencies from the coupling resonance (in regions 1 and 3), we have $Re\,y_{12}>0$ and $Im\,y_{12}\neq 0$. As noted above, for a given sign of the real part of the coupling conductivity, antiphase oscillations should be stable.

The behavior of the imaginary part of the bond conductance ($Im\, y_{12}(j\omega)$) in the series resonance region is similar to the example considered above. The parameter C_{12} of criterion (3) for the resonance frequency ω_0 is determined analytically $C_{12} = L/r^2$, that up to a coefficient coincides with the expression given in [5] as a particular example.

Thus, the criterion for the instability of the in-phase mode in the system of self-oscillators connected through a circuit in Figure 2 has the form:

$$\frac{L}{r^2} > C_{os} \qquad (6)$$

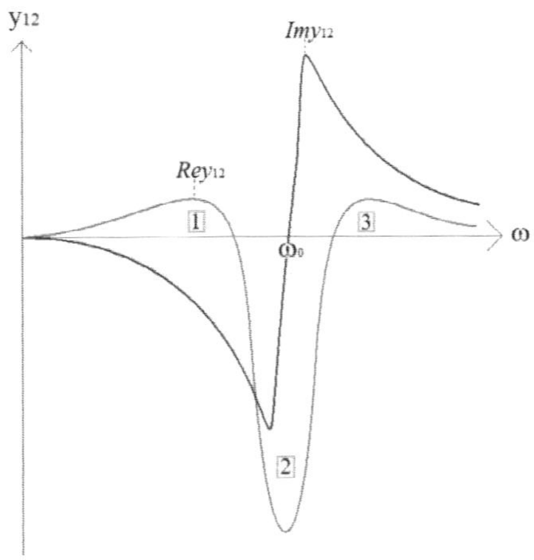

Figure 3. Circuit frequency characteristics.

Inequality (6) determines the way to control the stability of the in-phase mode with resonant tuning of the coupling circuit. Indeed, stability can be obtained for any L by choosing a sufficiently large value of the ohmic loss r. At sufficiently small losses, criterion (3) is satisfied and the in-phase mode becomes unstable. It should be noted that the parameter C_{12} remains positive in the frequency range, which is approximately estimated as

$$\omega_e \simeq \omega_0 \pm \frac{r}{2L}. \quad (7)$$

Thus, at typical Q-factors of 100–200, the frequency interval is very narrow.

At the same time, as follows from (7), with an increase in inductance, the width of the specified interval decreases in proportion to $1/L$, while the position of the resonance is proportional to $1/\sqrt{L}$. In this case, one can expect that a small (within the resonant tuning) shift of the resonance downward will be accompanied by an expansion of the frequency range in which the common-mode oscillations are unstable.

3. Numerical study of the stability of synchronous modes

The investigated circuit (Figure 4) represents two transistor self-oscillators connected to each other through a resonant coupling circuit. The self-oscillators are assembled according to the capacitive three-point scheme (the power supply circuits of the transistors are not shown in Figure 4). The resonant bipolar coupling circuit consists of sequential oscillatory circuits *r*, *C*, *L* and fully corresponds to Figure 2. The system is implemented in the NI Multisim program. The operating frequency of autonomous self-oscillators is of the order of 540 kHz, which is set by the elements of the oscillatory circuits C_1,$C_{,2}$. and L_1,$L_{,2}$. To study the temporal dynamics of oscillations in the in-phase and antiphase modes, step voltages of the same (+ +) or different (+ –) signs were introduced into the base circuit as initial conditions. The characteristics of oscillatory processes of self-oscillators at the stages of excitation, mutual transitions between types of synchronous oscillations and their suppression were observed and evaluated on two parallel oscillographic channels.

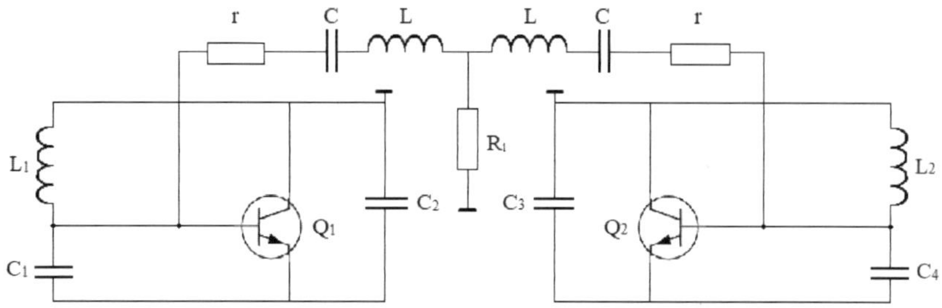

Figure 4. High-frequency circuit of a system of coupled self-oscillators with resonant coupling.

The main synchronous mode is the in-phase mode, corresponding to the summation of the power of self-oscillators in the total load R_l. Its stability is realized when tuning the coupling circuit to the resonance region 2 (Figure 3), where Re $y_{12}(j\omega)<0$ and only at sufficiently large values of r. In the case of complete symmetry of the circuit and with the identity of the oscillators, the frequency of stationary synchronous oscillations in the fundamental mode is equal to the partial frequencies of the oscillators. The antiphase mode at resonance tuning cannot exist due to large insertion losses (from the side of r). The conditions for the stability of antiphase oscillations are realized in regions 1 and 3 (Figure 3) where $Re\ y_{12}(j\omega)>0$, that is, with a significant detuning of the coupling resonance. On the one hand, such a detuning can be set by shifting the resonant frequency of the coupling circuits relative to the frequencies of the generators. On the other hand, detuning from resonance in antiphase mode occurs automatically due to shunting of resonant systems of generators with coupling circuit elements during dynamic short circuit of load resistance R_l. This situation always arises when the system goes into antiphase mode in the case of instability of in-phase oscillations; as a rule, such a transition is accompanied by increasing modulation and complete suppression of in-phase oscillations (Figures 5–7).

The conditions for stability of in-phase and antiphase oscillations are fulfilled at the edges of region 2 where где $Re\ y_{12}(j\omega)\approx 0$ and the coupling conductivity characterizes the reactive component. In a numerical experiment, these oscillations are realized by changing the initial conditions (+ +) or (+ –).

The main task of the research was to prove the plausibility of the criterion for the destruction of in-phase oscillations (6) at a high Q-factor of the coupling resonance. The following circumstance had to be taken into account when performing the experiment. The resonant frequency f_0 the coupling circuit r, C, L is uniquely determined, however, the complete coupling circuit also includes the elements of the output circuits of the oscillators themselves and the power supply decoupling elements. Since the estimate of the parameter $C_{12} = L/r^2$is valid only for the resonant frequency, at the preliminary stage it was necessary to localize the coupling resonance by selecting the values of the circuit elements. As it turned out, such a setting is not unambiguous, since a change in the control inductance L causes a change in the frequency f_0 and the parameter C_{12}. In addition, the width of the resonance region at small r is extremely small. Therefore, the resonance adjustment was proposed to be made on indirect reason related to the depth of inequality (6). Indeed, criterion (3) was formulated as a result of solving the problem of local stability of synchronous oscillations. Within the framework of this problem, the roots of the characteristic equation were found whose sign corresponds to phase stability or instability, and the quantity describes the initial rate of decrease or increase of small disturbances. Therefore, the stronger inequality (6) is, the faster the instability of the in-phase regime develops and vice versa with weakening of inequality. In the case of "neutral" stability, that is, immediately before the change in the sign of inequality, the rate of destruction of the in-phase mode should be very small.

In accordance with what has been said, in the model experiment, the time of restructuring of competing synchronous regimes into each other was estimated. This technique significantly complicated the study, but it was clear that only when tuning the coupling circuit to resonance with the frequencies of the oscillators is the possibility of unambiguous control of the parameter C_{12} realized with the help of ohmic losses r.

As a result of the described procedure, were selected L=1020 nH C=80 nF. Further, a series of experiments was carried out with a sequential change in resistance r in the range of 0.0001-0.01 Ohm. In this case, by direct observation of the oscillograms, the time of transition from an unstable in-phase mode to a stable antiphase one was estimated. In the range of 0.0001 - 0.008 Ohm, in-phase oscillations excited by the initial conditions (+ +) are unstable and the system goes into antiphase mode with a frequency of about 690 kHz. Phase restructuring is accompanied by characteristic self-modulation, which begins to develop after the active elements (transistors) enter the nonlinear mode (Figure 5). Further, the "modulation depth" increases, reaches 100% (Figure 6) and asymptotically decreases to zero. It should be noted a characteristic sign of the transient process: the envelopes of the oscillations are in antiphase. We can say that one oscillation alternately suppresses another.

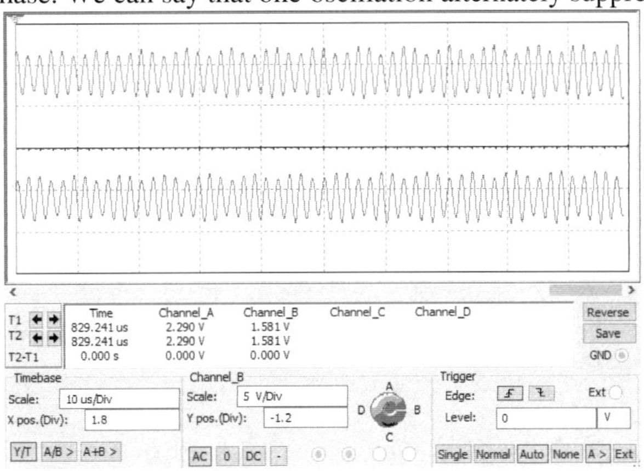

Figure 5. Oscillograms of oscillations of self-oscillators;start of modulation: L = 1020 nH, C = 80 nF.

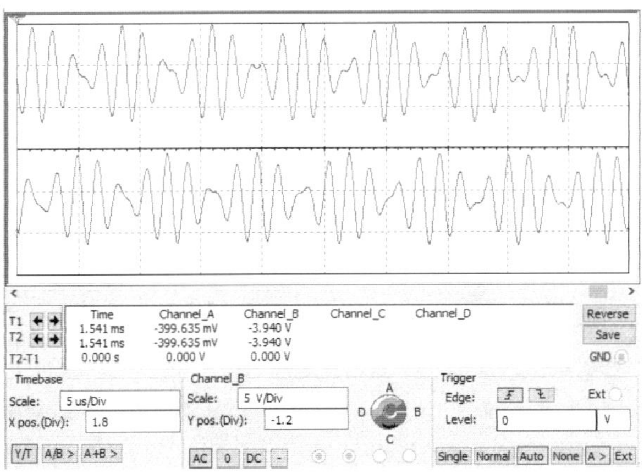

Figure 6. Oscillograms of oscillations of self-oscillators; modulation process:| L = 1020 nH, C = 80 nF.

The transition time in the specified interval varies from 1.5 ms to approximately 50 - 60 ms. On this basis, it can be assumed that near the upper limit r ≈ 0.008 Ohm, $C_{12} = L/r^2 \geq C$ takes place and the system is at the instability border. A further slight increase in r brings the system to stable in-phase oscillations; in this case, antiphase oscillations, initiated by the initial conditions (+ –), pass after a short self-modulation process into in-phase oscillations, or damp at a low level.

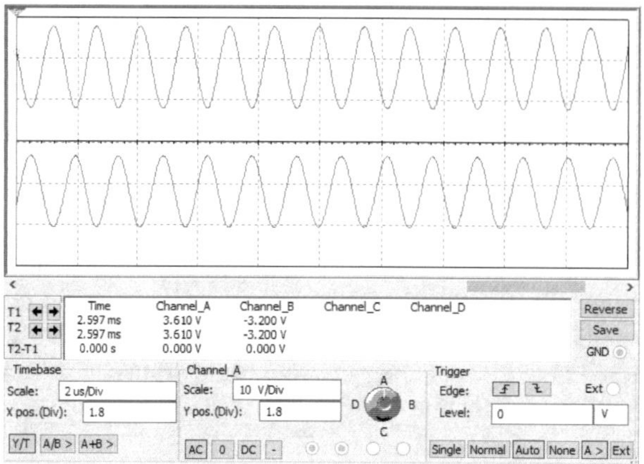

Figure 7. Oscillograms of oscillations; antiphase mode of operation: L = 1020 nH, C = 80 nF.

Further, the values of the inductances of the coupling circuit were increased four times (L = 4080 nH), and the values of the capacitances were reduced by the same amount (20 nF), which, by assumption, does not change the resonant tuning of the coupling circuit. As follows from (6), the boundary of the instability of the in-phase mode with respect to the parameter r should now be near $r \approx 0.016$ Ohm. Indeed, in a similar series of experiments with a sequential increase in resistance r, an extremely long (~ 60 ms) transition to a stable antiphase regime occurred at $r \approx 0.019$ Ohm. A fragment of the transient process for this case is shown in Figure 8.

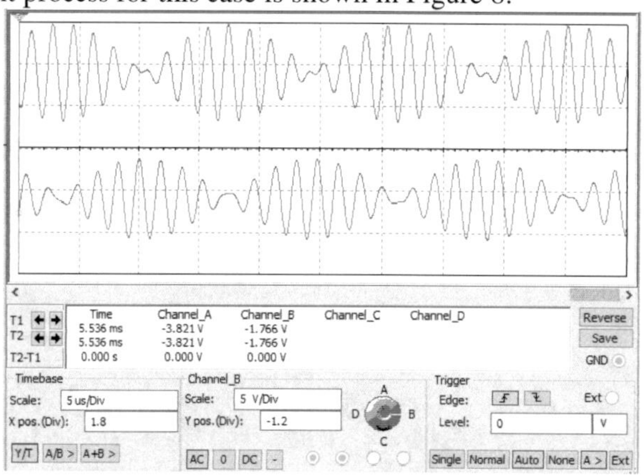

Figure 8. Oscillograms of oscillations of self-oscillators; transition to antiphase mode: L = 4080 nH, C = 20 nF.

Thus, the performed numerical experiment confirms the possibility of destruction of coherence in a system of coupled self-oscillators with strong resonant properties of the coupling channel. To assess these properties, an analytical criterion can be used that determines the phenomenological dependence of the stability condition on the parameters of the coupling chain.

4. Conclusion

In this paper, a system of two self-oscillators connected with each other and with a common load through series resonant circuits is considered. An analysis of the frequency characteristics of the interconnection parameter is carried out and an analytical criterion for the instability of in-phase oscillations is formulated. The character of dynamic processes of development of instabilities of synchronous regimes has been studied by the method of numerical modeling; it is shown that with resonance tuning of the coupling, both the conditions for the stability of in-phase oscillations and the conditions for their violation can be realized. The boundaries of instability of synchronous regimes are estimated and their compliance with the analytical criterion is shown. The results obtained can be useful in constructing systems with switching of coherent modes.

References

[1] Novikov S S, Usyukevich A A 2013 *Russian Physics Journal* **55**(11) 1296–1302.
[2] Novikov S S, Maidanovsky S A 2003 *Radio engineering and electronics* **48**(5) 595 – 600
[3] Vintizenko I I, Novikov C C 2009 *Relativistic magnetron microwave oscillators* (Tomsk: STL) p 430
[4] Shigeji N, Jenshan L, Tatsuo I 1993 *IEEE transactions on microwave theory and techniques* **41**(10) 1827–1837
[5] Vladimirov S N, Maidanovskii A S, Novikov S S 1993 *Nonlinear Oscillations of Multifrequency, Naturally Oscillatory Systems* (Tomsk: Tomsk State University Publishing House) p 203
[6] Novikov S S, Usyukevich A A 2012 *Applied Nonlinear Dynamics Izvestiya vuz.* **20**(5) 16–30
[7] Novikov S S, Usyukevich A A 2014 *Izvestiya vuz.. Physics* **57**(9/2)16–20
[8] Novikov S S 2017 *Progress In Electromagnetics Research Symposium* 1065–1069 doi: 10.1109/PIERS.2017.8261904.
[9] Novikov S S 2019 *Proceedings of the conference 8th International Conference APR collection* 200–205
[10] Novikov S S, Usjukevitch A A 2014 *Proceedings of the conference 24th International Crimean Conference Microwave & Telecommunication Technology* 109–110 doi:10.1109/CRMICO.2014.6959312
[11] Novikov S S, Usjukevitch A A 2015 *Proceedings of the conference 25th International Crimean Conference Microwave & Telecommunication Technology* **1** 525–526

Journal of Physics: Conference Series **2140** (2021) 012022 doi:10.1088/1742-6596/2140/1/012022

Flat UWB antenna with optimized ground plate

A Khalil and D Sukhanov

Tomsk State University, Tomsk, Russia

E-mail: sdy@mail.tsu.ru

Abstract. The design of a broadband antenna based on a combination of electric and magnetic emitters is proposed. Antenna size ratios are proposed that provide a wide operating frequency band. The results of numerical modeling of the standing wave ratio and radiation patterns for a particular case with a matching range from 13 GHz to 27 GHz are presented.

1. Introduction

Ultra-wideband (UWB) antennas are capable of transmitting and receiving ultra-short electromagnetic pulses. Usually wide bandwidth and low average power affect negatively on signal detection process; To overcome this, UWB antennas must receive all components of the signal spectrum with equal efficiency and without introducing significant phase distortion of these frequency components. Although, it is possible to implement compensating measures to counteract the effects of distortion, it is much preferable to look for antenna designs that introduce the lowest possible distortion in the phase response in order to simplify the overall system design [1–4].

Antenna development usually begins with defining the requirements for its characteristics and application features. We consider the following requirements for the ultra wide band (UWB) antenna:

1. 10 GHz bandwidth from 12 GHz to 22 GHz, where reasonable efficiency and satisfactory omnidirectional radiation pattern are required.
2. In the considered bandwidth, the highest level of radiation power should be provided.
3. The antenna transmits a signal with short pulses with minimal distortion over the entire frequency range.

The first point is the most important for antenna designers, which is expressed in the requirement that the antenna has a wide matching band, in which VSWR ≤ 2 [1,3,5–8]. This high impedance bandwidth is only realized when using a multi-resonant monopole printed antenna, which is usually exhibited by the characteristics of the high pass impedance [5,6]. For such a broadband antenna, in contrast to dipole or monopole antennas tuned to a single resonance, it is necessary to take into account some design features. The design parameters of these printed monopole antennas are not the resonant on operating frequency, but the lower cutoff frequency and the total bandwidth. The frequency of the lower band depends primarily on the maximum monopole height, while the antenna bandwidth depends on how well the impedance of the various modes is matched to the microstrip or coplanar feed line.

Dipole and monopole antennas are among the most popular antennas used in modern broadband wireless systems [3, 9]. These elements have a wide impedance band that can be expanded by increasing

Content from this work may be used under the terms of the Creative Commons Attribution 3.0 licence. Any further distribution of this work must maintain attribution to the author(s) and the title of the work, journal citation and DOI.
Published under licence by IOP Publishing Ltd

the radius of the cylindrical element. This is true up to the point where the stepped radius from the feed probe to the cylindrical member becomes abrupt. Narrowing this transition is useful and is used in broadband elements such as biconical dipoles, but the cost is limited in laboratory applications. However, it has been shown that replacing a cylindrical monopole or dipole element with flat elements provides a wider impedance bandwidth and, with some simple adjustment, a UWB bandwidth [6,3,7,8,10,11].

2. Antenna design

Consider the class of combined antennas [12]. Based on a combination of electric and magnetic emitters, it is possible to achieve matching over a wide frequency band. The geometry of the proposed planar monopole antenna with ground is shown in Figure 1. This antenna consists of two main parts: the first is a printed monopole of various widths, the second is an L matching loop. The length of the first part is $l_2 + l_3 + l_4$, it has an input matching line of l_1 length with variable width. The second part consists of 1) a line with a length of $l_5 + l_6$ and 2) a radial stub with an angle 90° and a radius equal to l_7 to match the impedance with the maximum available bandwidth. Both parts together form a small frame. The antenna has an incomplete ground plane, which has a height h_g and a width w_g.

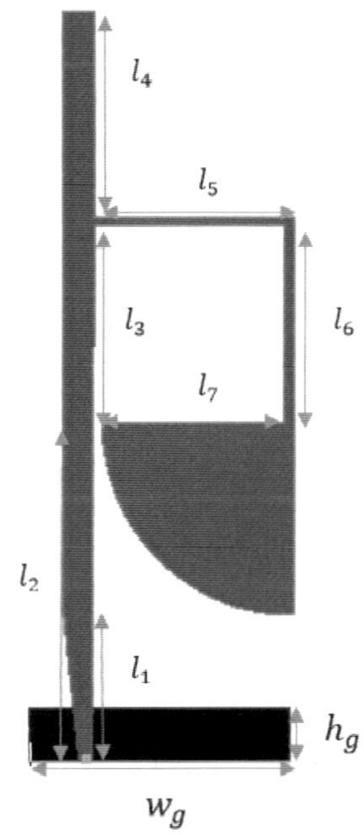

Figure 1. The geometry of the proposed UWB antenna.

For VSWR <2 in the selected 12-22 GHz frequency band, take into account some restrictions:

$2l_4 = l_5 + l_6$; $l_3 = l_6$; $l_5 = l_7 + \delta, 0 < \delta$; $l_2 = l_7 + l_1$.

Set $l_3 = 5$ mm; $l_4 = 6$ mm; $l_5 = 8.5$ mm; $l_6 = 5$ mm; $l_1 = 1$ mm; $l_2 = 5.5$ mm; $l_7 = 4.5$ mm; $h_g = 2$ mm; $w_g = 12$ mm. These values were obtained by selection in numerical modelling in order to expand the matching band.

To test the high performance of the proposed antenna VSWR simulation was carried out using the 4NEC2 program. Figure 2 shows the wired model of the antenna for 4NEC2 software. Dimensions and position of the antenna ground plate was optimized to ensure widest frequency range.

Figure 2. Wire antenna design for 4NEC2.

The simulation results are shown in Figures 2 and 3. It can be seen from the Figures 2-3 that the result meets the requirements. The antenna has an ultra bandwidth from 13 GHz to 27 GHz, which meets the system requirements of UWB and VSWR <2.

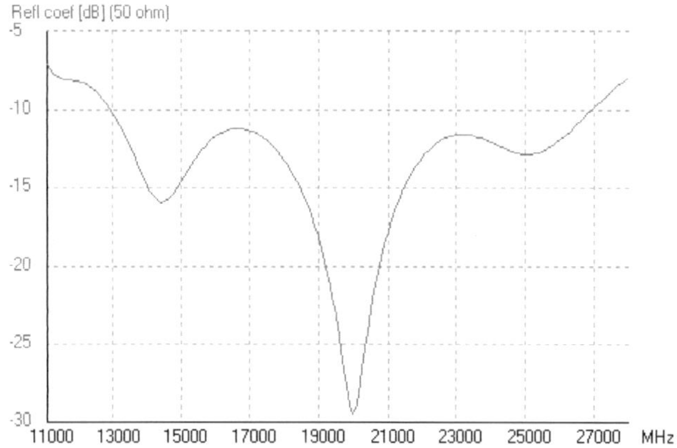

Figure 3. Reflection coefficient magnitude of proposed UWB antenna.

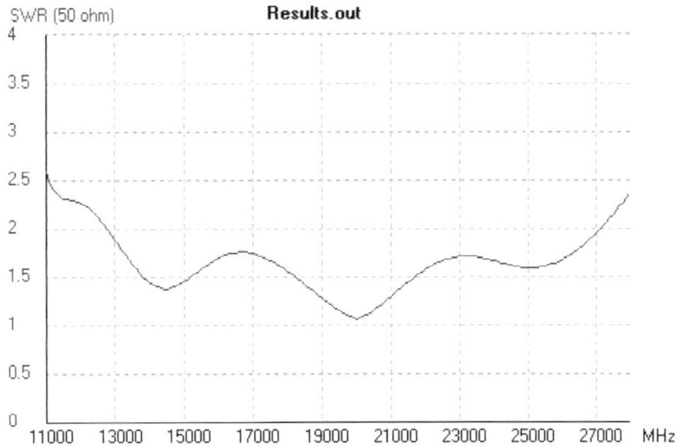

Figure 4. Voltage standing wave ratio of the proposed UWB antenna.

Simulation of directivity pattern of the antenna shows that the antenna is almost omni directional even on high operation frequency 26 GHz (Figure 5). At all other frequencies, the antenna is also omnidirectional. Omnidirectional antennas are in demand when implementing aperture synthesis technology or when creating antenna arrays. The orientation of the maximum of the radiation pattern shows that it is possible to manufacture an array of antennas on a printed circuit board, which should simplify the technology of manufacturing phased array systems.

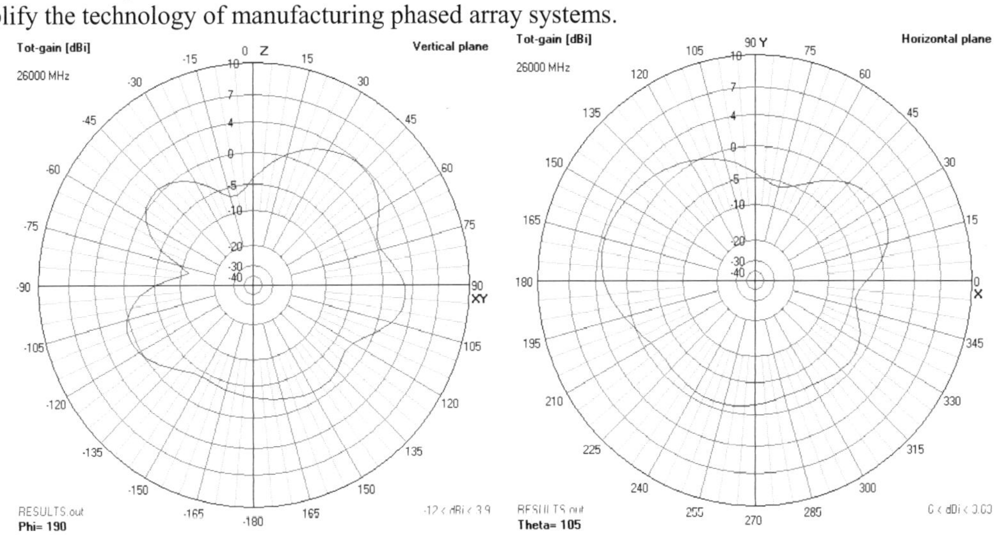

Figure 5. Directivity diagrams of the antenna on frequency 26 GHz.

3. Conclusion

A new monopole antenna was proposed for ultra wide band applications. It consists of a monopole and an L matching loop. The combination of electric and magnetic sources allows balancing the electrical and magnetic energy in the near field of the antenna and ensuring match with free space. A square frame was chosen as a magnetic source. Based on the requirements for the phasing of the two types of radiators, the requirements for the ratio of the antenna dimensions are specified. Further, by selecting the size and displacement of the ground plate, the optimal values for the widest antenna bandwidth are achieved.

Optimization was carried out using numerical simulation in the 4NEC2 software package. As a result of numerical simulation, the dependences of VSWR and Reflection coefficient on frequency were obtained. The resistance of the supply line of 50 ohms was considered. It has return loss less than -10 dB and VSWR less than 2 in the 14 GHz range from 13 GHz to 27 GHz. The constructed radiation patterns showed that this antenna is omnidirectional, which gives prospects for its application within the framework of aperture synthesis technology. Since the conductors in the antenna are oriented in the plane, this antenna can be manufactured using printed circuit board technology. The presence of a dielectric substrate will require re-optimization of the grounded plate.

Acknowledgments
The results were obtained as part of the implementation of the state task of the Ministry of Education and Science of Russia, project № 0721-2020-0038.

References
[1] Buyanov Yu I, Shipilov S E, Balzovsky E V 2020 *J. Phys.: Conf. Ser.* **1499** 012007 doi: 10.1088/1742-6596/1499/1/012007
[2] Galvan-Tejada G M, Peyrot-Solis M A, Aguilar H J 2015 *Ultra Wideband Antennas: Design, methodologies, and performance* (Florida: CRC Press)
[3] Kumar G, Kumar R 2019 *AEU - International Journal of Electronics and Communications* **109** 76-98 doi:10.1016/j.aeue.2019.07.004
[4] Satarov R N *et al* 2013 *Russian Physics Journal* **55**(8) 884–889 doi: 10.1007/s11182-013-9896-8
[5] Balzovsky E V, Buyanov Yu I, Koshelev V I, Nekrasov E S 2018 *IOP Conf. Ser.: Mater. Sci. Eng.* **363** 012002 doi: 10.1088/1757-899X/363/1/012002
[6] Ray K P 2008 *International Journal of Antennas and Propagation* **2008** 713858 doi: 10.1155/2008/713858
[7] Saeidi T, Ismail I, Wen W P, Alhawari A R, Mohammadi A 2019 *International Journal of Antennas and Propagation* **2019** 7918765 doi:10.1155/2019/7918765
[8] Allen B, Dohler M, Okon E, Malik W, Brown A, Edwards D 2007 *Ultra-wideband: Antennas and propagation for communications, radar and imaging* (USA: John Wiley & Sons)
[9] Balzovsky E V, Buyanov Yu I, Koshelev V I, Nekrasov E S 2017 *Journal of Physics: Conf. Ser.* **881** 012001 doi: 10.1088/1742-6596/881/1/012001
[10] Chen Z N, Liu D, Nakano H, Qing X, Zwick T 2016 *Handbook of Antenna Technologie.* (Singapore: Springer Singapore)
[11] Abdelraheem A M, Abdalla M A, Elregily H A, Mitkees A A 2012 *International Conference on Engineering and Technology (ICET)* **2012** 13270578 doi: 10.1109/ICEngTechnol.2012.6396155
[12] Balzovsky E V, Buyanov Yu I, Koshelev V I 2019 *IOP Conf. Ser.: Mater. Sci. Eng.* **516** 012043 doi:10.1088/1757-899X/516/1/012043

Journal of Physics: Conference Series **2140** (2021) 012023 doi:10.1088/1742-6596/2140/1/012023

Galvanic pH sensor for continuous monitoring of soil parameters in agriculture

M S Yuzhakov[1], D I Filchenko[1], A V Badin[1] and A K Berzin[2]

[1] National Research Tomsk State University, Tomsk, Russia

[2] National Research Novosibirsk State University, Novosibirsk, Russia

E-mail: miha.yuzhakov@gmail.com

Abstract. The article discusses trends in agriculture development in digitalization direction, substantiates need for monitoring pH level throughout year. The main solutions existing on market at moment are given, as well as need to develop a pH sensor for USKD-Agro hardware and software complex. The article discusses in detail the ongoing physical and chemical processes, on them basis pH level is determined, circuitry solutions used, as well as pH sensor implementation. The article describes the sensor calibration process, methodology and test results.

1. Introduction

Agriculture digitalization is an important issue around world. With increase in world's population, food products consumption increases. To increase food production, it is necessary not only to cultivate new territories, but also to monitor land plots that have been in use for a long time and have developed their potential. Digital and precision farming allows you to increase harvested crops amount without harming environment by rationalizing already plowed land.

This issue is especially relevant in Russia. This is confirmed by development he departmental Russian Federation Agriculture Ministry project "Digital Agriculture", the purpose of which is the project: agriculture digital transformation through the digital technologies

introduction of digital technologies and platform solutions to ensure a technological breakthrough in the agro-industrial complex and achieve a 2-fold increase in labor productivity at "digital" agricultural enterprises by 2024 [1]. The industry digital transformation trajectory at the agribusiness level is the massive implementation of digital Agro solutions [2].

In development course of hardware and software complex to monitoring climatic parameters, it became necessary to expand functionality by developing soil acidity sensor. Soil quality assessment is one of the most important preparatory stages before using the soil for agricultural needs. Timely and soil acidity state qualitative assessment will make it possible to make necessary changes to soil before sowing season beginning, in order to increase fertility for required crop.

Market research showed complete ready-made digital sensors lack with ability to microcontroller connect with further data transmission to network. At the moment, market for soil pH sensors can only offer independent devices, both analog and digital [3, 4]. Such separate devices are not suitable for recording soil pH levels dynamics of over a long time period.

To solve this problem, it was decided to study soil acidity sensors operation principle, develop its own sensor, carry out calibration and testing.

Content from this work may be used under the terms of the Creative Commons Attribution 3.0 licence. Any further distribution of this work must maintain attribution to the author(s) and the title of the work, journal citation and DOI.
Published under licence by IOP Publishing Ltd

2. Theory

In practice, potentiometric method for determining pH is common. This is a physicochemical method for measuring hydrogen ions activity based on measuring determination circuit consisting EMF and comparison electrodes immersed in test solution. A silver chloride is used for reference electrode, and a glass electrode is used for determination electrode. In this case, a change in hydrogen ions concentration leads to change in determination electrode potential, which leads to change in system EMF.

A system consisting of an aluminum and copper electrode is used to measure soil acidity. In chemical terms, it is interesting to understand the standard electrode potential, determine reaction course, and determine electrodes resulting system EMF. Determine reaction direction.

$$Cu^0 + 2Al^{3+} \leftrightarrow 3Cu^{2+} + Al^0, \tag{1}$$

compare copper and aluminum electrode potentials

$$Cu^{2+} + 2e = Cu^0 E^0_{\frac{Cu^{2+}}{Cu}} = +0.34 \text{ B}, \tag{2}$$

$$Al^{3+} + 3e = Zn^0 E^0_{\frac{Al^{2+}}{Al}} = -1.66 \text{ B}. \tag{3}$$

Since copper electrode potential is greater than aluminum potential. We come to conclusion that under standard conditions copper ion is a stronger oxidant. [5]

$$\text{ЭДС} = E^0_{\frac{Cu^{2+}}{Cu}} - E^0_{\frac{Zn^{2+}}{Zn}} = 0.34 - (-1.66) = 2 \text{ B}, \tag{4}$$

in this case, copper ion is an oxidizing agent, and aluminum is a reducing agent.

When soil acidity changes, EMF changes. The normal aluminum electrode potential an acidic medium is -1.66 V, while in an alkaline medium this potential is -2.4 V. In this case, aluminum electrode potential values of the aluminum electrode potential of in solutions 3 <pH <7 tends to zero. This phenomenon is explained by an oxide film formed on electrode surface. Good solubility of aluminum oxides is observed in acidic solutions with pH <3, in alkaline solutions with pH> 7 and determines metal chemical activity in these ranges. It is possible to increase metal activity range from 3 to 7 pH by removing the oxide film mechanically [6]. Table 1 shows standard electrode potentials values for various metals.

Table 1. Standard electrode potentials.

Electrode reaction	EMF, V	Electrode reaction	EMF, V
$Li^+ + e = Li$	-3.045	$Cr^{3+} + 3e = Cr$	-0.74
$Rb^+ + e = Rb$	-2.92	$Fe^{2+} + 2e = Fe$	-0.44
$K^+ + e = K$	-2.92	$Cd^{2+} + 2e = Cd$	-0.402
$Ba^{2+} + 2e = Ba$	-2.90	$2H^{2+} + 2e = H_2$	±0.0
$Sr^{2+} + 2e = Sr$	-2.89	$Cu^{2+} + 2e = Cu$	+0.337
$Ca^{2+} + 2e = Ca$	-2.87	$Ag^{2+} + e = Ag$	+0.799
$Al^{3+} + 3e = Al$	-1.66	$Pt^{2+} + 2e = Pt$	+1.20
$Zn^{2+} + 2e = Zn$	-0.763	$Au^{2+} + e = Au$	+1.68

3. Calibration and measurement methods

Potential difference measurement across electrodes with a conventional voltmeter can only be performed with a measurement error due to electrodes polarization and other effects. Potential difference measuring accuracy is achieved by compensation method. The connection diagram is shown in Figure 1.

Journal of Physics: Conference Series **2140** (2021) 012023 doi:10.1088/1742-6596/2140/1/012023

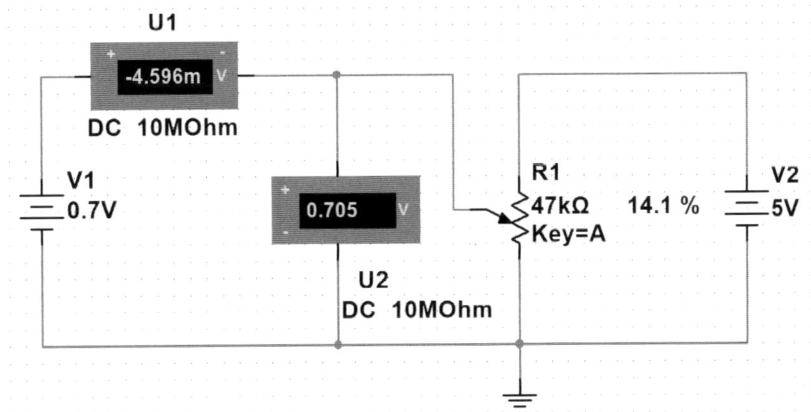

Figure 1. Compensation method diagram for measuring electrodes EMF.

For Figure 1, V1 – simulates electrodes with EMF 0.7V, V2 – power supply, R1 – potentiometer for voltage compensation, U1 and U2 voltmeters for measuring voltage difference and voltage across potentiometer. External installation view for measuring EMF by compensation method is shown in Figure 2.

Method physical meaning consists in potential difference equation across electrodes with a potential at a previously known resistance using an external power source, thus creating an equilibrium state. When galvanic cell EMF is infinitely small higher than external EMF, then conditions necessary for reversible processes occurrence are created, and cell electrical work at a given time will be maximum. The voltage measured at this moment by a high-resistance voltmeter at galvanic cell terminals is equal to its EMF. This scheme was used to calibrate sensor.

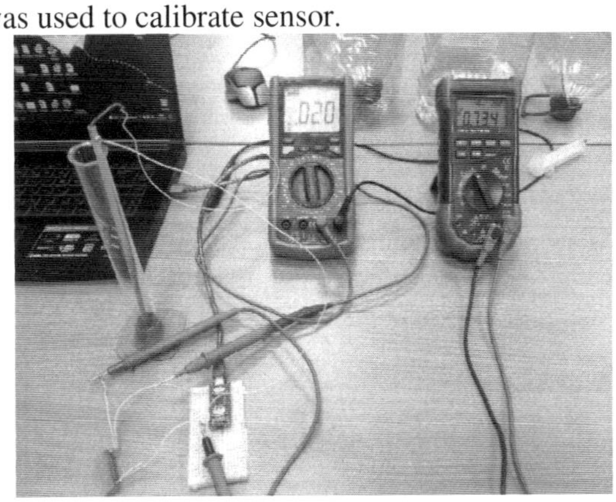

Figure 2. Installation external view for measuring by the compensation method.

Such a measurement scheme is inconvenient and requires automation. Figure 3 shows automated measurement diagram by compensation method using the MCU.

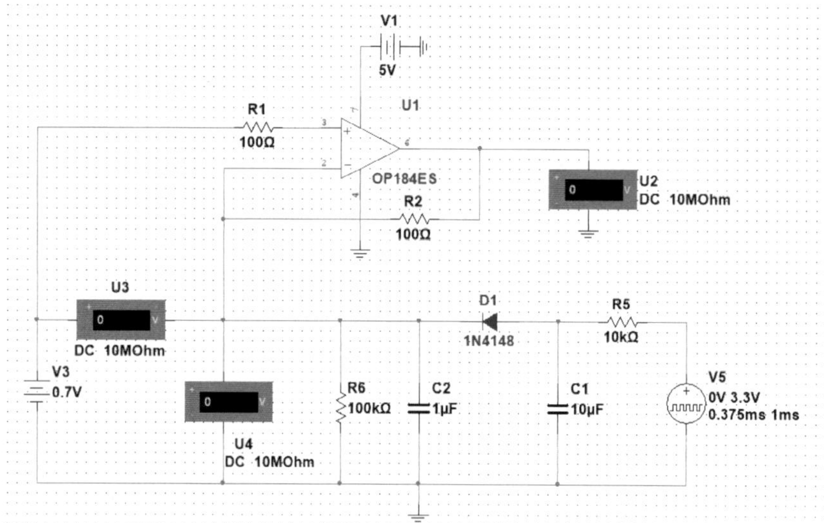

Figure 3. Measurement of EMF of electrodes by the compensation method:V5 – simulate PWM signal from MCU, R5 and C1 – form a low-pass filter for filtering PWM signal RF harmonics, D1, C2 and R6 – form an AM detector, V3 – simulate electrodes, U1 – an operational amplifier connected according to differential amplifier circuit, R1 and R2 – set gain, U2 – simulates ADC MCU.

To adjust compensation voltage, instead of potentiometer, this circuit uses a PWM signal from an MCU and a circuit for converting a PWM signal into a constant voltage. With differential amplifier help, a voltage equal to difference between compensation and electrode voltages is applied to MCU. Program code logic - PWM signal duty cycle increases until signal on microcontroller ADC reaches specified minimum level. However, this scheme did not justify itself due to measurement algorithm complexity and error in determining amplitude at AM detector output based on PWM signa duty cycle data.

A simpler measurement circuit was assembled and tested. Diagram is shown in Figure 4. In it, voltage at AM detector output is summed with electrodes EMF and read out using an ADC. Electrodes EMF must be shifted upward by a certain constant component so that MCU ADC channel can correctly process received signal. Such a circuit, in contrast to compensation circuit, cannot show EMF actual value on electrodes, but this is not required. It is enough to calibrate sensor using buffer solutions.

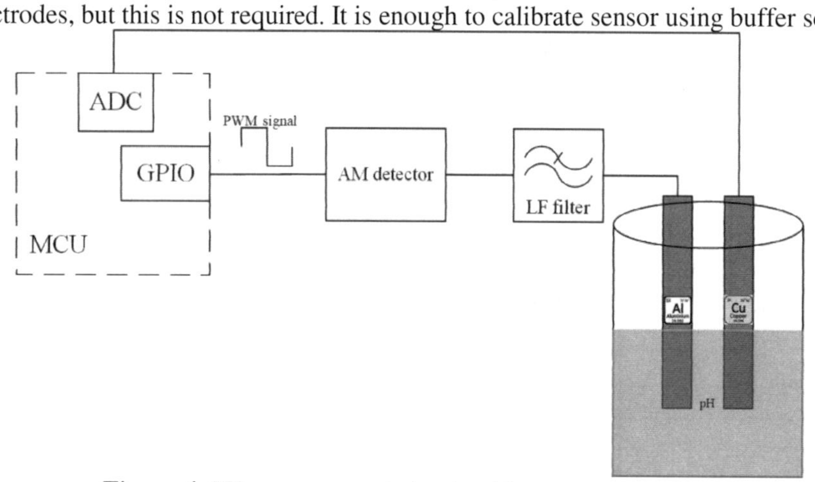

Figure 4. PH measurement circuit without compensation.

Soil acidity sensor two versions have been developed. For electrodes manufacture, mounting wires copper and aluminum core with a 3 mm cross section was used. Developed sensors differed in distance between electrodes and electrodes themselves length.

During tests, developed sensors were calibrated and investigated. Sensor for measuring soil acidity was calibrated on three solutions of known pH concentration. For this, buffer solutions were prepared with a 4pH, 6.86pH, 9.18pH concentration at a 25 °C temperature. Table 2 shows EMF dependence on electrodes depending on pH level for electrodes different configurations.

Table 2. Electrodes EMF dependence on pH level for electrodes different configurations.

Electrode length 8 cm, distance between electrodes 2 cm			Length of electrodes 15 cm, distance between electrodes 1 cm	
pH	**EMF, mV**	**ΔEMF, mV**	**EMF, mV**	**ΔEMF, mV**
4	605		484	-
6.86	667	62	590	106
9.18	853	186	837	247

Table 2 shows that longer electrodes and smaller distance between them, greater difference in EMF for different pH levels. Further research will describe 15 cm long probes spaced 1 cm apart.

The obtained values were used to calibrate sensor. Calibration curve is shown in Figure 5. According to resulting equation, conventional units are converted to pH readings.

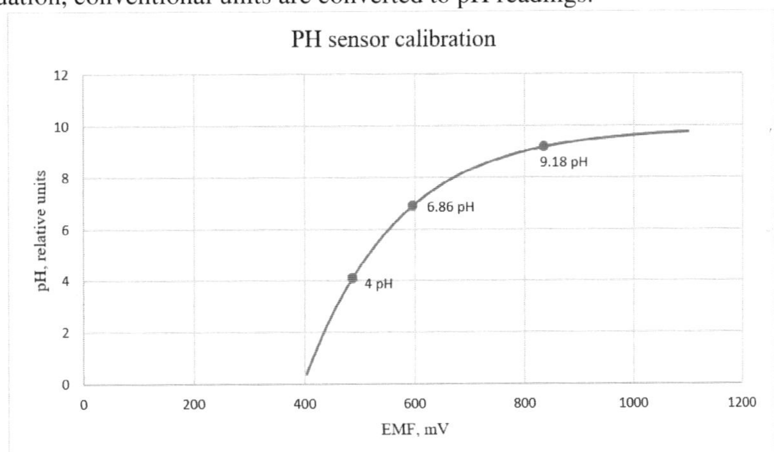

Figure 5. Calibration curve.

$$pH = A - B \cdot C^{ADC}, \tag{5}$$

where pH – acidity level, A = 9.89432, B = 106.4398, C = 0.99404 – coefficients, ADC – ADC readings in arbitrary units.

4. Soil acidity sensor test

To sensor test, a simple logger based on ATmega328P microcontroller was assembled. The block diagram is shown in Figure 6*a*.

To track measured parameters in real time, a program was developed in LabVIEW environment. The MCU is connected to PC via USB and transmits parameters to LabVIEW via a serial port. Then the received data are processed and graphically visualized. The program interface is shown in Figure 6*b*.

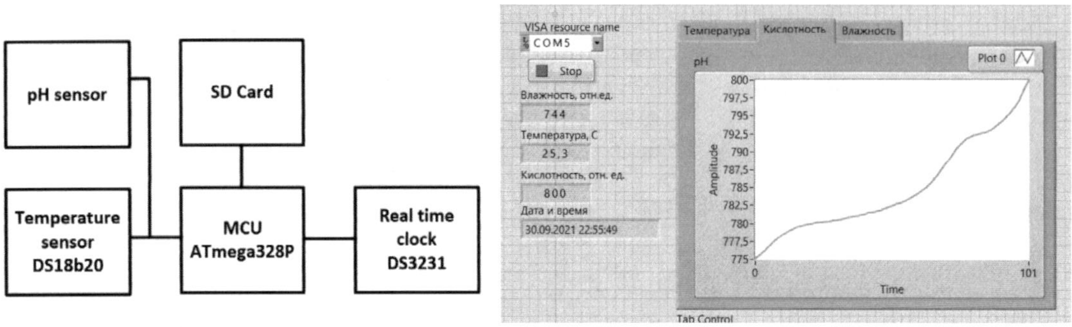

a b

Figure 6. Testing system hardware and software part: **a.** Logger block diagram for testing. **b.** Program interface.

Figure 7 shows a program code block diagram. Data from microcontroller is fed through serial port to LabVIEW programming environment. The NI VISA library is used to work with serial port. The data is transmitted in 18 characters string format. From general line, readings of moisture, temperature and soil acidity are highlighted. For plotting, the string data type is converted to integer.

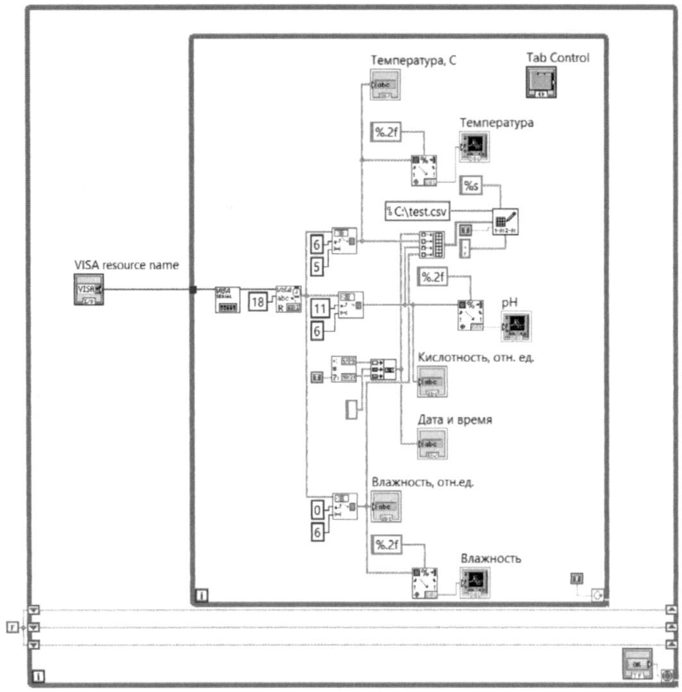

Figure 7. Program code block diagram.

Soil for testing sensor was purchased from a gardening store. Soil acidity 5.5 - 7.0 pH. An acidity sensor, a moisture sensor, a soil temperature sensor, and an analog acidity sensor from a store were installed in the soil. The testing lasted 2 days. The test setup is shown in Figure 8.

Figure 8. Testing pH sensor in soil.

5. Test results

Figure 9 shows pH sensor testing results. During testing, initially water with 5.5 – 7.0 pH was poured with a buffer solution with a pH = 9. Then at 22.07 at 14:24 a buffer solution with a pH = 4 was poured into a container with earth. , in turn, the acidity graph shows that the pH level has dropped. Over time, the pH should rise, as shown in the graph.

For comparison, an analog soil acidity sensor was used, which showed values similar to tested sensor values within permissible 1 pH error.

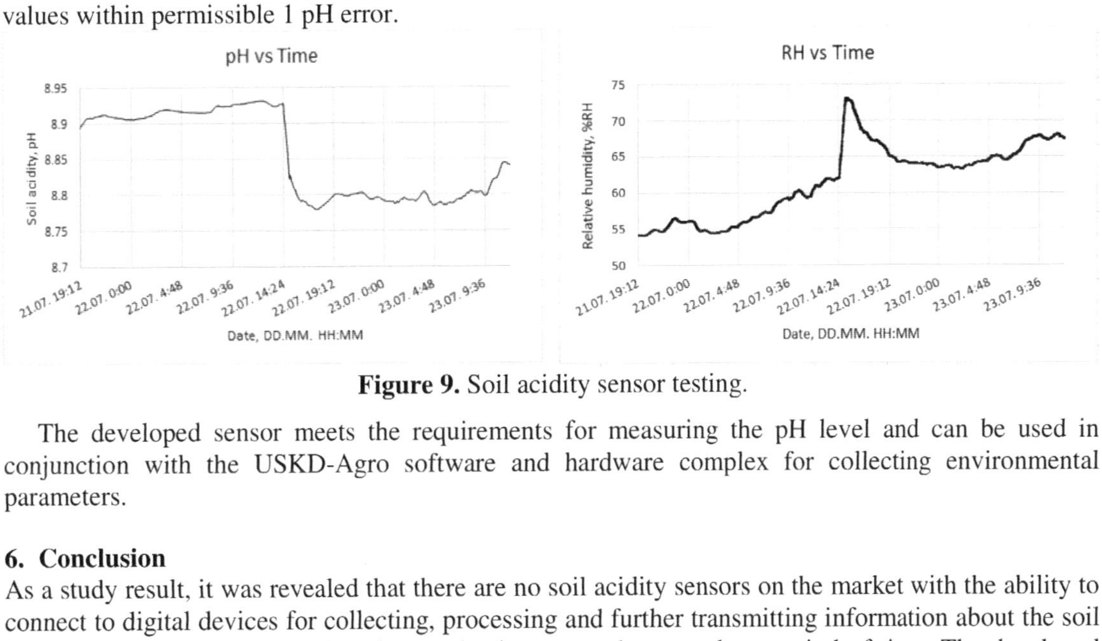

Figure 9. Soil acidity sensor testing.

The developed sensor meets the requirements for measuring the pH level and can be used in conjunction with the USKD-Agro software and hardware complex for collecting environmental parameters.

6. Conclusion

As a study result, it was revealed that there are no soil acidity sensors on the market with the ability to connect to digital devices for collecting, processing and further transmitting information about the soil acidity, which is a huge problem for conducting research over a long period of time. The developed sensor prototype was calibrated and tested for 2 days in the soil, during which the data was checked against an analog acidity sensor. The sensor needs further testing and improvement.

Acknowledgments

The project is carried out with support of Innovation Promotion Fund under UMNIK-Digital Russia program. Contract No. 730GUCES8-D3/63815.

Journal of Physics: Conference Series **2140** (2021) 012023 doi:10.1088/1742-6596/2140/1/012023

References

[1] On the creation and digital platforms maintenance for agro-industrial complex 2017 *Russian Federation Agriculture Ministry.* Available at: https://digital.gov.ru/uploaded/files/o-sozdanii-i-obespechenii-funktsionirovaniya-tsifrovyih-platform-agropromyishlennogo-kompleksa.pdf

[2] Departmental project "Digital Agriculture" 2019 *Russian Federation Agriculture Ministry.* Available at: https://mcx.gov.ru/upload/iblock/900/900863fae06c026826a9ee43e124d058.pdf

[3] Professional soil PH meter 2021 *Aliexpress.* Available at: https://clck.ru/Xtowf

[4] pH meter for soil 3 in 1 MEGEON 35280 2021 *Ozon.* Available at: https://clck.ru/XtpCB

[5] Putilova I N 1967 *General chemistry course* (Moscow: Higher School) p 244

[6] Sedikin F V 1976 *Dimensional electrochemical machining of machine parts* (Moscow: Mechanical Engineering) p 55

[7] Lemos S, Nogueira A, Torre-Neto A *et al* 2007 *Journal of Agricultural and Food Chemistry* **55**(12) 4658–4663 doi: 10.1021/jf063746a

[8] Singh M, Patka R, Vinchurkar M *et al* 2020 *IEEE Sensors Journal* **20**(1) 47–54 doi: 10.1109/JSEN.2019.2941695

[9] Oberoi A, Basavaraju S, Lekshmi S 2018 *Proceeding of the 2017 IEEE International Conference on Computational Intelligence and Computing Research (ICCIC)* doi: 10.1109/ICCIC.2017.8524170

[10] Staggenborg S, Carignano M, Haag L 2007 *Agronomy Journal* **99**(3) 854–861 doi: 10.2134/agronj2006.0254

Refraction of electromagnetic waves in pseudo-passive media

V V Fisanov

National Research Tomsk State University, Tomsk, Russia

E-mail: fisanov@mail.tsu.ru

Abstract. The problem of reflection and refraction of a plane monochromatic wave is considered in the most general formulation for a flat interface of two isotropic pseudo-passive media using the vector covariant approach of F.I. Fedorov. The generalized Fresnel coefficients for the perpendicular and parallel polarizations turn out to be complex values, except for cases when the phase difference of the permeabilities of the two media is zero or 180 degrees. The critical Brewster angles also turn out to be complex quantities. The phenomenon of total reflection, in contrast to passive transparent media, is not realized in pseudo-passive media. The energy coefficients of reflection and transmission are calculated and analyzed. Restrictions imposed on the values of the material parameters of the problem are indicated, which should be taken into account when studying refractive wave phenomena involving pseudo-passive metamaterials.

1. Introduction

Electromagnetic metamaterials are characterized by almost unlimited values of effective material parameters, but their diverse practical application is hampered by large associated losses. In this regard, active metamaterials are promising (see, for example, Ref. [1]), in which there is a gain (negative loss), which allows, in principle, to reduce or even completely eliminate unwanted absorption. In pseudo-passive isotropic media, the refractive index n is a real value. This is achieved under the condition that the effective complex dielectric permittivity and magnetic permeability have equal in absolute value, but opposite in sign of equiphases: $\mu=|\mu|\exp(\mathrm{i}\alpha)$, $\varepsilon=|\varepsilon|\exp(-\mathrm{i}\alpha)$ (here $\mathrm{i}=\sqrt{-1}$, $-\pi<\alpha\leq\pi$). When the permittivity modules are equal, they are complex-conjugate quantities. The refraction of waves for pseudo-passive structures surrounded by a transparent semi-infinite isorefractive dielectric [2] and a vacuum [3] has already been studied in the literature.

In this paper, the problems of reflection and refraction of a plane wave are considered in a more general formulation for the boundary of two pseudo-passive media using the vector covariant approach of F.I. Fedorov [4], rather than the coordinate approach, as in [2, 3].

2. Plane harmonic waves

Traveling plane harmonic waves with circular frequency ω in the complex representation are characterized by the strengths of the electromagnetic field of the form $\boldsymbol{E}=\boldsymbol{E}_1+\mathrm{i}\boldsymbol{E}_2=\boldsymbol{E}_0\mathrm{e}^{\mathrm{i}\varphi}$, $\boldsymbol{H}=\boldsymbol{H}_1+\mathrm{i}\boldsymbol{H}_2=\boldsymbol{H}_0\mathrm{e}^{\mathrm{i}\varphi}$, where $\boldsymbol{E}_0=\boldsymbol{E}_0'+\mathrm{i}\boldsymbol{E}_0''$, $\boldsymbol{H}_0=\boldsymbol{H}_0'+\mathrm{i}\boldsymbol{H}_0''$ are the complex vector amplitudes, the phase is a function of spatial coordinates and time and is written as $\varphi=k_0\boldsymbol{m}\cdot\boldsymbol{r}-\omega t$ (the notation is used here: $k_0=\omega/c$ is the wave number of free space (vacuum), $\boldsymbol{r}$ is the radius vector and $\boldsymbol{m}=n\hat{\boldsymbol{n}}=\boldsymbol{k}/k_0$ is the

Content from this work may be used under the terms of the Creative Commons Attribution 3.0 licence. Any further distribution of this work must maintain attribution to the author(s) and the title of the work, journal citation and DOI.
Published under licence by IOP Publishing Ltd

refraction vector proportional to the wave vector $\boldsymbol{k}$). Maxwell's equations in relation to plane waves are reduced to a system of vector algebraic relations

$$\boldsymbol{D}=-\boldsymbol{m}\times\boldsymbol{H}\,,\ \boldsymbol{B}=\boldsymbol{m}\times\boldsymbol{E}\,,\ \boldsymbol{m}\cdot\boldsymbol{D}=0\,,\ \boldsymbol{m}\cdot\boldsymbol{B}=0\,. \tag{1}$$

The constitutive relations for pseudo-passive media $\boldsymbol{D}=\varepsilon\boldsymbol{E}$, $\boldsymbol{B}=\mu\boldsymbol{H}$ contain complex permittivity and permeability; for the time dependence of the type of $\exp(-i\omega t)$ and ε-active medium adopted here, we should take $\varepsilon=\varepsilon'-i\,\varepsilon''$, $\mu=\mu'+i\,\mu''$ ($\varepsilon''>0$, $\mu''>0$). Excluding the vector $\boldsymbol{E}$ or $\boldsymbol{H}$ from Maxwell's equations, we come to the dispersion equation $\boldsymbol{m}^2=n^2=\varepsilon\mu$ with a positive refractive index $n=\sqrt{|\varepsilon||\mu|}$. In formulas (1), it is advisable to separate the real and imaginary parts:

$$\begin{cases}\mathrm{Re}(\varepsilon\boldsymbol{E})=-\boldsymbol{m}\times\boldsymbol{H}_1,\ \mathrm{Re}(\mu\boldsymbol{H})=\boldsymbol{m}\times\boldsymbol{E}_1,\\ \mathrm{Re}(\varepsilon\boldsymbol{E})\cdot\boldsymbol{m}=0,\ \mathrm{Re}(\mu\boldsymbol{H})\cdot\boldsymbol{m}=0;\end{cases},\ \begin{cases}\mathrm{Im}(\varepsilon\boldsymbol{E})=-\boldsymbol{m}\times\boldsymbol{H}_2,\ \mathrm{Im}(\mu\boldsymbol{H})=\boldsymbol{m}\times\boldsymbol{E}_2,\\ \mathrm{Im}(\varepsilon\boldsymbol{E})\cdot\boldsymbol{m}=0,\ \mathrm{Im}(\mu\boldsymbol{H})\cdot\boldsymbol{m}=0.\end{cases}. \tag{2}$$

Combining formulas (2), we find

$$\boldsymbol{H}_1=|\mu|^{-2}\boldsymbol{m}\times(\mu'\boldsymbol{E}_1+\mu''\boldsymbol{E}_2),\ \boldsymbol{H}_2=|\mu|^{-2}\boldsymbol{m}\times(\mu'\boldsymbol{E}_2-\mu''\boldsymbol{E}_1),\ \boldsymbol{m}\cdot\boldsymbol{H}_1=\boldsymbol{m}\cdot\boldsymbol{H}_2=0\,,\ \boldsymbol{m}\cdot\boldsymbol{E}_1=\boldsymbol{m}\cdot\boldsymbol{E}_2=0\,. \tag{3}$$

3. Fresnel coefficients

Let us consider the incidence of a plane homogeneous wave on a plane interface between two pseudo-passive media. The unit vector $\hat{\boldsymbol{q}}$ of the normal to the boundary is directed from the first medium to the second (in which a uniform refracted wave propagates). The laws of reflection and refraction in vector form are expressed in the form $\boldsymbol{m}_i\times\hat{\boldsymbol{q}}=\boldsymbol{a}$, where the index $i=0$ is applied to the incident wave, $i=1$ – for the reflected wave, and $i=2$ – for the refracted wave. This implies the scalar form of these laws: $n_i\sin\psi_i=|\boldsymbol{a}|$, where $n_0=n_1$ is the refractive index of the first medium, $\psi_i=\angle(\hat{\boldsymbol{n}}_i,\hat{\boldsymbol{q}})$ are the angles of plane waves that form the unit vectors $\hat{\boldsymbol{n}}_i$ of the refraction vectors with the normal $\hat{\boldsymbol{q}}$ (in this case $\psi_1=\pi-\psi_0$). The vectors of refraction of the reflected and refracted waves are calculated by the well-known formulas: $\boldsymbol{m}_1=\hat{\boldsymbol{q}}\times\boldsymbol{a}-\boldsymbol{m}_0\cdot\hat{\boldsymbol{q}}\hat{\boldsymbol{q}}$, $\boldsymbol{m}_2=\hat{\boldsymbol{q}}\times\boldsymbol{a}+\sqrt{\boldsymbol{m}_2^2-\boldsymbol{a}^2}\hat{\boldsymbol{q}}$. Before the square root, the sign "$+$" should be changed to "$-$" if a backward wave propagates in the second medium. The rule of distinguishing between forward and backward waves in media with losses and gain is known [5], but the presence of backward waves in one of the media does not change the Fresnel formulas [6]. To find the Fresnel coefficients, the boundary conditions of continuity of the fields tangent to the boundary are used: $\left(\boldsymbol{E}^{(0)}+\boldsymbol{E}^{(1)}-\boldsymbol{E}^{(2)}\right)\times\hat{\boldsymbol{q}}=0$, $\left(\boldsymbol{H}^{(0)}+\boldsymbol{H}^{(1)}-\boldsymbol{H}^{(2)}\right)\times\hat{\boldsymbol{q}}=0$. Since the boundary conditions contain only complete complex vectors $\boldsymbol{E}^{(i)}$, $\boldsymbol{H}^{(i)}$ and there are no material parameters, the process of obtaining the Fresnel coefficients turns out to be identical to that in the case of a transparent passive medium. First, the fields are decomposed into orthogonal components: $\boldsymbol{E}^{(i)}=A_i\boldsymbol{a}+B_i\hat{\boldsymbol{n}}_i\times\boldsymbol{a}$, $\boldsymbol{H}^{(i)}=(n_i/\mu_i)(B_i\boldsymbol{a}+A_i\hat{\boldsymbol{n}}_i\times\boldsymbol{a})$, and then boundary conditions are applied, which are conveniently represented as a pair of expressions $\boldsymbol{E}^{(0)}+\boldsymbol{E}^{(1)}-\boldsymbol{E}^{(2)}=\beta\hat{\boldsymbol{q}}$, $\boldsymbol{H}^{(0)}+\boldsymbol{H}^{(1)}-\boldsymbol{H}^{(2)}=\gamma\hat{\boldsymbol{q}}$ (symbols β and γ denote some scalars). After scalar multiplication of the first of them by $\boldsymbol{a}$, and the second by $\hat{\boldsymbol{q}}\times\boldsymbol{a}$, systems of algebraic equations for the coefficients A_i (for perpendicular polarization) and B_i (for parallel polarization) are obtained. Their solutions can be presented in the form [4]

$$\frac{A_0}{\frac{n_1}{\mu_1}\hat{\boldsymbol{n}}_1\cdot\hat{\boldsymbol{q}}-\frac{n_2}{\mu_2}\hat{\boldsymbol{n}}_2\cdot\hat{\boldsymbol{q}}}=\frac{A_1}{\frac{n_2}{\mu_2}\hat{\boldsymbol{n}}_2\cdot\hat{\boldsymbol{q}}-\frac{n_0}{\mu_0}\hat{\boldsymbol{n}}_0\cdot\hat{\boldsymbol{q}}}=\frac{-A_2}{\frac{n_0}{\mu_0}\hat{\boldsymbol{n}}_0\cdot\hat{\boldsymbol{q}}-\frac{n_1}{\mu_1}\hat{\boldsymbol{n}}_1\cdot\hat{\boldsymbol{q}}}, \tag{4}$$

$$\frac{B_0}{\frac{n_1}{\mu_1}\hat{\boldsymbol{n}}_2\cdot\hat{\boldsymbol{q}}-\frac{n_2}{\mu_2}\hat{\boldsymbol{n}}_1\cdot\hat{\boldsymbol{q}}}=\frac{B_1}{\frac{n_2}{\mu_2}\hat{\boldsymbol{n}}_0\cdot\hat{\boldsymbol{q}}-\frac{n_0}{\mu_0}\hat{\boldsymbol{n}}_2\cdot\hat{\boldsymbol{q}}}=\frac{-B_2}{\frac{n_0}{\mu_0}\hat{\boldsymbol{n}}_1\cdot\hat{\boldsymbol{q}}-\frac{n_1}{\mu_1}\hat{\boldsymbol{n}}_0\cdot\hat{\boldsymbol{q}}}\,. \tag{5}$$

Fresnel coefficients are introduced by the formulas $\rho^{\perp}=A_1/A_0$, $\tau^{\perp}=A_2/A_0$, $\rho^{\|}=B_1/B_0$, $\tau^{\|}=B_2/B_0$ taking into account the ratios $n_0=n_1$, $\mu_0=\mu_1$, $\hat{\boldsymbol{n}}_i\cdot\hat{\boldsymbol{q}}=\cos\psi_i$ and are reduced to a standard form using trigonometric functions:

$$\rho^{\perp}=\frac{1-\mu_1\,\mathrm{tg}\psi_0/(\mu_2\,\mathrm{tg}\psi_2)}{1+\mu_1\,\mathrm{tg}\psi_0/(\mu_2\,\mathrm{tg}\psi_2)},\quad \tau^{\perp}=\frac{2}{1+\mu_1\,\mathrm{tg}\psi_0/(\mu_2\,\mathrm{tg}\psi_2)},\tag{6}$$

$$\rho^{\|}=\frac{(\mu_1/\mu_2)\sin 2\psi_0-\sin 2\psi_2}{(\mu_1/\mu_2)\sin 2\psi_0+\sin 2\psi_2},\quad \tau^{\|}=\frac{2(n_1/n_2)\sin 2\psi_0}{(\mu_1/\mu_2)\sin 2\psi_0+\sin 2\psi_2}.\tag{7}$$

Formulas (6) and (7) are universal, since they cover all possible combinations of material parameters in pairs of bordering pseudo-passive and truly passive transparent media. Considered in [2], the case of a wave incidence from an ordinary to a pseudo-passive medium corresponds to a phase parameter $\arg(\mu_1)=\alpha_1=0$. Fresnel coefficients in pseudo-passive media are complex quantities, except when the phase difference $\alpha_1-\alpha_2$ is zero or π. Critical Brewster angles [7]

$$\psi_{\mathrm{Br}}^{\perp}=\mathrm{arctg}\sqrt{\frac{1-(Y_1/Y_2)^2}{(n_1/n_2)^2-1}},\quad \psi_{\mathrm{Br}}^{\|}=\mathrm{arctg}\sqrt{\frac{(Y_2/Y_1)^2-1}{1-(n_1/n_2)^2}},$$

where $Y_i=\sqrt{\varepsilon_i/\mu_i}=|Y_i|\exp(-i\alpha_i)$, are the wave admittances of the media, turn out to be also complex quantities, with the indicated exception. A paradoxical result takes place in the case $\alpha_1-\alpha_2=\pm\pi/2$: $|\rho^{\perp}|=|\rho^{\|}|=1$ as if there is a total reflection at all angles of incidence. In addition, in contrast to conventional transparent contacting media, under conditions when the angle of incidence exceeds the critical angle of total reflection ($\psi_0>\arcsin(n_2/n_1)$), equality $|\rho^{\perp}|=|\rho^{\|}|=1$ for pseudo-passive media is not realized. As is known [8], in a medium with purely imaginary permeabilities ($\alpha=\pm\pi/2$), the average flux of the Poynting vector is zero, therefore, for $\alpha_1=0$ and $\alpha_2=\pm\pi/2$, there is a complete reflection from the second medium. Apparently, a similar phenomenon takes place in the more general case ($\alpha_1-\alpha_2=\pm\pi/2$), due to the appropriate phase matching of the permeabilities of both media.

4. The energy coefficients of reflection and transmission

For what follows, it is of interest to link the Fresnel coefficients with the energy coefficients of reflection and transmission. Using the relationship between the Fresnel coefficients $1+\rho^{\perp}=\tau^{\perp}$, $Y_1(1-\rho^{\perp})\cos\psi_0=Y_2\tau^{\perp}\cos\psi_2$ and $Y_1(1+\rho^{\|})=Y_2\tau^{\|}$, $(1-\rho^{\|})\cos\psi_0=\tau^{\|}\cos\psi_2$ [9], one can come to a universal expression $(\rho^{\perp,\|})^2+(\tau^{\perp,\|})^2Y_2\cos\psi_2/(Y_1\cos\psi_0)=1$. It is rethought for positive permeabilities as $r^{\perp,\|}+d^{\perp,\|}=1$, where $r^{\perp,\|}$ and $d^{\perp,\|}$ are the reflection and transmission coefficients, respectively. For a pseudo-passive medium with its complex permeabilities and wave admittance, the reflection coefficient is defined as $r^{\perp,\|}=|\rho^{\perp,\|}|^2$, and the transmission coefficient is defined as $d^{\perp,\|}=1-r^{\perp,\|}$.

On the other hand, to consider the wave intensities, one should turn to the law of conservation of energy during the interaction of waves at the interface between two media, which states that the normal component of the energy flux density vector is continuous on the boundary surface of these media: $(\boldsymbol{P}_0+\boldsymbol{P}_1-\boldsymbol{P}_2)\cdot\hat{\boldsymbol{q}}=0$, where the Poynting vectors are given by the formula $\boldsymbol{P}_i=(c/4\pi)\mathrm{Re}\boldsymbol{E}^{(i)}\times\mathrm{Re}\boldsymbol{H}^{(i)}$. The originality of the calculation of the Poynting vector as applied to a pseudo-passive medium in comparison with a transparent passive medium is due to the differences between the systems of equations (1) and (2). Using the expression for $\mathrm{Re}\boldsymbol{H}^{(i)}=\boldsymbol{H}_1^{(i)}$ from (3), we find

$$\boldsymbol{P}_i = \frac{c}{4\pi|\mu_i|^2}\boldsymbol{E}_1^{(i)} \times \left\{\boldsymbol{m}_i \times \left(\mu_i'\boldsymbol{E}_1^{(i)} + \mu_i''\boldsymbol{E}_2^{(i)}\right)\right\} = \frac{c}{8\pi|\mu_i|^2}\left\{\mathrm{Re}\left(\mu_i^*\boldsymbol{E}^{(i)2}\right) + \mu_i'\left|\boldsymbol{E}^{(i)}\right|^2\right\}\boldsymbol{m}_i =$$

$$= \frac{c}{8\pi|\mu_i|^2}\left\{\mu_i'\left(\boldsymbol{E}_1^{(i)2} - \boldsymbol{E}_2^{(i)2} + \left|\boldsymbol{E}^{(i)}\right|^2\right) + 2\mu_i''\boldsymbol{E}_1^{(i)}\cdot\boldsymbol{E}_2^{(i)}\right\}\boldsymbol{m}_i . \quad (8)$$

When deriving formula (8), relations of the form $4\boldsymbol{E}_1^2 = \boldsymbol{E}^2 + \boldsymbol{E}^{*2} + 2|\boldsymbol{E}|^2$, $4\mathrm{i}\boldsymbol{E}_1\cdot\boldsymbol{E}_2 = \boldsymbol{E}^2 - \boldsymbol{E}^{*2}$ were used. In formula (8), it is assumed that the full vectors $\boldsymbol{E}^{(i)}$ contain a common phase factor $\exp(\mathrm{i}\varphi)$, and therefore, as a result of averaging over the oscillation period, we obtain $\bar{\boldsymbol{P}}_i == c\mu_i'\left|\boldsymbol{E}^{(i)}\right|^2\boldsymbol{m}_i \Big/ \left(8\pi|\mu_i|^2\right)$.

The law of conservation of energy allows for the separation of polarizations. Therefore, we assume $\left(\bar{\boldsymbol{P}}_0^{\perp} + \bar{\boldsymbol{P}}_1^{\perp} - \bar{\boldsymbol{P}}_2^{\perp}\right)\cdot\hat{\boldsymbol{q}} = 0$, $\left(\bar{\boldsymbol{P}}_0^{||} + \bar{\boldsymbol{P}}_1^{||} - \bar{\boldsymbol{P}}_2^{||}\right)\cdot\hat{\boldsymbol{q}} = 0$ and after that we introduce the average coefficients of reflection $\bar{r}^{\perp,||} = -\left(\bar{\boldsymbol{P}}_1^{\perp,||}\cdot\hat{\boldsymbol{q}}\right)\Big/\left(\bar{\boldsymbol{P}}_0^{\perp,||}\cdot\hat{\boldsymbol{q}}\right)$ and transmission $\bar{d}^{\perp,||} = \left(\bar{\boldsymbol{P}}_2^{\perp,||}\cdot\hat{\boldsymbol{q}}\right)\Big/\left(\bar{\boldsymbol{P}}_0^{\perp,||}\cdot\hat{\boldsymbol{q}}\right)$, at that $\bar{r}^{\perp,||} + \bar{d}^{\perp,||} = 1$. In expanded form, these coefficients are given by the formulas

$$\bar{r}^{\perp} = \frac{|A_1|^2}{|A_0|^2} = \left|\rho^{\perp}\right|^2, \ \bar{d}^{\perp} = \frac{\mu_2'}{\mu_1'}\frac{|\mu_1|^2}{|\mu_2|^2}\frac{\eta_2}{\eta_0}\frac{|A_2|^2}{|A_0|^2} = \frac{\mu_2'}{\mu_1'}\frac{|\mu_1|^2}{|\mu_2|^2}\frac{\eta_2}{\eta_0}\left|\tau^{\perp}\right|^2, \quad (9)$$

$$\bar{r}^{||} = \frac{|B_1|^2}{|B_0|^2} = \left|\rho^{||}\right|^2, \ \bar{d}^{||} = \frac{\mu_2'}{\mu_1'}\frac{|\mu_1|^2}{|\mu_2|^2}\frac{\eta_2}{\eta_0}\frac{|B_2|^2}{|B_0|^2} = \frac{\mu_2'}{\mu_1'}\frac{|\mu_1|^2}{|\mu_2|^2}\frac{\eta_2}{\eta_0}\left|\tau^{||}\right|^2, \quad (10)$$

where $\eta_i = n_i\cos\psi_i$. The reflection coefficients in formulas (9) and (10) are identical to the coefficients obtained above from the Fresnel formulas. It is easy to see that they have the following analytical structure:

$$r^{\perp,||} = \left|\rho^{\perp,||}\right|^2 = \left|\frac{1 - M^{\perp,||}e^{\mathrm{i}\delta}}{1 + M^{\perp,||}e^{\mathrm{i}\delta}}\right|^2 = \frac{1 + \left(M^{\perp,||}\right)^2 - 2M^{\perp,||}\cos\delta}{1 + \left(M^{\perp,||}\right)^2 + 2M^{\perp,||}\cos\delta}, \quad (11)$$

where the notation $M^{\perp} = |\mu_1|\mathrm{tg}\psi_0/\left(|\mu_2|\mathrm{tg}\psi_2\right)$, $M^{||} = |\mu_1|\sin 2\psi_0/\left(|\mu_2|\sin 2\psi_2\right)$, $\delta = \alpha_1 - \alpha_2$ is introduced. Formula (11) shows that the values of the reflection coefficient exceed one, that is, they have no physical meaning if $\cos\delta < 0$ ($\pi/2 < |\delta| < \pi$).

5. Conclusion

Thus, the classical model problem of the incidence of a plane homogeneous wave on a plane interface between two transparent isotropic media can be generalized to the case of pseudo-passive electromagnetic media, but with some restrictions. These restrictions impose a prohibition on the mismatch of the arguments of the material parameters of the contacting media, which exceeds the right angle, but less than the deployed angle. With refraction in pseudo-passive media, Brewster's polarization law is not fulfilled, and the phenomenon of total reflection does not occur at angles of incidence exceeding the critical angle. For a more complete study of the phenomena of refraction of plane waves, as already noted [4], it is necessary to take into account the limitedness of the wave in space or in time.

Acknowledgments

This work was partially supported within the scope of State Assignment of the Russian Ministry of Education and Science (Project No. 0336-2019-0008).

References

[1] Wuestner S, Pusch A, Tsakmakidis K L *et al.* 2011 *Transactions of the Royal Society A: Mathematical, Physical and Engineering Sciences* **369**(1950) 3525–3550 doi:10.1098/rsta.2011.0140
[2] Dragoman D 2011 *Optics Communications* **284**(8) 2095–2098 doi: 10.1016/j.optcom.2010.12.069
[3] Basiri A, Vitebskiy I, Kottos T 2015 *Physical Review A* **91**(6) 063843 doi: 10.1103/PhysRevA.91.063843
[4] Fedorov F I 1958 *Optics of Anisotropic Media* (Minsk: Publishing House of the Academy of Sciences of the BSSR) 380 p [In Russian]
[5] Fisanov V V 2017 *Russian Physics Journal* **60**(5) 797–802 doi: 10.1007/s11182-017-1141-4
[6] Fisanov V V 2017 *Russian Physics Journal* **59**(9) 1395–1400 doi: 10.1007/s11182-017-0922-0
[7] Kizel' V A 1973 *Reflection of Light* (Moscow: Nauka) 352 pages [In Russian]
[8] Bai P, Ding K, Wang G *et al.* 2016 *Physical Review A* **94**(6) 063841 doi: 10.1103/PhysRevA.94.063841
[9] Nikol'skiy V V 1973 *Electrodynamics and Propagation of Radio Waves* (Moscow: Nauka) 608 p [In Russian]

Modeling of electrotransport properties of Li-intercalated graphene film

D Sergeyev[1,2], A Duisenova[1] and Zh Embergenov[1]

[1]K. Zhubanov Aktobe Regional University, Aktobe, Republic of Kazakhstan
[2]T. Begeldinov Aktobe Aviation Institute, Aktobe, Republic of Kazakhstan

E-mail: serdau@mail.ru

Abstract. In this work, within the framework of density functional theory combined with the method of nonequilibrium Green's functions the density of states, transmission spectrum, current-voltage characteristics, and differential conductivity of Li-intercalated graphene (LiC_6) have been determined. It is shown that in the energy range of -1.3÷-1.05 eV the quasiparticle transport through the nanostructure is disable. The features of IV- and dI/dV-characteristics of LiC_6 in the form of decreasing of resistance in the range of -0.4÷0.4 V were revealed, and in the interval of 0.4÷1.4 V formation of negative differential resistance area, related to scattering of quasiparticles. It is established, that LiC_6 nanodevice has 12÷13 ballistic channels and has the maximum amount of conductance ~12÷13G_0, where G_0 is the conductance quantum.

1. Introduction

Recently, to reduce the size, mass, and energy consumption of electronic techniques, electrophysical properties of different nanoscale contacts are being investigated to be used as active and passive elements in electric circuits and to replace traditional elements (see, for example, [1, 2]). New directions of electronics based on exotic materials are being developed to create a new generation of electronic devices and appliances, such as superconductor [3–5], organic [6], single-electronics [7, 8], spintronics [9] and others. In this respect, one of the most promising nanomaterials for such nanocontacts is graphene, which is an allotropic modification of carbon in the form of a two-dimensional monatomic layer [10,11]. After the discovery of graphene, the search for new two-dimensional materials was started, and some quasi-two-dimensional materials with unusual electronic properties were created [12–14]. In theoretical works [15–17], the possible superconducting properties of the calcium- and lithium-doped graphene film have been considered. However, there is considerable disagreement in explaining the electronic properties of these two-dimensional materials. In work [18] it is noted that for the appearance of superconductivity is responsible non-phonon mechanism of formation of Coopers pairs of electrons due to the weak repulsion between them, leading to spiral type d-pairing, and in work [19] usual phonon mechanism of electron pairing in s-channel is considered, and within the framework of Eliashberg theory is estimated the critical temperature of LiC_6 nanostructure, equal to ~8 K.

Even though the superconductivity in the LiC_6 monolayer has not been confirmed experimentally yet, a significant change in the electronic properties of the Li-intercalated graphene compared to the "pure" graphene are observed. In this work, the electrotransport characteristics of Li-intercalated graphene are simulated and analyzed within the framework of the electron density functional theory (DFT) in combination with the non-equilibrium Green's function (NEGF) method.

Content from this work may be used under the terms of the Creative Commons Attribution 3.0 licence. Any further distribution of this work must maintain attribution to the author(s) and the title of the work, journal citation and DOI.
Published under licence by IOP Publishing Ltd

2. Geometry

The geometry of the Li-intercalated graphene nanodevice is shown in Figure 1. The length of the nanodevice is ~51.15 Å. The electrodes are an extension of the Li-intercalated graphene with a length of ~8.525 Å. The Li-graphene surface with a size of ~34.1×17.23 Å consists of 255 carbon atoms and 40 lithium atoms. To describe the interatomic interaction and optimize the nanostructure, Brenner [20] and ReaxFF_CHONSSiLi_2013 [21] potentials have been used. During the nanodevice geometry optimization, the parameters of the atomic configuration were relaxed until the forces at all atoms became lower than a given threshold value of 0.05 eV/Å.

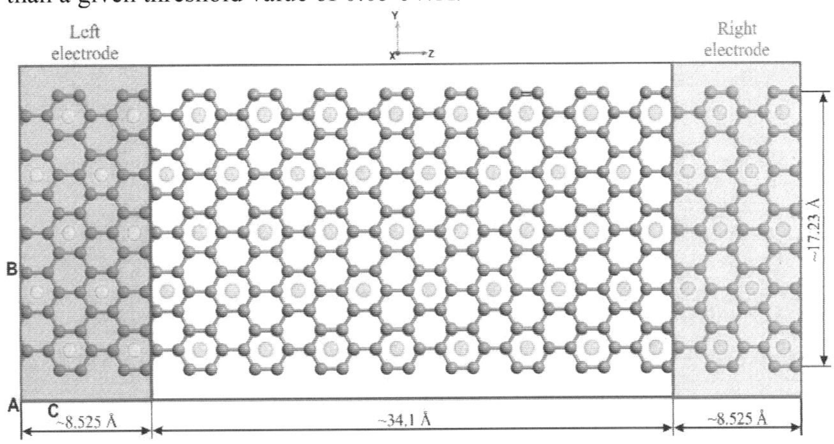

Figure 1. The geometry of a Li intercalated graphene nanodevice.

3. Fundamental equations

Computer simulation of electrotransport characteristics of Li-graphene nanodevice was carried out in the framework of DFT + NEGF in local density approximation (LDA). Modeling of electrotransport characteristics of the nanodevice was realized in program Atomistix ToolKit with Virtual NanoLab [21]. (Fundamental equations of this method are described in detail in our previous works [22, 23]). The current-voltage-characteristic (CVC) of the nanostructure was determined by solving the Landauer equation which indicates the fundamental coupling of the electric current I with the transmittance spectrum $T(E)$:

$$I(V_L, V_R, T_L, T_R) = \frac{2e}{h}\int_{-\infty}^{+\infty} T(E)\left[f\left(\frac{E-\mu_R}{k_B T_R}\right) - f\left(\frac{E-\mu_L}{k_B T_L}\right)\right]dE, \tag{1}$$

where e is electron charge, h is Planck's constant, E is energy, $f(\varepsilon)$ is Fermi energy distribution function of quasiparticles, k_B is the Boltzmann's constant, T_L, T_R are current temperatures and μ_R, μ_L are electrochemical potentials of the right and left electrodes. $T(E) = \mathrm{tr}\left[\Gamma^L G \Gamma^R G^\dagger\right] = \mathrm{tr}\left[\Gamma^R G \Gamma^L G^\dagger\right]$ is transmission spectrum (function), Γ^L is broadening matrix, G, $G^\dagger$ is lagged and leading Green's functions. The differential conductivity of the nanostructure was obtained by calculating the self-consistent current at a number of applied biases and performing numerical differentiation.

4. Results

The density of states (DOS) of the LiC_6 nanodevice is shown in Figure 2. The DOS of the LiC_6 nanostructure near the Fermi energy ($\varepsilon_F \approx 0$) is 35–52 eV^{-1}, when in "pure" graphene it is 20 eV^{-1}, i.e., the DOS of LiC_6 increases in 1.75–2.6 times relative from graphene. This means that the number of energy levels in the LiC_6 nanodevice near the Fermi energy increases, which should lead to the strengthening of the electron transport of quasiparticles in low bias voltage.

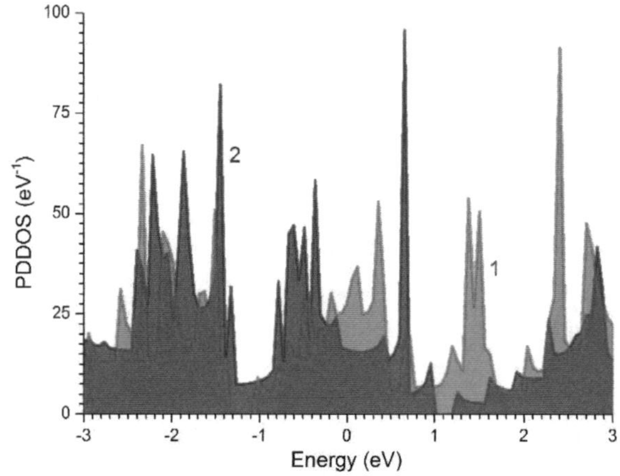

Figure 2. The density of states of LiC_6 nanodevices: 1 - LiC_6, 2 - graphene (for comparison).

The evolution of the transmission function (spectrum) $T(E)$ of the LiC_6 nanodevice with increasing bias voltage from -2 V to 2 V in 0.2 V steps is shown in Figure 3*a,b*. As the bias voltage increases, the intensity of the transmission spectrum increases. At positive energy, the transmission capacity of quasiparticles through the nanostructure in consideration is higher than at negative energy. In the energy range -1.3÷ -1.05 eV the probability of quasiparticles passing through the nanostructure is small. As can be seen, the features of DOS appear in the transmission spectrum of the considered nanostructure at the same energy values, as these values are directly proportional $T(\varepsilon) = D(\varepsilon - U)2\pi\gamma_1\gamma_2/\gamma$ (here U is self-consistent potential, γ is Luttinger parameter) [24].

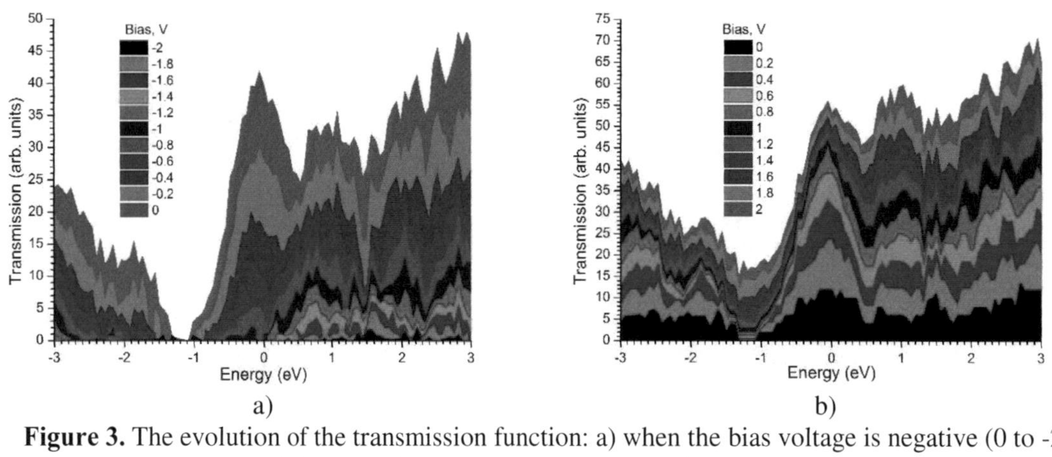

a) b)

Figure 3. The evolution of the transmission function: a) when the bias voltage is negative (0 to -2 V); b) when the bias voltage is positive (0 to 2 V).

The current voltage characteristic (CVC) and the differential conductivity of the LiC_6 nanodevice are shown in Figure 4*a,b*. In the voltage range of -0.4÷0.4 V in LiC_6 nanodevice a rapid linear increase of current to 0.11 mA is observed. It is supposed that the decreasing of the film resistance in the stated interval of voltage appears because of the increasing of number of energy levels near to the Fermi energy. Further increasing of bias voltage from 0.4 to 1.4 V will lead to quasi-linear decrease of current due to scattering of quasiparticles. Notice that a significant decrease in current in this voltage range forms a negative differential resistance (NDR) region on the CVC. These changes are reflected in the differential conductivity of the LiC_6 nanodevice (Figure 4*b*). The LiC_6 nanodevice demonstrates 12÷13 ballistic channels and has a maximum conductance value of ~12÷13G_0, where $G_0 = e^2/h$ is the conductance quantum.

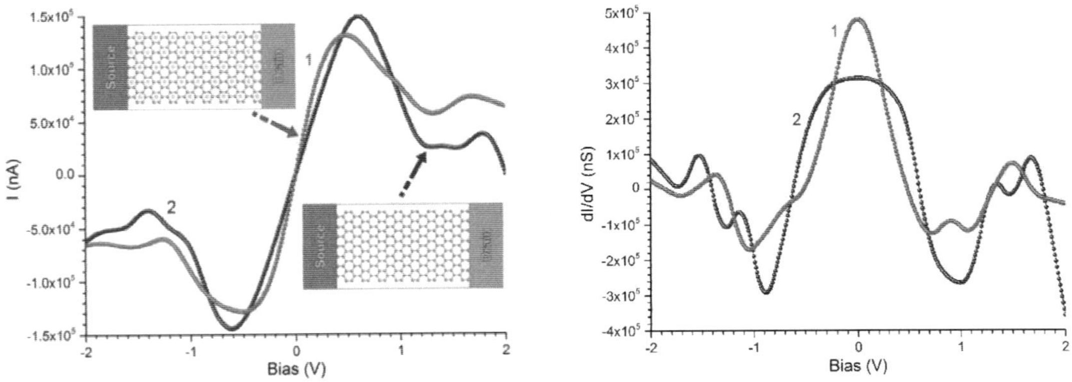

a) b)

Figure 4. The current -voltage - (a) and dI/dV-characteristics (b) of LiC_6 nanodevices: 1 – LiC_6, 2 – graphene (for comparison).

5. Conclusion

In summary, the main electrotransport characteristics of Li-intercalated graphene (DOS, transmission spectrum, CVC) and differential conductivity have been determined within the framework of DFT + NEGF. It is found that in the energy interval -1.3÷-1.05 eV the transport of quasiparticles through the nanostructure is disable. The features of IV- and dI/dV-characteristics are determined: decreasing of resistance in the range of -0.4÷0.4V leading to the linear increase of current, appearance of NDR region in the range of 0.4÷1.4V because of quasiparticles scattering. It is established that the electronic conductivity of LiC_6 near the Fermi energy is higher in comparison with graphene.

Acknowledgments

The research has been supported by the grant of the Ministry of Education and Science of the Republic of Kazakhstan AP08052562.

References

[1] Cuevas J C, Scheer E 2017 *Molecular Electronics: an Introduction to Theory and Experiment* (World Scientific Publishers) p 846
[2] Dragoman M, Dragoman D 2017 *2D Nanoelectronics: Physics and Devices of Atomically Thin Materials* (Switzerland: Springer International Publishing,) p 199
[3] Likharev K K 2012 *Physica C* **482** 6–18 doi: 10.1016/j.physc.2012.05.016
[4] Hinken J H 1989 *Superconductor Electronics: Fundamentals and Microwave Applications* (Berlin: Springer-Verlag Heidelberg) p 158
[5] Sergeyev D M 2012 *Russian Physics Journal* **55** 84–91 DOI: 10.1007/s11182-012-9779-4
[6] Kanbur Y, Irimia-Vladu M, Głowacki E D , Voss G, Baumgartner M , Schwabegger G *et* al 2012 *Organic Electronics* **13**(5) 919–924 doi: 10.1016/j.orgel.2012.02.006
[7] Averin D V, Likharev K K 1986 *Journal of Low Temperature Physics* **62** 345–373 doi: 10.1007/BF00683469
[8] Patel R, Agrawal Y, Parekh R 2021 *Microsystem Technologies-Micro-And Nanosystems-Information Storage and Processing Systems* 27(5) 1863–1875 doi: 10.1007/s00542-020-05002-5

[9] Baranov P G, Kalashnikova A M, Kozub V I, Korenev V L *et* al 2019 *Phys. Usp.* **62** 795–822 doi: 10.3367/UFNe.2018.11.038486

[10] Novoselov K S, Geim A K, Morozov S V, Jiang D *et* al 2004 *Science* **306**(5696) 666–669 doi: 10.1126/science.1102896

[11] Ratnikov P V, Silin A P 2018 *Phys. Usp.* **61** 1139–1174 doi: 10.3367/UFNe.2017.11.038231

[12] Liu N, Bo G, Liu Y, Xu X *et al* 2019 *Small* **15**(32) 1805147 doi: 10.1002/smll.201805147

[13] Kiraly B, Liu X, Wang L, Zhang Zh, Mannix A J *et al* 2019 *ACS Nano* **13**(4) 3816–3822 doi: 10.1021/acsnano.8b09339

[14] Sahoo S K, Wei K H 2019 *Advanced materials interfaces* **6**(18) 1900752 doi: 10.1002/admi.201900752

[15] Emery N, Herold C, d'Astuto M, Garcia V *et al* 2005 *Phys Rev Lett* **95**(8) 087003 doi: 10.1103/PhysRevLett.95.087003

[16] Guzman D M, Alyahyaei H M, Jishi R A 2014 *2D Materials* **1**(2) 021005 doi: 10.1088/2053-1583/1/2/021005

[17] Gholami R, Moradian R, Moradian S, Pickett W E 2018 *Scientific Reports* **8** 13795 doi: 10.1038/s41598-018-32050-9

[18] Nandkishore R, Levitov L, Chubukov A 2012 *Nature Phys* **8** 158–163 doi: 10.1038/nphys2208

[19] Profeta G, Calandra M, Mauri F 2012 *Nature Phys* **8** 131-134 doi: 10.1038/nphys2181

[20] Brenner D W, Shenderova O A, Harrison J A *et* al 2002 *J. Phys.: Condens. Matter.* **14** 783-802 doi: 10.1088/0953-8984/14/4/312

[21] Brandbyge M, Mozos J L, Ordejon P, Taylor J, Stokbro K 2002 *Phys. Rev. B* **65** 165401 DOI: 10.1103/PhysRevB.65.165401

[22] Sergeyev D, Zhanturina N 2019 *Radioengineering* **28**(4) 714-720 doi: 10.13164/re.2019.0714

[23] Sergeyev D M 2018 *Journal of Nano- and Electronic Physics* **10**(3) 03018 doi: 10.21272/jnep.10(3).03018

[24] Datta S 2005 *Quantum Transport: Atom to Transistor* (Cambridge: Univ. Press) p 404

Investigation of the parameters of the reflected wave near the Brewster angle

V P Krylov

ONPP "Technology" named after A. G. Romashina, Obninsk, Russia

E-mail: info@technologiya.ru

Abstract. In free space, the relative permittivity is determined by the Brewster formula without taking into account dielectric and magnetic losses. In experimental studies, discrepancies in the angular position of the minimum of the reflected wave from dielectric materials are observed in comparison with calculations, which are known as deviations from Fresnel's laws. By solving the task of inclined falling wave on an plate made of a dielectric material with complex of the dielectric and magnetic permittivity, the parameters of the reflected wave were calculated, according to which the angles corresponding to the minimum reflection were determined, depending on the dielectric losses of the material. From the condition that the reflected wave is equal to zero, a formula for determining the Brewster angle for a material with dielectric and magnetic losses was analytically obtained, the results of calculations for which coincided with the calculations for the reflected wave in the context of geometric optics. It is determined that in the general case, the conditions for determining the position of the minimum of the complex amplitude and the phase jump by 180° of electromagnetic waves do not coincide and can be found only when solving the task an falling wave on a plate with complex electrodynamic parameters of the material in the context of geometric optics.

1. Introduction

For the solution of the main objectives of radio engineering during creation of artificial composite materials with variable electrodynamic parameters, it is necessary application of methods of measurement in free space in wide strip for definition of their properties. To solve this problem, it is of interest to study the reflected wave from a dielectric plate with variable electromagnetic parameters in the region of the Brewster angle.

2. Design model

To describe this problem in the context of geometric optics, calculations are performed for a plate made of a homogeneous material with an effective permittivity, but the results of these calculations using classical formulas when studying the reflected wave in the Brewster angle region do not correspond to experimental data, which leads to assumptions about deviations in the Fresnel laws [1]. The reasons for the deviations are associated with the presence of losses in the material, the presence of an intermediate layer at the plate boundaries, but experiments show that this is not entirely true [2].

The coefficient of reflection TM wave, when the vector of electric field lies in plane of incident, is from expression [3]:

$$\dot{r} = \frac{(M_{11}+M_{12}\cdot q_1)\cdot q_1-(M_{21}+M_{22}\cdot q_1)}{(M_{11}+M_{12}\cdot q_1)\cdot q_1+(M_{21}+M_{22}\cdot q_1)} \tag{1}$$

Content from this work may be used under the terms of the Creative Commons Attribution 3.0 licence. Any further distribution of this work must maintain attribution to the author(s) and the title of the work, journal citation and DOI.
Published under licence by IOP Publishing Ltd

where: i – is the imaginary unit, $M_{11} = M_{22} = \cos\beta_2$; $M_{12} = \frac{-i}{q_2}\cdot\sin\beta_2$; $M_{21} = -i\cdot q_2\cdot\sin\beta_2$;
$\beta_2 = \frac{2\cdot\pi}{\lambda_0}\cdot\sqrt{\dot{\varepsilon}_2\cdot\dot{\mu}_2}\cdot d_2\cdot\cos\alpha_2$ – parameter, corresponding to the plate thickness d_2,
$q_1 = \cos\alpha_1$; $q_2 = \sqrt{\frac{\dot{\mu}_2}{\dot{\varepsilon}_2}}\cdot\cos\alpha_2$; λ_0 – the wavelength, α_1 – the angle of inclined falling wave on the plate, α_2 – the angle of refraction,

$$\dot{\varepsilon}_2 = \varepsilon_2' + i\cdot\varepsilon_2'' = \varepsilon_2'\cdot(1 + tg\delta_2^{э}); \ \dot{\mu}_2 = \mu_2' + i\cdot\mu_2'' = \mu_2'\cdot(1 + tg\delta_2^{м}),$$

$tg\delta_2^{э}$ - dielectric losses in the material of the plate, `
$tg\delta_2^{м}$– magnetic losses in the material of the plate.

The phase of the reflected TM wave for a material without dielectric losses and at $\dot{\mu}_2 = 1$ is equal to:

$$\delta_{rs} = -\,arctg\frac{2\cdot\varepsilon_2'\cdot\sqrt{\varepsilon_2' - \sin^2\alpha_1}\cdot\cos\alpha_1}{\operatorname{tg}\beta\cdot(\varepsilon_2'^2\cdot\cos^2\alpha_1 - \varepsilon_2' + \sin^2\alpha_1)} \tag{2}$$

3. The received results

Two consequences follow from Brewster's law:

-the angle between the directions of the reflected and refracted waves is $\alpha_0 = 90°$,

-$\alpha_{Бр}$ angle corresponding to the minimum of the modulus of the amplitude and corresponding phase 90° the reflected wave is calculated from the formula (1).

As can be seen from the expression (2), in the phase of the reflected TM wave jump occurs at the point of uncertainty functions arctg provided that:

$$\varepsilon_2{}^2\cdot\cos^2\alpha_1 - \varepsilon_2' + \sin^2\alpha_1 = 0 \tag{3}$$

which is fulfilled under the condition $\sqrt{\varepsilon_2'} = \operatorname{tg}\alpha_1$ and corresponds to the Brewster angle when the TM wave is reflected from the dielectric plate and is derived from the condition of the perpendicular of the reflected and transmitted waves.

The second jump of the phase shift of the reflected wave corresponds to the condition:

$$\operatorname{tg}\beta = 0 \tag{4}$$

which is performed for the phase of the reflected wave at the break points of the function $arctg = \pm\frac{\pi}{2}$, then its argument is equal to ±∞: $\tan\beta = 0$ or $\beta = \pi\cdot n$, n=0,1,...N, is performed for the plate thickness multiple of half the wavelength in the material at the angle of inclined incident wave α_1:

$$d_2 = \frac{\lambda_0}{2\cdot\sqrt{\varepsilon_2' - \sin^2\alpha_1}}\cdot n \tag{5}$$

thickness of the plate for both polarizations of TE and TM waves. The results of experimental measurements of the frequency dependences of the amplitude (1st) and phase shift (2nd) of the reflected wave at the normal angle incident wave on the half-wave plate taken from the screen of the vector circuit analyzer are shown in Figure 1.

Consider the fulfillment of the consequences of Brewster's law for materials with losses. To do this, we determine the angle with the minimum amplitude of the reflected wave from the calculated angular dependence of the modulus of the reflected wave amplitude on the tangent of the dielectric loss angle of the material $tg\delta_2^{э}$ at a constant value of the permittivity from the condition: $\min(|\dot{r}(\alpha_1)|)$ and the phase under the condition: $\delta = -\operatorname{arctg}\frac{Im(\dot{r})}{Re(\dot{r})} = 90°$.

The Brewster angle is calculated by the well-known formula:

$$\operatorname{tg}\alpha_{Br1} = \sqrt{\varepsilon_2'} \tag{6}$$

We will calculate the angles corresponding to the consequences of Brewster's law, depending on the dielectric losses of the plate material.

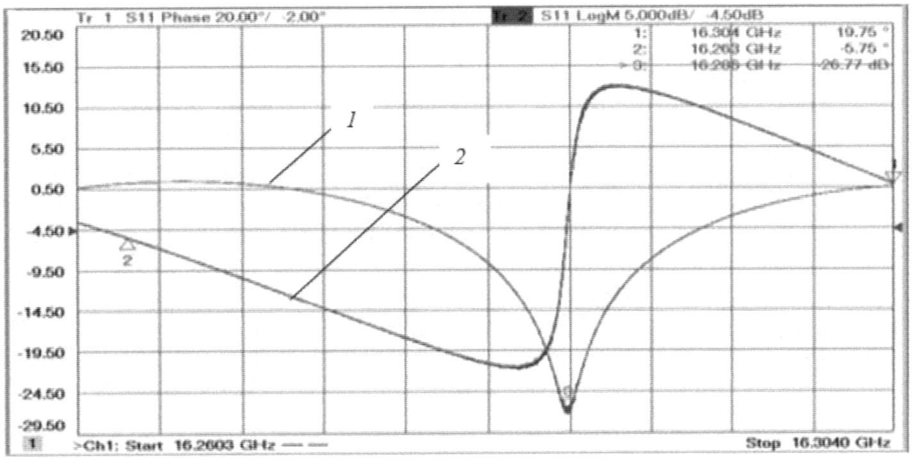

Figure 1. Experimental frequency dependence of amplitude and phase of the reflected wave on plate of half-wave thickness.

In the Figure 2 the calculated dependences on the loss tangent of a dielectric, corresponding are presented: for curve 1nd provision of corners corresponding to reflection minimum, for curve 2nd phase of coefficient of reflection equal the 90°, for curve 3rd for corner equal the 90° between the directions the incident and reflected waves, at reflection from plate with ε_2=3,81 quarter of wave electric thickness.

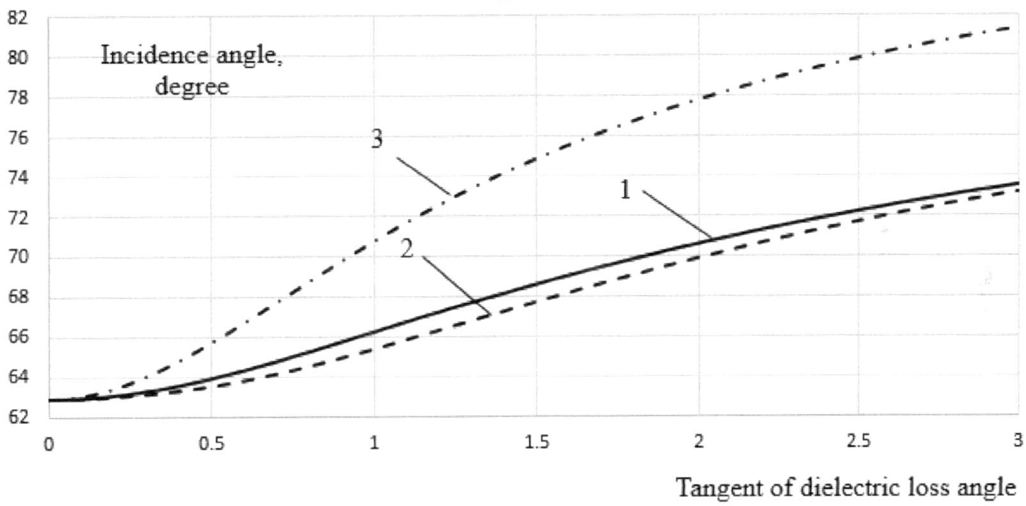

Figure 2. The provision of the angles depending on the tangent of dielectric loss angle, corresponding to minimum of amplitude (1), jump of phase (2) and for corner of 90 degrees between the passed and reflected waves (3).

From the comparison of the curves presented in Figure 2, it can be seen that with an increase in losses in the plate material, the difference between curves 1st and 2nd increases, that is, the coincidence condition, as according to formula (3), between the angles at which the minimum amplitude is observed and at which the phase is 90°, in the presence of losses in the plate material, is not fulfilled and corresponds only to the condition for the plate material without losses.

Figure 2 also shows that the angle at which the direction of the reflected wave is observed is 90 degrees with the direction of the transmitted wave. as the losses in the plate material increase, it differs more and more from the one calculated by the formula (3) for the plate material without losses.

From the analysis of the received rated dependences which modeling did not go beyond geometrical optics, and deviation from Fresnel's laws assumed in works [1,2] are not observed, and calculations match experimental observations that speaks about not applicability of formula of Brewster in the form of (3).

4. Discussion of results

Let us consider the condition under which is fulfilled by one of the investigation from of Brewster's law: the equality of the reflected wave to zero when solving the problem of the inclined incident of a plane wave on an infinite plate of material. Having carried out analytical calculations from the expression (1), the authors have established that the equality $|\dot{r}| = 0$ is fulfilled for the angle of the falling wave on the plate $\alpha_1 = \alpha_{Br2}$ under condition:

$$\cos \alpha_{Br2} = \sqrt{\frac{\dot{\mu}_2 - \dot{\varepsilon}_2 \cdot \dot{\mu}_2^{\,2}}{\dot{\mu}_2 - \dot{\varepsilon}_2^{\,2} \cdot \dot{\mu}_2}} \tag{7}$$

Accordingly, the Brewster angle for a material with losses is equal to:

$$\alpha_{Br2} = arccos\left|\sqrt{\frac{1 - \dot{\varepsilon}_2 \cdot \dot{\mu}_2}{1 - \dot{\varepsilon}_2^{\,2}}}\right| \tag{8}$$

If there is no magnetic component in the formula (5): $\mu_2' = 1;\ tg\delta_2^{м} = 0{,}0$, the angle corresponding to the minimum of the reflected wave amplitude modulus for a plate made of a material with dielectric losses is determined from the expression:

$$\alpha_{Br2} = arccos\left|\sqrt{\frac{1}{1 + \dot{\varepsilon}_2}}\right| \tag{9}$$

Accordingly, we change the formula (3), representing it as:

$$\alpha_{Br3} = arctg\left|\sqrt{\dot{\varepsilon}_2}\right| \tag{10}$$

A comparison of the formulas for determining the Brewster angle value shows the difference in determining it for a dielectric material with losses (9) and without them (3).

We will calculate the value of Brewster angles using various criteria following from Brewster's law and compare them with calculations using formulas 9 and 10, depending on the loss value of the plate material.

The calculation results are shown in Figure 3 for the plate material with $\varepsilon_2' = 3.81$, a quarter of the wave electrical thickness. Curve 1st is the dependence of the angle corresponding to the minimum of the reflected wave amplitude modulus on the tangent of the dielectric loss angle. Curve 2nd is the dependence of the position of the angle corresponding to the value of the reflected wave phase equal to 90° on the tangent of the dielectric loss angle. Curve 3rd and 4 is the dependence of the position of the Brewster angle calculated by formulas (9) and (10), depending on the tangent of the dielectric loss angle.

It can be seen from Figure 3 that the results of calculations according to formula (9) (curve 3rd) coincided with the results of numerical model calculation of the angles corresponding to the minimum of the reflection coefficient modulus (curve 1st), and the calculations according to formula (10) do not coincide with either curve 1st or curve 2nd, moreover, with an increase in losses in the material, an increase in $tg\delta_2^{э}$, this difference increases.

The calculations were carried out using the Microsoft Visual Studio software platform of the FORTRAN 2008 standard with the use of additional options for floating-point numbers to reduce calculation errors, both according to formulas (9, 10) and when modeling the problem of an of the inclined falling electromagnetic wave on an infinite plate made of a homogeneous dielectric material with losses.

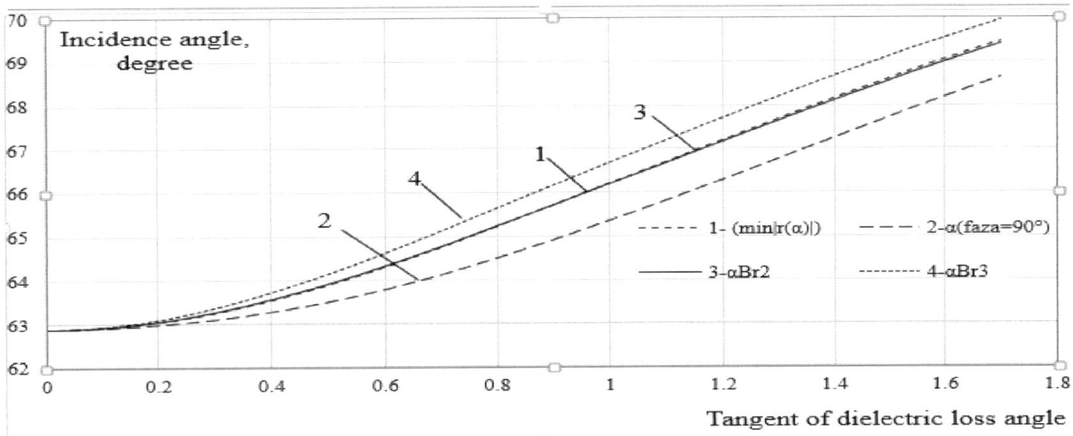

Figure 3. Calculated dependences of the Brewster angle on the tangent of the dielectric loss angle of the plate material according to various criteria and formulas: curve 1nd - according to the minimum of the amplitude modulus, curve 2nd-according to the phase value of 90°, curve 3rd- according to formula 9, curve 4nd - according to formula 10.

Summarizing the obtained results of theoretical and experimental studies and analyzing mathematical formulas for determining the phase and amplitude of electromagnetic waves in the form:

$$\delta_r = -arctg\frac{Im(\dot{r})}{Re(\dot{r})} \quad (11)$$

$$|\dot{r}| = \sqrt{Re(\dot{r})^2 + Im(\dot{r})^2} \quad (12)$$

it can be concluded that the phase jump of the electromagnetic wave by 180° occurs when the condition that the real part of the amplitude is equal to zero is fulfilled:

$$Re(\dot{r}) = 0 \quad (13)$$

and the condition of the minimum amplitude when the amplitude modulus is equal to zero:

$$|\dot{r}| = 0 \quad (14)$$

It is performed only if the real and imaginary parts of the amplitude of the electromagnetic wave are equal to zero:

$$\begin{cases} Re(\dot{r}) = 0 \\ Im(\dot{r}) = 0 \end{cases} \quad (15)$$

therefore, as can be seen from the comparison of conditions (13) and (15), in general, the points of the minimum amplitude and the phase jump by 180° should not coincide, but since the mathematical description of electromagnetic amplitudes is based on harmonic functions, it is obvious that for a medium with electromagnetic parameters with insignificant dielectric and magnetic losses, the coincidence of conditions (13) and (15) is quite often observed for frequency dependencies and less often for angular dependencies, both for reflected and transmitted waves.

Therefore, it is obvious that the difference in the angular positions of the minimum and the phase jump by 180° of the reflected wave in the region of the Brewster angle, observed in experiments, is not due to deviations in the Fresnel laws, but to the difference in the conditions of their natural position, which can be seen when calculating the reflected waves in the context of geometric optics, taking into account the complex values of the electrodynamic parameters of the medium.

5. Conclusion

Thus, it is determined that the formula (3) presented in the literature is applicable for calculating the Brewster angle value only for lossless materials.

Since all known materials have dielectric and magnetic losses, it is necessary to use the formulas (9,10) to find the value of the Brewster angle.

When an electromagnetic linearly polarized wave with an electric field vector lying in the plane of incidence falling at a Brewster angle on a plate of material with losses, the angle between the reflected and transmitted waves differs from 90° and increases with increasing dielectric and magnetic losses.

From the analysis of the calculated dependences obtained, the modeling of which did not go beyond the limits of geometric optics, it is clear that there are no deviations from the Fresnel laws assumed in [1, 2, 4], and the results of calculations of the angular position of the minimum of the reflected wave module coincide with the results of calculations according to formula (9), which coincides with experimental observations and indicates that the Brewster formula is not applicable for determining angles corresponding to its consequences.

It is obvious that the use of formula (3) to determine the permittivity of materials when measuring in free space is impossible, because the existing losses in the material will be interpreted as wrong data on the permittivity, and the use of formula (9) for these purposes requires additional measurement and methodological procedures.

References

[1] Mandelstam L I 1972 *Lectures on optics, relativity theory and quantum mechanics* (Moscow: Nauka)
[2] Azzam R M A, Basharaa N M 1977 *Ellipsometry and polarized light Amsterdam (*N-Y, Oxford: North-Holland Publishing company)
[3] Bourne M, Wolf E 1968 *Principles of optics* (Oxford: Pergamon Press)
[4] Paniagua-Domı´nguez R *et al* 2016 *Nature Communication* 1-9 doi: 10.1038/ncomms10362

Application of the migration method for radiotomography of breast cancer

A I Eremeev[1], V V Vasin[1], R N Satarov[1], I S Tseplyaev[1] and S S Shipilova[2]

[1] Tomsk State University, Tomsk, Russia
[2] Tomsk Polytechnic University, Tomsk, Russia

E-mail: aleksandr_eremeev_93@mail.ru

Abstract. A new method to visualize microwave image is presented to early non-invasive detection of breast cancer tumors. A new processing method to compute microwave images of heterogeneity in a biological environment is described here, as well as a new algorithm for accelerating the calculation of three-dimensional radio images. Sounding of synthetic phantoms with dielectric properties of breast tissue was carried out in the range of 2–8 GHz using a special radar system developed by the authors. Results show that it is possible to use this microwave imaging method to form 3D accurate images using hemispherical scan Images of tumor phantoms were obtained during probing in the 2–8 GHz range with a resolution of about 5–7 mm.
According to the results of the reconstruction of three-dimensional radio images, it was revealed that the calculation time can be reduced by at least 2 times with an insignificant loss of quality.

1. Introduction

This article proposes the approach to the development of a microwave visualization method for solving the problem of early non-invasive diagnosis of breast neoplasms. This direction is actively developing in the world, since breast cancer is the second most common cancer in the world after lung cancer and, along with cardiovascular and respiratory diseases, is a socially significant disease. Survival for cancer primarily depends on the stage at which treatment began, so they need to be diagnosed as early as possible (the probability of complete recovery reaches 93% in the case of early diagnosis of breast cancer (stages I-II)) [1, 2]. Non-invasive and comfortable for the patient (as opposed to painful X-ray mammography), but at the same time fast diagnosis is a long-standing medical request, despite the existing methods of X-ray and ultrasound diagnostics. At the same time, each of the methods is not absolute and sufficient and has its own limitations associated with the following factors: inaccurate location and size of tumors, long scanning time, the requirement for a homogeneous background environment, ambiguity of the images obtained, the presence of ionizing radiation, high cost of installations, low probability of detecting small forms of benign and malignant neoplasms. Most of these limitations are associated with the characteristics of oncological tumors. On the one hand, the density practically does not change in tumors at an early stage, but these changes are recorded during X-ray and ultrasound examinations. On the other hand, the dielectric constant changes in the area of tumor formation due to the proliferation of blood vessels that provide blood flow to new tumor cells in this area of space. Such change in the electrophysical characteristics can be recorded by radio wave sounding. Thus, the tumor detection can be based on the fact that the dielectric properties of a healthy

Content from this work may be used under the terms of the Creative Commons Attribution 3.0 licence. Any further distribution of this work must maintain attribution to the author(s) and the title of the work, journal citation and DOI.
Published under licence by IOP Publishing Ltd

tissue and a malignant tissue are different. This is confirmed by large-scale studies that have shown that the values of the dielectric constant of the tumor and healthy tissue are different [3, 4].

It follows from here that methods of detecting neoplasms are needed at present, that will improve the diagnosis of breast cancer and allow starting treatment as early as possible, thereby significantly increasing the chances of recovery. Use of microwave electromagnetic radiation is the most promising non-invasive method for searching for neoplasms in a biological environment. Microwave radio waves penetrate deep enough into biological tissues that can be used for mammography purposes. The advantages of radio wave methods are that they are harmless to humans and do not require the special conditions for working with the equipment, unlike X-ray installations.

Scientists around the world have shown promising results in the development of a microwave tomograph for the diagnosis of breast cancer. However, long scan time and low resolution are still the main unsolved problems.

2. Method for calculating microwave images

The radiolocation approach to obtaining the information about inhomogeneities in the biological environment is used in this investigation. It is known that the use of focusing radiation significantly reduces the influence of multiple interactions and diffraction effects so that an adequate description of the dominant wave interactions can be described in the single interaction approximation.

The main idea of this operation is to sum all received signals with calculated delay for a given focus point $\mathbf{r}_F$. Scanning the dedicated sounding area by focus point allows getting the spatial distribution of inhomogeneities. If sounding is carried out from a large number of angles, then the superposition (sum) of a plurality of interference patterns of the received signals decreases the level of side lobes and increases the value of the main maximum at the focal point. In the case of a monostatic locating scheme, the result of signal focusing is written as

$$U(\mathbf{r}_F) = \sum_n S(\mathbf{r}_n, \tau_n), \ \tau_n = \frac{2\sqrt{\varepsilon}}{c}\left|\mathbf{r}_F - \mathbf{r}_n\right| \tag{1}$$

where S is the received and transmitted signal at the point $\mathbf{r}_n$, ε is the average background medium dielectric constant, τ_n is the signal delay when reflected from heterogeneity.

For studies of volumetric objects, it is convenient to scan along a hemispherical surface with the radius B and angular coordinates of the azimuth and zenith (φ_n, θ_m), surrounding the object under study. In this case, the expression (1) takes the form:

$$U(\mathbf{r}_F) = \sum_n \sum_m S(\varphi_n, \theta_m, B, \tau_{n,m}), \tag{2}$$

where the delay for focus point with the coordinates $\mathbf{r}_F = (x_F, y_F, z_F)$ is calculated by the formulas

$$\tau_{n,m} = \frac{2\sqrt{\varepsilon}}{c}\sqrt{\left(x_F - B\sin(\theta_m)\cos(\varphi_n)\right)^2 + \left(y_F - B\sin(\theta_m)\sin(\varphi_n)\right)^2 + \left(z_F - B\cos(\theta_m)\right)^2}. \tag{3}$$

The problem of eliminating the additive interference that occurs during reflection from the scanner design, internal re-reflections in the channels of microwave paths and receiving-transmitting antennas, arises when processing the received signals. There are a number of methods for eliminating this interference. In some cases, the differential calibration is used [5], in which the previously obtained signals from the healthy breast of the same patient, previously measured at periodic examinations, are subtracted. The disadvantage of this approach is the need for long-term observation of the patient and regular sounding procedure. Another approach uses the technology of rotating the antenna array [6]. When the grating is rotated, two signals are recorded, and then they are subtracted. This assumes that the chest has cylindrical symmetry, however, this approximation is too rough. The standard filtering method is to subtract the DC component of the received signal P for each time point t_i. Mathematically, this operation is described by the following expression.

$$U(\mathbf{r}_F)=\sum_n\left(S(\mathbf{r}_n,\tau_n)-P_{f(\mathrm{n})}\right),\ P_i=\frac{1}{N}\sum_n S(\mathbf{r}_n,t_i),\ f(\mathrm{n})=\left[\frac{\tau_n}{\Delta t}\right]. \quad (4)$$

Here the operator $f(\mathrm{n})$ calculates the integer part of the number, and Δt is time sampling step. This operation assumes that at each sounding point the noise has the same temporal shape. In the case of sounding on a hemispherical surface, the interference associated with multiple reflections in the microwave channels and antennas is the same, and the interference associated with reflections from the scanner design is different.

To eliminate the interference in this work, it is proposed to use the special filtering of the constant components of the signal Q at each of the angular latitudes of sounding θ_m using the expression

$$U(\mathbf{r}_F)=\sum_n\sum_m\left(S(\varphi_n,\theta_m,B,\tau_{n,m})-Q_{m,f(\mathrm{n})}\right),\ Q_{m,i}=\frac{1}{N}\sum_n S(\varphi_n,\theta_m,B,t_i). \quad (5)$$

The proposed original solution will be applied further to obtain three-dimensional radio images of the phantom breast.

3. Accelerated algorithm for calculating 3D microwave image

To develop an algorithm for increasing the speed of calculating the radio image, we consider the situation (Figure 1). The area is divided into nodal and small grids. The grid step is determined by the formula (6):

$$N=\frac{D}{d}, \quad (6)$$

where D is the scale of the nodal grid, d is scale of the small grid. The nodes that correspond to the elements of the small grid are highlighted in red. The elements of the small grid are highlighted in blue. Receiving and transmitting antennas are highlighted in yellow.

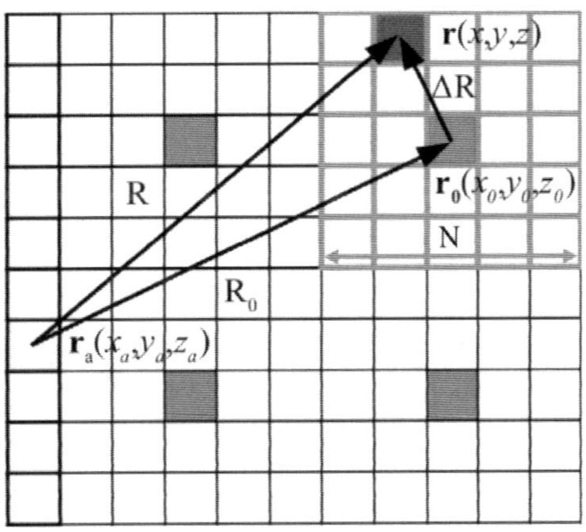

Figure 1. Dividing the space into nodal and small grids.

The algorithm calculates the distance for small elements of the space relative to the nodal grid. The distance for nodes is calculated without any approximations. Vector $\mathbf{r}$ is the vector between the node and the element of small grid, vector $\mathbf{r}_0$ is the vector between antenna and the node, $\mathbf{r}_a$ is the vector between antenna and the element of small grid. R is the distance between antenna and the element of small grid, ΔR is the distance between the node and the element of small grid, R_0 is the distance between antenna and the node. The distance between the nodes, expressed in the number of elements of the small grid, is called the step of the nodal grid. To calculate the distance using the formula (7) and expanding it to Taylor series we obtain the formula (8), which allows to calculate the distance approximately necessary for calculating the radio image.

$$R_0 = \sqrt{\sum_{i=1}^{3}(\mathbf{r}_i - \mathbf{r0}_i)^2}, \tag{7}$$

$$R \approx R_0 - \frac{1}{R_0}\sum_{i=1}^{3}(\Delta\mathbf{r}_i(\mathbf{r}_i - \mathbf{r0}_i)) \tag{8}$$

A program in the C++ language has been created. The program allows to compute the distance with approximation and without it. The calculation of the radio image was carried out on the configuration I7-3635QM 2.4 GHz, 8 GB DDR3. Two material points were modeled in the space and the inverse problem of building a radio image is solved. 3D Radio Image Resolution is 200×200×200. Figure 2 shows the dependences in red of the time of solving the inverse problem with increasing the step of the nodal grid and the difference between the images with increasing the step of the nodal grid (in blue). The Figure 2 shows that use the nodal grid with the smallest possible step reduced the computation time of the radio image by two times, and the quality of the radio image changed by several percent.

With an increase in the step of the nodal grid, the time for calculating the radio image decreases to a certain value, while the quality of the calculated radio image decreases. The optimal step of the nodal grid is chosen to be 20.

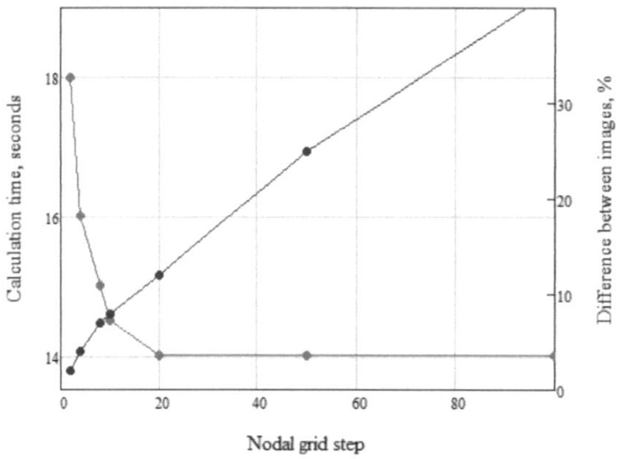

Figure 2. Calculated radio image at different step of the nodal grid.

4. Hardware for sounding the model environments

To test the microwave tomography method, a laboratory model in the form of a scanner was developed (Figure 3).

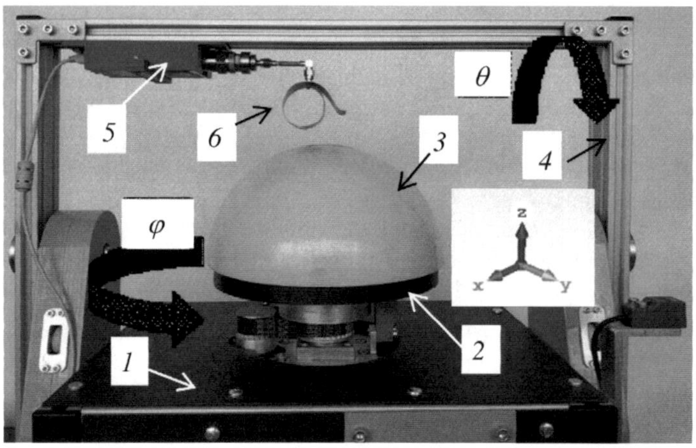

Figure 3. Hemispherical scanner. 1 – body; 2 – rotating platform; 3- object of investigation; 4 – rotation frame; 5 – VNA; 6 – antenna.

Scanner allows to scan in the spherical system by rotating the table on the angle φ in the diapason of 0–360 degrees and rotating frame on the angle θ in the diapason of 0–90 degrees. VNA CABAN R140 by "Planar" firm with the working frequency range of 0.3–14 GHz uses as RF generator. The vector reflectometer was attached to the moving frame of the scanner. An ultra-wideband antenna was connected to the reflectometer. A wide antenna frequency band and a stable directional pattern allow providing an operating sounding frequency band in the 2–8 GHz range.

5. Results

The results of detecting inhomogeneities in a homogeneous medium were obtained earlier for the case of scanning along the plane [7]. In this study, a breast phantom similar in electrophysical properties to a real breast was developed in order to fully test the system for detecting the heterogeneities. During the development of the breast phantom, polyurethane mixtures with the various graphite contents were used [8]. With the help of calculations and experiments, the proportions of polyurethane and graphite were selected to simulate human fat, skin and glands. Figure 4 shows the photo of the phantom and its cross section.

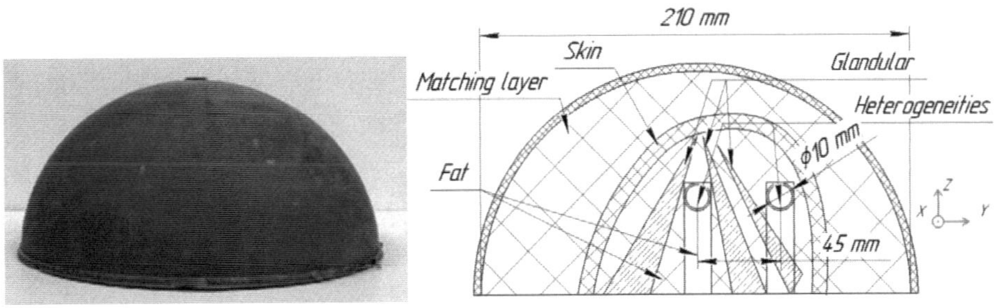

Figure 4. Breast phantom and its cross section.

The developed mammary gland phantom was used to test the algorithms for calculating the radio image. Multi-angle sounding was provided using a spherical scanner. Figure 5 shows the results of experiments with the phantom. Radio image resolution is 100×100×20. The computation time without optimization was 47 seconds, and the use a nodal grid with the smallest possible step reduced the computation time to 37 seconds. The use of a nodal grid with the maximum possible step has reduced the time for calculating the radio image to 27 seconds. In this case, it is possible to determine the location and size of the tumors. Without optimization, the calculated radio image quality improved slightly.

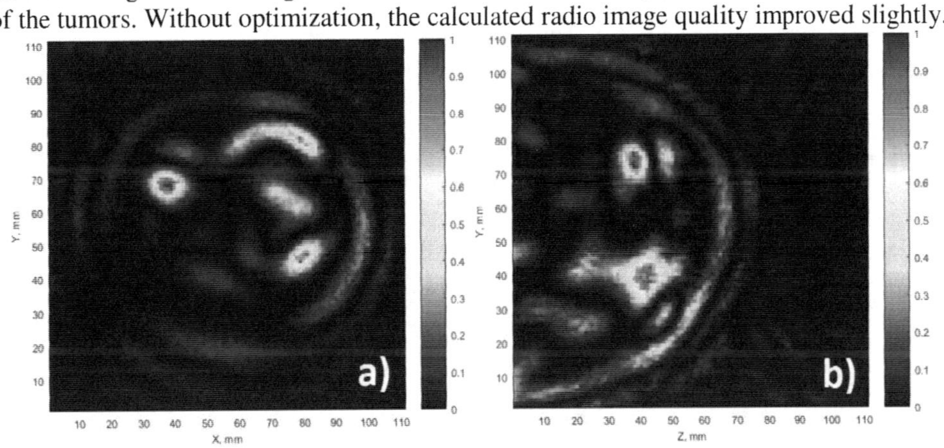

Figure 5. Radio images of heterogeneities in the coronal plane (a) and in the sagittal plane (b).

Also, analyzing the obtained results, it is possible to estimate the resolution of the proposed approach, which does not exceed the dimensions of the heterogeneities themselves and is about 5–7 mm.

The obtained result allows us to conclude about the advantages of the proposed approach for calculating the radio image in comparison with the standard methods.

6. Conclusion

This paper proposes the approach to the development of a microwave imaging method for solving the problem of early non-invasive diagnosis of breast tumors. The method of data processing for obtaining radio images of heterogeneities in a biological environment is described, as well as the algorithm for accelerating the calculation of a three-dimensional radio image is described. The results obtained by scanning on the hemisphere of the breast phantom using the proposed method show that this microwave visualization method is able to form 3D accurate images.

Acknowledgments

This research was supported by Ministry of Science and Higher Education of the Russian Federation, project № FSWM-2020-0038.

References

[1] Onitilo A A *et al.* 2009 *Clinical medicine and research* **7**(1–2) 4–13 doi: 10.3121/cmr.2008.825
[2] Cancer: Facts and Figures 2019–2020 2020 *American Cancer Society.* Available at: https://www.cancer.org/research/cancer-facts-statistics/all-cancer-facts-figures/cancer-facts-figures-2020.html
[3] Lazebnik M *et al.* 2007 *Phys. Med. Biol.* **52** 6093–6115 doi: 10.1088/0031-9155/52/20/002
[4] Lazebnik M *et al.* 2007 *Phys. Med. Biol.* **52** 2637–2656 doi: 10.1088/0031-9155/52/10/001
[5] Porter E *et al.* 2013 *IEEE Antennas and Wireless Propagation Letters* **12** 229–232 doi: 10.1109/LAWP.2013.2247374
[6] Klemm M *et al.* 2010 *IEEE Transactions on Antennas and Propagation* **58** (7) 2337–2344 doi: 10.1109/TAP.2010.2048860
[7] Shipilov S *et al.* 2018 *MATEC Web of Conferences* **155** 01025 doi: 10.1051/matecconf/201815501025
[8] Shipilov S *et al.* 2020 *Medical Physics* **47** (10) 5147–5157 doi: 10.1002/mp.14408

Quantum-mechanical model of dielectric losses in nanometer layers of solid dielectrics with hydrogen bonds at ultra-low temperatures

V A Kalytka[1], A D Mekhtiev[2,3], P Sh Madi[1,3] and A V Bashirov[4]

[1]Karaganda Technical University, Karaganda, Republic of Kazakhstan
[2]Kazakh Agrotechnical University name of Saken Seifullin, Nur-Sultan, Republic of Kazakhstan
[3]Tomsk Polytechnic University, Tomsk, Russia
[4]Karaganda Economic University of Kazpotrebsoyuz, Karaganda, Republic of Kazakhstan

E-mail: valerii.kalytka@gmail.com

Abstract. Upon based the finite difference methods construct the solutions for Liouville quantum kinetic equation linearized by the external field, in complex with the stationary Schrodinger equation and the Poisson operator equation, for an ensemble of non-interacting hydrogen ions (protons) migrating in the field of a crystal lattice perturbed by a variable polarizing field. The influence of the phonon subsystem is not taken into account. The equilibrium (non-balanced) proton density matrix is calculated using quantum Boltzmann statistics. The temperature spectra of dielectric losses tangent angle for hydrogen bonded crystals (HBC) in a wide temperature range (50–550 K) are calculated. At the theoretical level detected the effects of nano-crystalline states (1–10 nm) during the polarization of HBC in the region of ultra-low temperatures (4–25 K).

1. Introduction

In the last two decades, composite materials based on layered dielectrics with an ion-molecular chemical bond, capable of exhibiting high ionic conductivity in certain ranges of field parameters and temperatures, have played an essential role in modern industry [1,2].

A separate category of ion-molecular dielectrics are the hydrogen bonded crystals (HBC), used in electronics (elements of electronically controlled systems, microwave devices [1]), optoelectronics (nonlinear transducers of optical signals; fiber optic sensors) [3–9], laser technology (regulators of radiation parameters and electric gates (KDP) [10–12]), microelectronics (field-effect transistors, resonant tunneling diodes, TIR-structure), power generation and high voltage equipment (insulating cover collector items generators in Thermal Power Plants (TPP)) [12].

According to their electrophysical properties, HBC are classified as proton semiconductors and dielectrics (PSCD) and exhibit the property of proton conductivity, which is reduced to the diffusive transfer of hydrogen ions (protons) along hydrogen bonds in the direction of the lines of force of an external electric field [1].

The manifestation of a rectangular hysteresis loop (RHL) with an abnormally long relaxation time of residual polarization (up to 10 years) in ferroelectrics of the HBC class (KDP, DKDP [1], triglycine

Content from this work may be used under the terms of the Creative Commons Attribution 3.0 licence. Any further distribution of this work must maintain attribution to the author(s) and the title of the work, journal citation and DOI.
Published under licence by IOP Publishing Ltd

sulfate (TGS) [10], ferrotic salt, etc.) allows using these materials in capacitors of non-volatile high-speed storage devices (memory cells such as DRAM, FeRAM, etc.) of electronic computing devices [1,11,12].

2. Quantum mechanical model of dielectric losses in proton semiconductors and dielectrics

In the high temperature range (100-550 K), the main contribution to the dielectric relaxation in HBC is made by thermally activated (classical) transitions of relaxators (protons) along hydrogen bonds. In this case, theoretical studies of the mechanisms of polarization and conduction are carried out by methods of quasi-classical kinetic theory, based on the joint solution of a system of nonlinear Fokker-Planck and Poisson equations, in an arbitrary approximation k for a dimensionless small parameter of the perturbation theory $(\gamma \approx 0.001 \div 0.01)$ [1,11,12] and, already in a linear approximation for the comparison parameter, they $(\gamma \approx 0.01 \div 0.1)$ agree well with the experiment [1].

In the low temperature range (50–100 K), the dominant contribution to the relaxation processes in the HBC is made by tunneling (quantum) transitions of protons through the potential barrier, and the decomposition parameter increases markedly $(\gamma < 0.1)$ [1,11,12], which requires more strict consideration of the influence of nonlinear effects on polarization. At the same time, the energy spectrum of protons in the potential field of the crystal lattice becomes quasi-discrete and the statistical properties of the proton subsystem in the equilibrium state are described by the density matrix calculated by the Gibbs quantum canonical distribution [11].

Direct quantum-mechanical studies of nonequilibrium electrophysical processes in a dielectric will be based on the solution of the quantum kinetic Liouville equation [2], which allows, together with the Poisson operator equation and the Schrodinger equation, to construct a nonequilibrium density matrix and, on this basis, proceed to calculations of the quantities measured in the experiment (the current of thermally stimulated depolarization, the tangent of the dielectric loss angle, etc.).

In the absence of an external disturbance, the simulated system (a crystal of the HBC class) is described by the Liouville equation [12]

$$\frac{\partial \hat{\rho}_C^{(0)}}{\partial t} + \frac{1}{i\hbar}\left[\hat{\rho}_C^{(0)}, \hat{H}_C^{(0)}\right] = 0 \tag{1}$$

The Hamiltonian of an undisturbed system (crystal) is represented as [12]

$$\hat{H}_C^{(0)} = \hat{H}_{pr,C}^{(0)} + \hat{H}_{ph} + \hat{H}_{pr,ph} \tag{2}$$

In (2) $\hat{H}_{pr,C}^{(0)}$ – The Hamiltonian of the proton subsystem; $\hat{H}_{ph}$ – the Hamiltonian of the phonon subsystem; $\hat{H}_{pr,ph}$ – the operator of the proton-phonon interaction.

According to the assumptions experimentally established for the HBC [2], neglecting the proton-phonon interaction $\hat{H}_{pr,ph} \to 0$, taking the Hamiltonian of the phonon subsystem as a numerical operator $\hat{H}_{ph} \to \mathrm{const}$ and limiting ourselves to the model of an ideal proton gas, we transform equation (1) to the Liouville equation for the proton subsystem [2, 12].

$$\frac{\partial \hat{\rho}_{pr}^{(0)}}{\partial t} + \frac{1}{i\hbar}\left[\hat{\rho}_{pr}^{(0)}, \hat{H}_{pr}^{(0)}\right] = 0 \tag{3}$$

Here, the $\hat{H}_{pr}^{(0)}$ – Hamiltonian of an individual proton is undisturbed by the external field.

Since the Fermi energy of an ideal gas of protons in the HBC tends to zero, the solution of equation (3) is written according to Boltzmann quantum statistics for protons moving in a one-dimensional potential crystal $W_{pr,C}^{(0)}(x)$ field (the hydrogen bond field) and distributed over undisturbed energy levels $E_n^{(0)}$ [2, 12].

$$\hat{\rho}_{pr}^{(0)} = N_{pr.F}\left\{\sum_{n=0}^{\infty}\exp\left(-\frac{E_n^{(0)}}{k_B T}\right)\right\}^{-1} \times \exp\left(-\frac{\hat{H}_{pr}^{(0)}}{k_B T}\right) \quad (4)$$

The energy levels are $E_n^{(0)}$ calculated from the solution of the stationary Schrodinger equation of an individual proton $\hat{H}_{pr}^{(0)}\psi_n = E_n^{(0)}\psi_n$, where $\hat{H}_{pr}^{(0)} = -\frac{\hbar^2}{2m}\hat{\nabla}^2 + \hat{W}_{pr.C}^{(0)}(x)$, and the ψ_n – wave function of the nth stationary state of the proton, calculated by the quasi-classical method [2, 12]; m is the mass of the proton.

The statistical operator of a proton subsystem perturbed by an external polarizing field of intensity is $E(x;t) = E_0\exp(i\omega t)$ calculated from equation [2]

$$\frac{\partial\hat{\rho}_{pr}}{\partial t} + \frac{1}{i\hbar}\left[\hat{\rho}_{pr}, \hat{H}_{pr}\right] = 0 \quad (5)$$

where $\hat{H}_{pr} = \hat{H}_{pr}^{(0)} + \hat{W}_{el}(\vec{r};t)$ is the proton Hamiltonian perturbed by the external field; $\hat{W}_{el}$ is the perturbing correction to the operator $\hat{H}_{pr}^{(0)}$; is the $\vec{r}$ – radius vector of the proton; is the $\hat{\rho}_{pr} = \hat{\rho}_{pr}^{(0)} + \delta\hat{\rho}_{pr}$ – nonequilibrium statistical operator of protons; is the $\delta\hat{\rho}_{pr}$ – nonequilibrium correction to the unperturbed statistical operator (4). Acting by the operators $\frac{\partial\hat{\rho}_{pr}}{\partial t}$ and $\frac{1}{i\hbar}\left[\hat{\rho}_{pr}, \hat{H}_{pr}\right]$ on the proton wave function in the nth stationary state and excluding the second-order terms in the field $\delta\hat{\rho}_{pr} \times \left(\hat{W}_{el}\psi_n\right)$, we $\hat{W}_{el} \times \left(\delta\hat{\rho}_{pr}\psi_n\right)$ have

$$\frac{\partial\left(\delta\hat{\rho}_{pr}\psi_n\right)}{\partial t} + \frac{1}{i\hbar}\left\{\delta\hat{\rho}_{pr}\left(\hat{H}_{pr}^{(0)}\psi_n\right) - \hat{H}_{pr}^{(0)}\left(\delta\hat{\rho}_{pr}\psi_n\right) + \hat{\rho}_{pr}^{(0)}\left(\hat{W}_{el}\psi_n\right) - \hat{W}_{el}\left(\hat{\rho}_{pr}^{(0)}\psi_n\right)\right\} = 0 \quad (6)$$

Introducing a scalar function into (6) $u_n(\vec{r};t) = \delta\hat{\rho}_{pr}\psi_n(\vec{r};t)$, we have

$$\frac{\partial u_n}{\partial t} + \frac{E_n^{(0)}}{i\hbar}u_n - \frac{1}{i\hbar}\hat{H}_{pr}^{(0)}u_n + \frac{1}{i\hbar}\left[\hat{\rho}_{pr}^{(0)}; \hat{W}_{el}\right]\psi_n = 0 \quad (7)$$

For the model of blocking electrodes [2,12]

$$\hat{j}_{pr}(d;t) = \hat{j}_{pr}(0;t) = 0 \quad (8)$$

where d is the thickness of the dielectric.

There is no polarization at the initial moment of time [2]

$$u_n(\vec{r};0) = 0 \quad (9)$$

We construct the solution of equation (7) together with the Poisson operator equation [12]

$$\frac{\partial\hat{E}}{\partial x} = \frac{q}{\varepsilon_0\varepsilon_\infty}\delta\hat{n}(\vec{r};t) \quad (10)$$

taking into account the boundary condition [2,12]

$$\int_0^d \hat{E}(x;t)\,dx = V_0\exp(i\omega t) \quad (11)$$

where q is the proton charge; is the high ε_∞ – frequency permittivity; is the V_0 ω amplitude and circular frequency of the external field, respectively.

Quantum-mechanical expression for excessive proton concentration [12]

$$\delta\hat{n}(\vec{r};t) = \sum_{n=0}^{\infty}\psi_n^*\,\delta\hat{\rho}_{pr}(\vec{r};t)\,\psi_n = \sum_{n=0}^{\infty}\psi_n^* u_n(\vec{r};t) \quad (12)$$

allows you to calculate the polarization operator of the proton subsystem

$$\hat{P}(\vec{r};t) = q\sum_{n=0}^{\infty}\psi_n^* \,\vec{r}\times u_n(\vec{r};t) \tag{13}$$

The averaging of operator (13) by coordinates and energies is carried out using a non-equilibrium polarization density matrix

$$\langle\overline{P}\rangle \equiv Sp(\hat{\rho}_{pr}\times\hat{P}) = \int_V \Psi^*\hat{P}(\vec{r},t)\Psi dV \tag{14}$$

Crystal polarization in an alternating field [12]

$$\langle\hat{P}\rangle = \varepsilon_0(\varepsilon^* - \varepsilon_\infty)E_0\exp(i\,\omega t) \tag{15}$$

Based on (14), (15), we write down the theoretical spectrum of the complex permittivity

$$\varepsilon^*(\omega,T) = \varepsilon_\infty + \frac{1}{\varepsilon_0 E_0}\exp(-i\,\omega t)\int_V \Psi^*\hat{P}(\vec{r},t)\Psi dV \tag{16}$$

The degree of accuracy of the calculation results according to the formula (16) will be estimated by comparing with the experimental data on measuring the tangent spectra of the dielectric loss angle [2,12]

$$tg\delta(\omega,T) = \frac{Im[\varepsilon^*(\omega,T)]}{Re[\varepsilon^*(\omega,T)]} \tag{17}$$

3. Numerical study of the quantum mechanical model of dielectric losses

Numerical calculations of the temperature spectra of the tangent of the dielectric loss angle $tg\delta_{th}^{(v)}(T)$ [12] were performed according to formulas (16), (17) and compared with the results of experimental measurements in Onot talc and gypsum in the temperature range $T = 50-550$ K, at the frequency of an external alternating electric field $\nu_1 = 7\cdot10^6$ Hz [12,13]. The thickness of the crystal in the experiment was 30 microns [12].

According to the result of numerical calculation, in the low-temperature region (50-100 K), in Onmot talc, at the crystal thickness d=30 microns, at the frequency of the alternating polarizing field $7MHz$ detected the theoretical maximum of the temperature tangent angle of the dielectric loss $tg\delta_{th}^{(v)}(T)$ with the amplitude $tg\delta(\nu_1, T_{max}) = 0.5\cdot10^{-4}$ at the temperature $T_{max} = 105$ *K* due to the tunneling (quantum) motion of relaxators (protons) with an activation energy $U_0 = 0.045$ *eV*. In gypsum, accordingly, a maximum of this kind is revealed, also theoretically, under the same external conditions ($\nu_1 = 7\cdot10^6$ Hz; d=30 microns), but at a temperature of $T_{max} = 97\,K$, with an amplitude $tg\delta(\nu_1, T_{max}) = 0.9\cdot10^{-4}$ and an activation energy $U_0 = 0.05$ *Ev*.

According to Tables 1,2, the low-temperature maximum is $tg\delta(\nu_1, T_{max})$ most sensitive to changes in the thickness of the crystal layer, as a result, the theoretical maximum in talc with a decrease in the layer thickness from 30 microns to 3 nm shifts from low (105 K) to ultra-low temperatures (22 K), and the maximum amplitude decreases from $0.5\cdot10^{-4}$ to $0.39\cdot10^{-8}$. In gypsum, the maximum $tg\delta(\nu_1, T_{max})$ shifts from 97 K to 15 K with an increase in the amplitude from $0.9\cdot10^{-4}$ at 30 microns to $0.86\cdot10^{-8}$ at 3 nm.

Table 1. Temperature maxima of the tangent of the dielectric loss angle $tg\delta_{th}^{(\nu_1)}(T)$ in Onot talc, calculated at the frequency of the polarizing field 7 MHz, at different thicknesses of crystal layers.

Layer thickness, nm		**Amplitudes and temperature positions of theoretical maxima** $tg\delta(\nu_1, T_{max})$			
30000	$0.5 \cdot 10^{-4}$ (105)	$2.5 \cdot 10^{-3}$ (160)	$3.23 \cdot 10^{-3}$ (220)	$4.25 \cdot 10^{-3}$ (265)	$4.5 \cdot 10^{-3}$ (310)
3000	$0.4 \cdot 10^{-4}$ (96)	$2.2 \cdot 10^{-3}$ (157)	$2.14 \cdot 10^{-3}$ (219)	$4.25 \cdot 10^{-3}$ (265)	$4.5 \cdot 10^{-3}$ (310)
300	$0.3 \cdot 10^{-4}$ (88)	$2.4 \cdot 10^{-3}$ (146)	$1.17 \cdot 10^{-3}$ (215)	$4.25 \cdot 10^{-3}$ (265)	$4.5 \cdot 10^{-3}$ (310)
30	$4.6 \cdot 10^{-6}$ (50)	$6.8 \cdot 10^{-6}$ (127)	$1.05 \cdot 10^{-3}$ (207)	$4.25 \cdot 10^{-3}$ (265)	$4.5 \cdot 10^{-3}$ (310)
3	$0.39 \cdot 10^{-8}$ (22)	$5.8 \cdot 10^{-7}$ (103)	$0.54 \cdot 10^{-3}$ (190)	$3.3 \cdot 10^{-3}$ (259)	$4.4 \cdot 10^{-3}$ (308)

Table 2. Temperature maxima of the tangent of the dielectric loss angle $tg\delta_{th}^{(\nu_1)}(T)$ in gypsum, calculated at the frequency of the polarizing field 7 MHz, at different thicknesses of crystal layers.

Layer thickness, nm		**Amplitudes and temperature positions of theoretical maxima** $tg\delta(\nu_1, T_{max})$			
30000	$0.9 \cdot 10^{-4}$ (97)	$1.8 \cdot 10^{-3}$ (145)	$3.9 \cdot 10^{-3}$ (210)	$3.3 \cdot 10^{-3}$ (270)	$4.1 \cdot 10^{-3}$ (320)
3000	$0.3 \cdot 10^{-4}$ (89)	$1.3 \cdot 10^{-3}$ (142)	$3.24 \cdot 10^{-3}$ $3.3 \cdot 10^{-3}$ (209)	$3.3 \cdot 10^{-3}$ (270)	$4.1 \cdot 10^{-3}$ (320)
300	$0.1 \cdot 10^{-4}$ (80)	$0.4 \cdot 10^{-3}$ (138)	$2.9 \cdot 10^{-3}$ (200)	$3.3 \cdot 10^{-3}$ (270)	$4.1 \cdot 10^{-3}$ (320)
30	$1.6 \cdot 10^{-6}$ (47)	$7.4 \cdot 10^{-6}$ (130)	$1.6 \cdot 10^{-3}$ (190)	$3.2 \cdot 10^{-3}$ (267)	$4.1 \cdot 10^{-3}$ (320)
3	$0.86 \cdot 10^{-8}$ (15)	$3.4 \cdot 10^{-7}$ (102)	$5.1 \cdot 10^{-2}$ (183)	$2.9 \cdot 10^{-3}$ (264)	$3.5 \cdot 10^{-3}$ (317)

Low-temperature peaks (50-100 K) of the loss angle tangent in layered minerals are characterized by an anomalous dependence of the maximum amplitude $tg\delta(\nu_1, T_{max})$. It depends on the layer thickness in the range of 1-10 nm with a maximum shift towards helium temperatures (1–10 K) and a sharp drop in its amplitude by 3-4 orders of magnitude [14].

4. Scientific and practical significance of the research results

Currently, a fairly large amount of experimental data has been accumulated on the application of proton semiconductors and dielectrics (PSCD), mainly in the field of electrochemical technologies and physical chemistry (in the development of solid-state electrolytes (perovskites, orthoperiodates and biperiodates of alkali metals) [4–9]. At the same time, there are not so many applications of this phenomenon in the field of theoretical electrical engineering and physical electronics. Theoretical developments aimed at applying the quantum kinetic theory of proton conduction and polarization [13–17] to the issues of microelectronics (field-effect transistors, tunneling and resonant diodes, etc. [2]) and the quantum theory of high-temperature superconductivity [12] are relevant.

Semi-empirical studies of the migration of adsorbed protons over the surface of single-layer carbon nanotubes [18] they are not complete, due to the lack of a strict theoretical justification: 1) the relationship between the configuration of the tube surface and the dominant physical mechanism of proton transfer (tunneling or thermally activated proton transfer through a potential barrier); 2) the temperature dependence of the probability of proton tunnel transitions; 3) the shape of the potential relief and activation energy for protons. The methods of quantum kinetic theory, the foundations of which are laid by the authors [19–26], will allow us to consider in more detail, at a more rigorous theoretical level, using the density matrix apparatus, the quantum mechanism of proton tunneling transport in nanoscale materials with high proton conductivity.

The results of research in the future, will find application in microelectronics and nonlinear optics in particular, in studies of the effects on the nonlinear optical processes of the second order (second-harmonic generation, parametric oscillation and amplification of light, a mixture of frequencies, the electro-optic effect) nonlinearities of higher order (the effect of self-action of laser radiation) is important for technology femptoseconds lasers [27,28].

The scientific and practical significance of the methods developed in this article consists in the development of unified algorithms for modern software and hardware, which allows not only to analyze, but also to predict, with a high degree of accuracy, the results of scientific and industrial experiments [29,30].

5. Conclusion

1.The influences of the parameters of the polarizing field (frequency, EMF amplitude) and temperature on the mechanism of diffusive tunneling transport of hydrogen ions (protons) in the potential field of a crystal lattice perturbed by the polarizing field (quantum diffusion polarization) in crystals with hydrogen bonds (HBC) is studied by the methods of quasi-classical kinetic theory and non-relativistic quantum theory.

2.The calculating of the equilibrium (non-balanced) density matrix for the ensemble of the main relaxers (hydrogen ions) was carried out from the solutions of the Liouville equation (3), together with the stationary Schrodinger equation, which had not taken into account the proton-proton and proton-phonon interactions, and the Hamilton operator for the phonon subsystem was taken as a numerical constant and calculated by computer method as a parameter for comparing the theory with the experiment. The Hamilton of the proton subsystem was constructed for the model of an ideal proton gas in equilibrium with the ionic subsystem of the crystal lattice, and the unperturbed statistical operator of the proton subsystem was written with the help of Boltzmann quantum statistics (4).

3.Constructed the non-linear quantum kinetic equation of proton relaxation (7) solving in complex with the Poisson operator equation (10), by the finite difference method, for the model of blocking electrodes (8), (11).

4.Measuring in the experiments the value of dielectric polarization (15) was calculated by complete quantum mechanical averaging of the polarization operator (13), (14) using the density matrix (statistical matrix) apparatus. On this basis, the theoretical spectra of the complex permittivity (16), calculated over a wide range of field parameters (100 kV/m–1000 MV/m) and temperatures (0–1500 K).

5.An expression is written for calculating the dielectric loss tangent of angle (17), calculated taking into account the electrical conductivity due to the through motion of protons in the dielectric

6.At the theoretical level, dimensional effects in the HBC are found, which manifest themselves in shifts of low-temperature (50–100 K) maxima of the dielectric loss tangent of angle towards ultra-low temperatures (4–25 K) with a decrease in the amplitudes of the maxima by 3–4 orders of magnitude, in the case of a reduction in the thickness of the crystal layer from 1–10 microns to 1–10 nm. This effect is explained by changes in the configuration of the discrete energy spectrum of protons, when, at a film thickness of 1–10 nm, the activation energy of low-temperature relaxers decreases to 0.01–0.03 eV (versus 0.05–0.1 eV at 1–10 microns) and, for energy levels falling into an isolated potential well, abnormally high quantum transparencies of the potential barrier (0.8–0.9) are manifested. Thus, in HBC nanofilms, near ultra-low temperatures (4–25 K), the dielectric, due to the restructuring of the structure of the hydrogen sublattice, passes into a quasi-magnetoelectric state, characterized also by abnormally high values of the real component of the KDP (2.5–3.5 million)

References

[1] Kalytka V, Korovkin M, Mekhtiev A, Yurchenko A 2018 *Russian Physics Journal* **61**(4) 757–769 doi: 10.1007/s11182-018-1457-8
[2] Kalytka V, Korovkin M, Madi P, Kalacheva S, Sidorina Y 2020 *IOP Conf. Series: Journal of Physics: Conf. Series* **1499** 012046 doi:10.1088/1742-6596/1499/1/012046
[3] Tan Wei, Sun Yong, Chen Hong & Shen Shun-Qing 2014 *Scientific Reports* **4** 3842
[4] Wells B, Zayats A and Podolskiy V 2014 *Physical Review B* **89** 035111
[5] Slobozhanyuk A, Ginzburg P, Powell D, Iorsh I, Shalin A, Segovia P, Krasavin A, Wurtz G, Podolskiy V, Belov P, Zayats A 2015 *Physical Review B* **92** 195127
[6] Khanikaev A B *et al* 2013 *Nature Materials* **12** 233–239
[7] Kudyshev Zh *et al* 2017 *ACS Photonics* **4**(6) 1413–1420
[8] Zhaxylyk A *et al* 2017 *ACS Photonics* **4**(10) 2470–2478
[9] Antonova A M, Vorobyov A V, Lyalikov B A 2008 *Proceeding of the 14 All-Russian Scientific and Technical Conference Energy: Ecology, Reliability, Security: Materials Reports* 59–62
[10] Yatsenko O, Chudotortsev I, Stekhanova Zh, Milovidova S, Rogazinskaya O 2006 *Bulletin of the Voronezh State University (VSU). Series: Chemistry. Biology. Pharmacy* (2) 117–121
[11] Kalytka V *et al* 2019 *Journal of Computational and Theoretical Nanoscience* **16**(7) 2799–2804 doi:10.1166/jctn.2019.8130
[12] Kalytka V, Neshina Y, Madi P, Naboko Y 2019 *IOP Conf. Series: Materials Science and Engineering: Series* **698** 022002 doi:10.1088/1757-899X/698/2/022002
[13] Kalytka V, Korovkin M, Bashirov A, Madi P, Mekhtiyev A, Mekhtiyev R 2021 *IOP Conf. Series: Journal of Physics: Conf. Series* **1843** 012011doi:10.1088/1742-6596/1843/1/012011
[14] Nocedal J, Stephen W 2006 *Numerical Optimization* (Berlin: Springer) doi:10.1007/978-0-387-40065-5
[15] Kulagin I, Ganeev R, Tugushev R, Rasnianskii A, Usmanov T 2004 *Quantum Electronics* **34**(7) 657–662
[16] Belonenko M 1998 *Quantum Electronics* **25**(3) 255–258
[17] Pyatakov A P, Star A K 2012 *Successes of physical sciences* **182** 593–620 doi: 10.3367/UFNr.0182.201206b.0593
[18] Krotov K 2016 *Bulletin of the NSU. Series: Information Technologies* **14**(1) 39–60
[19] Patent na izobretenie № 6. 27.02.2009, Patent holder: FGBOU HPE "Russian Naval State Academy named after Admiral F. F. Ushakov". Multifunctional device for the study of physical and technical characteristics of semiconductors, dielectrics and electrical insulation materials
[20] Tihomirov P, Emelianov P, Plotnik N, Zirianov A 2014 *Bulletin of the NSU. Series: Information Technologies* **12**(4) 112–120
[21] Ivanov V 2015 *Bulletin of the Perm University. Series: Mathematics. Mechanics. Computer science* **31**(4) 16–27
[22] Andreev N F *et al* 2004 *Quantum electronics* **34**(4) 381–385
[23] Bespalov V I *et al* 1987 *News of the USSR Academy of Sciences. Physical series* **51**(8) 1354–1360
[24] Parafonova V 2003 *Science and Life* **2**(18) 174–182
[25] Aghamalyan N R, Hovsepyan R K and Petrosyan S I 2008 *J. Contemp. Phys.* **43**(3) 128–133
[26] Dhananjay A, Singh S, Nagaraju J and Krupanidhi S B 2007 *Appl. Phys. A* **88** 421–424
[27] Agamalan N R, Ovsepyan R to 2008 *News of National Academy of Armenia, Physics* **43**(4) 177–182.
[28] Poghosyan A R, Aghamalyan N R, Guo R, Hovsepyan R K, Vardanyan E S 2010 *SPIE* **7781** 43
[29] Bredichin V I *et al* 1982 *Quantum electronics* **3**(9) 1263–1265
[30] Kafadaryan E A, Petrosyan S I, Badalyan G R, Harutyunyan S R, Kuzanyan A S 2002 *Thin Solid Films* **32**(9) 1046–1049

Issues of technical implementation of ground-based microwave radiometric system calibration

I N Rostokin[1], E V Fedoseeva[1], E A Rostokina[1] and G G Shchukin[1, 2]

[1]Vladimir State University named after A.G. and N.G. Stoletovs Murom Institute, Murom, Russia
[2]Military Space Academy named after A.F. Mozhayskogo, St.-Petersburg, Russia

E-mail: Rostockin.Ilya@yandex.ru

Abstract. The theoretical bases of a method of carrying out calibration of microwave radiometric system on a noise signal of the external noise generator are analysed, the parameters setting value of gain of antenna temperature at introduction in structure of system of such external source of a calibration signal are determined. The questions of technical realization of an external source of a calibration signal for microwave radiometric system are considered, the structural scheme is presented, the place of its arrangement in structure of the antenna is determined. The necessity of using a horn antenna to create directional radiation towards the mirror antenna emitter is shown. The data on the design of such a calibration signal source for the case of its use in a three-band microwave radiometric system with the reception of a noise signal at the common antenna aperture and a sequential separation of the input signals of the three bands in the feeder path of the common mirror irradiator are presented. The influence of the design and, accordingly, the directional properties of the horn antenna of a microwave radiometric system calibration signal source on the coefficient characterizing the radiation divergence created by the noise generator horn, as well as the frequency properties of this coefficient are analysed. The results of measurements of radio-thermal radiation by a three-band microwave radiometric system when two calibration periods with reception of a radio-noise signal from an external calibration signal source located at the base of the antenna mirror are presented.

1. Introduction

Calibration of the microwave radiometric system is performed to obtain an unambiguous correspondence of the value of the system output signal to the radio brightness temperature of the probed area of space. To perform it, it is necessary to ensure the reception of radio-noise radiation from sources that create a certain level of noise signal at the antenna input [1, 2].

In practice, the sources of the test noise signal can be reference regions of the surrounding space, the level of radio-noise radiation of which can be set with a high degree of accuracy. In ground-based microwave system such region is a cloudless atmosphere, observed in the zenith. But to determine the parameters of the linear calibration characteristic of the microwave radiometric system, linking the value of the output signal and the radio brightness temperature, it is necessary to have a second source of the test noise signal [3, 4].

As such a source is known application of wide-aperture passive sources of own radio-thermal radiation, the characteristics of which are close to the characteristics of the "black body". Limitations for the use of such sources, due to the need to perform them with large enough linear dimensions for the possibility of overlapping the angular area of the main lobe of the antenna pattern in the far zone, which

Content from this work may be used under the terms of the Creative Commons Attribution 3.0 licence. Any further distribution of this work must maintain attribution to the author(s) and the title of the work, journal citation and DOI.
Published under licence by IOP Publishing Ltd

can not always be realized in practice, especially in conditions of mobile movement of the microwave system to the base location [5–7].

To implement the mobile version of the ground-based microwave radiometric system calibration a noise generator is introduced into the measuring complex, whose signal through a small horn, located in the base of the mirror, is radiated towards the mirror antenna of the system, creating a stable level of noise signal. In order to create several reference levels for determining the parameters of the calibration characteristic the noise generator is supplied with an attenuator.

In the proposed method of ground based microwave radiometric system calibration the external noise signal source is placed in the base of the mirror and the noise signal is received by the antenna illuminator. As a result, in the output signal of the system this noise signal passes with a gain corresponding to the gain of the antenna illuminator.

2. Theoretical basis of calibration of a microwave radiometric system by an external noise generator

The input signal of the microwave radiometric system is usually expressed through the value of the antenna temperature, proportional to the power of the received noise signal. The antenna temperature of the microwave radiometric system, the smoothed characteristic of the radio noise radiation surrounding the antenna space, is given by the expression:

$$T_a = [\bar{T}_b(1-\beta)\eta + \bar{T}_r\beta\eta + T_0(1-\eta)]k_{af} \quad , \tag{1}$$

where $\bar{T}_b$ and $\bar{T}_r$ – averaged values of radioburst temperature over the space area, corresponding to the angular area of the main lobe and dispersion area of the antenna patterns; β – antenna dispersion factor; η – antenna efficiency; T_0 – antenna thermodynamic temperature; k_{af} – antenna feeder transmission coefficient of the antenna radiation path in a given frequency range.

The first term in expression (1) defines the component of antenna temperature, conditioned by reception of radio noise radiation from the angular area of the main lobe of the antenna patterns, which characterizes the radio brightness temperature of the investigated area of space. To determine the numerical value of the measured brightness temperature for the value of the output signal of radiometer it is necessary to calibrate the system that is possible when measuring the radio noise radiation of the reference source with the known brightness temperature [8].

One of the ways to create a reference radio noise signal for a microwave radiometric system is to use a noise generator located outside the radiometer, creating a certain level of noise signal at the radiometer's input. For possibility to perform external calibration taking into account antenna's influence we are considering a noise generator placed at the base of the mirror and producing at the input of the antenna emitter a reference radio-noise signal, equivalent to the given noise temperature. Then in the composition of the antenna temperature there will be an additional component the value of which is proportional to the noise temperature of the source:

$$\Delta T_a = T_{ng}\gamma_t k_{af}, \tag{2}$$

where T_{ng}– radio brightness temperature of the noise generator; γ_t– coefficient taking into account the reduction of noise power at the input of the illuminator due to the spatial divergence of the radiation generated by the noise generator.

The presence of the coefficient γ_t as a part of expression (2) determines the necessity to evaluate the influence of constructive realization of a calibrated radio noise source on the value of the antenna temperature increment at the radiometer input. When performing measurements with reception at the common aperture of the mirror antenna in several frequency ranges, as suggested in [9], it is additionally necessary to estimate the change in the value of the coefficient γ_t at the specified location of the reference source of the noise signal.

Journal of Physics: Conference Series **2140** (2021) 012029 doi:10.1088/1742-6596/2140/1/012029

3. Technical issues of implementing the calibration of a microwave radiometric system by an external noise generator

The problem of calibration with an external noise generator was solved for a three-band microwave radiometric system with compensation of background noise when receiving radio noise radiation on a common mirror antenna aperture with frequency separation in an irradiator operating in two-mode mode (two modes of circular waveguide H_{11} and E_{01}), with a device for separation of mode signals - a mode separator for each frequency range. The sequential algorithm for frequency separation of input signals allowed the system to use a single mirror antenna illuminator for three frequency channels.

Figure 1 shows the schematic diagram and the appearance of the test noise source of the microwave radiometric system. As a reference noise signal source the microwave noise generator module M31305-4 (NG) on avalanche spanning diode. The generated noise power reaches 33 dB relative to kT_0, which allows including the noise generator in the measurement path through a controlled attenuator (ATT) with a large dynamic range of attenuation change (from 0 to -33 dB) with output to the transmitting corrugated horn with a radiation pattern width of 25° at a frequency of 22 GHz.

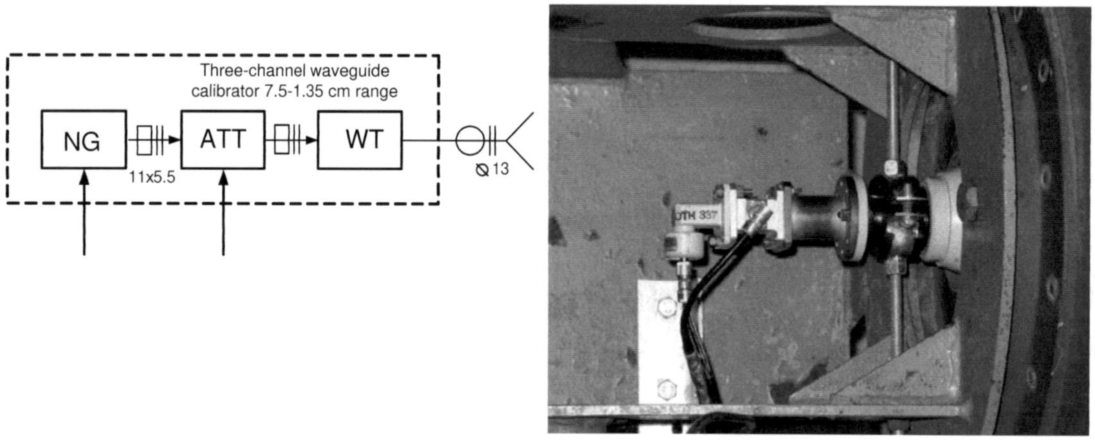

a) block diagram b) appearance

Figure 1. Wideband waveguide noise signal source: NG – noise generator module on avalanche spanning diode; ATT – controlled attenuator; WT – waveguide transition from rectangular to circular waveguide.

To control the noise temperature of the noise generator there is an absorption attenuator at its output. When the attenuator absorption coefficient is zero, the total intensity of the noise coming to the output is maximum and is equal to $T_{NG\ MAX} = T_{NG} = 33\ kT_0$. When the absorption factor is equal to one, the noise intensity will reach its minimum value equal to the noise intensity produced by the attenuator $T_{NG\ MIN} = T_{ATT} = 1\ kT_0$. As a result, all of the attenuator absorption values from 0 to 1 correspond to specific noise intensity values between 1 kT_0 and 33 kT_0. Therefore, the attenuator can be directly graduated in noise intensity units.

While adjusting and tuning the level of test noise signal a waveguide absorbing attenuator was used, which allows smooth adjustment of attenuation level from 1 dB to 35 dB; the attenuation setting error is 1.5 dB (to 19 dB) and 10% (to 35 dB); permissible maximum power 0.5 W; VSWR 1,08; microwave tract section 17×8 mm.

For research of possibility of practical realization of method of calibration of three-band microwave radiometric system with the central wavelengths of working bands 1.35 cm, 3.2 cm and 7.5 cm at reception on the common antenna emitter [10] on an external source of a noise signal measurements for a cloudless atmosphere are executed at two levels of power of a noise signal differing on level of noise power on 3 dB. The example of realization of calibration of the three band microwave system is shown in Figure 2.

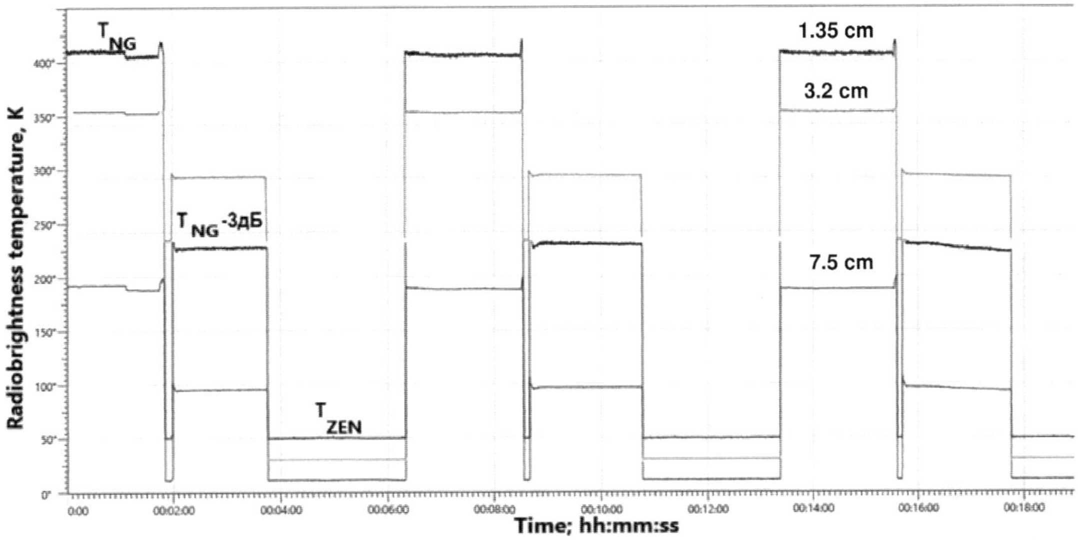

Figure 2. Time course of the results of measurements of radio-thermal radiation of the atmosphere when introducing the signal from an external noise generator for three bands 1.35 cm, 3.2 cm and 7.5 cm

Figure 2 shows the data of measurements in calibration mode by an external noise generator in the values of radio brightness temperatures for the three working frequency ranges with central wavelengths of 1.35 cm, 3.2 cm and 7.5 cm working ranges.

The obtained results showed the fundamental possibility of calibration of the three-band microwave radiometric system by the radio-noise emission of an external broadband source - noise generator simultaneously in the three working wavelength ranges when receiving a common mirror antenna irradiator with subsequent frequency and mode separation of the signals of two polarizations [11].

4.Conclusion

The executed analysis of features of technical realization of a method of calibration of the multiband microwave radiometric system by an external noise generator, located in the base of the antenna mirror, at reception of radio-noise radiation on a common antenna aperture and consecutive allocation of signals of working frequency ranges in an irradiator, has shown a necessity of careful study of a question of a choice of a source of a noise signal taking into account frequency dependence of parameters as radiating system - horn, and frequency dependence of noise temperature of noise generator in a frequency band.

The revealed features of realization of a source of a noise calibration signal for a multiband microwave radiometric system should be taken into account at the stage of development and technical setup. At the same time, the main advantage of this method of calibration with the use of an external generator is the possibility of its performance in any conditions of ground-based microwave radiometric system.

The obtained results of the research have shown the principal possibility of implementing the calibration of the three-band microwave radiometric system by the radio noise emission of an external broadband source - noise generator simultaneously in the three working wavelengths at the reception on a common mirror antenna irradiator with subsequent frequency and mode separation of the signals of two polarizations.

Acknowledgments

The research was carried out at the expense of a grant from the Russian Science Foundation №21-19-00378, https://rscf.ru/en/project/21-19-00378/

References

[1] Stepanenko V D, Shchukin G G, Bobylev L P *et al.* 1987 *Radiotepolocation in meteorology* (Leningrad: Gidrometeoizdat) p 283

[2] Falin V V 1997 *Microwave radiometric systems* (Moscow: Luch) p 440

[3] Grankov A, Milshin A 2016 *Problems of the environment and natural resources* **3** 3–29

[4] Martin-Neira M *et al* 2008 *IEEE Trans. Geosci. Remote Sens.* **46**(3) 613–620.

[5] Yong Han, Westwater Ed R 2018 *IEEE Trans. Geosci. Remote Sensing* **38**(3) 1260–1276

[6] Kaisti M, Altti M, Poutanen T 2014 *IEEE Transactions on Microwave Theory and Techniques* **62**(10) 2436–2446

[7] Butakova S 1984 *Proceedings of Higher Education Institutions. Radiophysics* **27**(11) 1431–1435

[8] Fedoseeva E, Rostokin I, Shchukin G 2020 *Measurement Techniques* **63**(4) 301–307 doi: 10.1007/s11018-020-01787-z

[9] Rostokin I, Fedoseyeva E, Fedoseyev A 2015 *Radiotechnical and telecommunication systems* **1** 5–11

[10] Rostokin I, Fedoseeva E, Rostokina E, Shchukin G 2020 *Journal of Physics: Conference Series* **1632** 012025 doi:10.1088//1742-6596/1632/1/012025

[11] Rostokin I, Fedoseeva E, Rostokina E, Shchukin G 2019 *Proceeding of the Russian Conference on Radio Wave Propagation (RWP)* 384–387 doi: 10.1109/RWP.2019.8810166

Journal of Physics: Conference Series **2140** (2021) 012030 doi:10.1088/1742-6596/2140/1/012030

Investigation of the parameters of a single photon detector for quantum communication

K I Khomiakova, A P Kokhanenko, A V Losev

Tomsk State University, Tomsk, Russia

E-mail: homiackowa.kristina@yandex.ru

Abstract. Nowadays the best single photon detectors from a practical view are those based on InGaAs/InP avalanche photodiodes, operating at a wavelength of 1.55 μm. The dependence of quantum efficiency and noise levels on the temperature and bias voltage of avalanche photodiodes were carried out.

1. Introduction

Many research groups engaged in the study of quantum cryptographic systems strive to create a system not only with absolute secrecy of information exchange, but also with a high data transfer rate. However, with an increase in the data transfer rate in quantum cryptographic systems, problems arise associated with the detection of single photons, which is associated with the limitation of the performance of single photon detectors (SPD) and an increase in noise [1]. Therefore, research is required in the field of increasing the speed of detectors of single photons and reducing their noise characteristics. Therefore, SPD are increasingly used in such areas as quantum key distribution (QKD), positron emission tomography, optical reflectometry, and biomedical research. Currently, various devices and methods for detecting single photons are known, for example, photomultipliers, single photon avalanche photodiodes (SPAD), detectors based on the sum-frequency generation, superconducting thin-film nanostructures (SSPD) and SPD based on quantum dots and semiconductor defects. The given classes of devices differ in design features, in the spectral range sensitivity, in quantum efficiency and in the ability to distinguish single photons.

Recently, the best single-photon detectors in the near infrared range for practical use in the field of quantum telecommunications are avalanche photodiodes based on InGaAs / InP. In the literature covering photodetectors, one finds the terminologies of avalanche photodiode (APD) and SPAD. Normally, a device is referred to as an APD when it operates below the breakdown voltage in the linear-mode, for which the output photocurrent is linearly proportional to the input optical power. The term SPAD refers to a device operated in Geiger mode, for which biasing above the breakdown voltage can result in a self-sustaining avalanche in response to the absorption of just a single photon. For a SPAD based detector system, there are two crucial parts: the SPAD device [2–4] and the quenching electronics [5–7]. Therefore, the performance of single-photon detector system depends both on the SPAD device itself and the quenching electronics as well.

The advantage of such SPADs is that they are able to register weak optical signals due to internal amplification due to the impact ionization process. For use in the field of quantum cryptography, SPD must have high detection efficiency, high signal-to-noise ratio, low dead time, low jitter and the ability to distinguish the number of photons is desirable as well [1, 8].

Content from this work may be used under the terms of the Creative Commons Attribution 3.0 licence. Any further distribution of this work must maintain attribution to the author(s) and the title of the work, journal citation and DOI.
Published under licence by IOP Publishing Ltd

This work is devoted to measuring the characteristics of SPD and finding the optimal parameters of their operation for quantum communications. There are many different stand layouts designed for automated measurements of the parameters of single photon detectors operating in standby and gated modes. There are two main modes of SPD operation: gated and waiting. In the gated mode, a periodic signal is supplied to the SPAD, which for a short period of time transfers the diode to the Geiger state - the photon is capable of generating an avalanche [1, 2]. If a constant bias voltage is applied to the SPAD above the breakdown voltage, that is, the diode is constantly in the Geiger state, then this mode is called the waiting mode. In gated mode, the dark count and the likelihood of afterpulses are lower than in standby mode, however, the design of the control electronics is more laborious.

2. Results and discussion

In this work, using the experimental stand of the Russian Quantum Center (RQC) and the experimental stand of the Department of Quantum Electronics and Photonics, we measured the parameters of SPD produced by the RQC, made on the basis of avalanche photodiodes with the InGaAs / InP structure Figure 1, for two operating modes: waiting and gated.

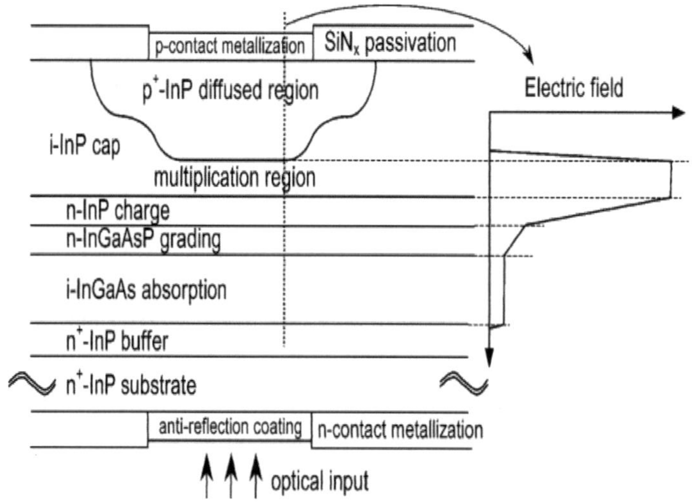

Figure 1. Cross section of SPAD InGaAs / InP with bottom illumination [3].

To measure the characteristics, pulses were applied to the SPD by a semiconductor laser with a wavelength of 1550 nm, with a repetition rate of 100 kHz, attenuated by 0.1 photons / pulse using variable and constant attenuators. Using special software, the temperature (from -40 to -60 °C) and the bias voltage of the APD were changed. The measurement of the characteristics of the SPD: quantum efficiency, dark noise (dark counting), was carried out using an oscilloscope and a frequency counter.

Quantum efficiency was calculated using the formula:

$$\eta = \frac{f_{\sin} - f_D}{\nu \cdot \mu}, \tag{1}$$

where ν is the laser pulse repetition rate, μ is the average number of photons per pulse, f_{sin} is the pulse frequency at the detector output when laser pulses with a different average number of photons are applied, f_D is the frequency of dark counts with the laser turned off.

Dark noise originates from thermal or tunnel excitation mechanisms. At sufficiently high temperatures of SPAD use, thermal excitation will make the main contribution to the DCR, while at low temperatures, but at a strong electric field, the main contribution to DCR will be made by tunnelling excitation caused by excess bias [9].

In this work, experimental temperature dependences of the SPAD dark count were obtained, and measurements of the SPAD dark count and quantum efficiency on the bias voltage were carried out for three waiting and two gated photodetectors.

In Figure 2 shows the temperature dependences of the dark count at a quantum efficiency of 10% and corresponding bias voltages for four detectors (D1, D2, D4, D5) operating in the standby mode, and one (D3) in the gated mode. It can be seen that the DCR value sharply decreases with decreasing temperature. It can be seen from the graph that, in comparison with the SPD operating in the standby mode, the dark count of the detector in the gated mode is much lower.

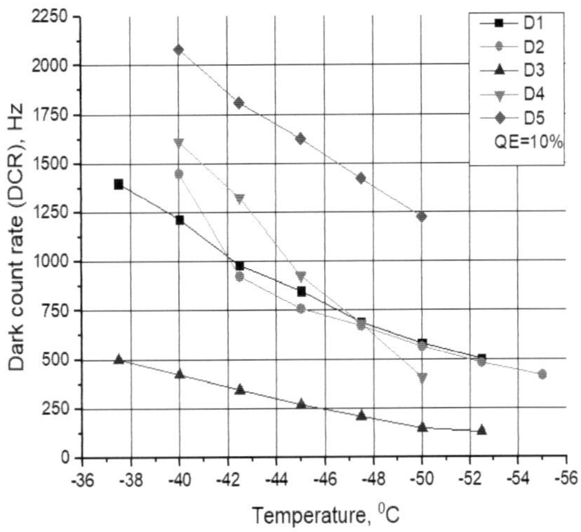

Figure 2. Temperature dependences of the dark count at quantum efficiency 10% and constant bias voltages for five SPDs.

Analyzing the dependence of the dark count on temperature for all SPDs, it is possible to observe the decrease in DCR from the decrease in temperature. Secondly, the dependence for D3 SPD can be distinguished, that the lower the temperature, the decrease in the DCR decay. The dark count at the same temperatures has a large scatter in values.

Figures 3 and 4 show the dependences of the dark count and quantum efficiency on the bias voltage for four detectors (D1, D2, D4, D5) operating in the standby mode, and for one (D3) in the gated mode. From the graph, you can see that each detector has its own operating bias voltage to obtain a quantum efficiency of 10%. The dark count and quantum efficiency increase with increasing bias voltage.

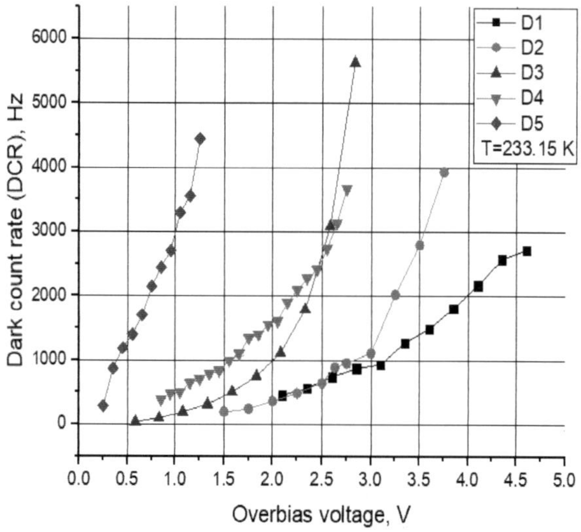

Figure 3. Temperature dependences of the dark count at quantum efficiency 10% and constant bias voltages for five SPDs.

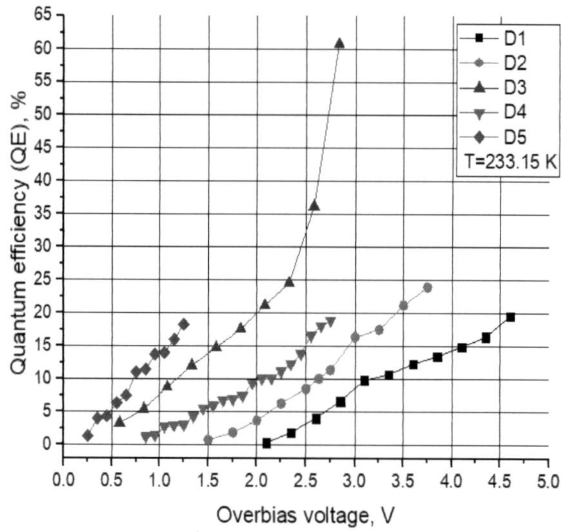

Figure 4. Temperature dependences of the dark count at quantum efficiency 10% and constant bias voltages for five SPDs.

According to Figure 4, it can be seen that the exaggeration of the bias voltage, the quantum efficiency increases, this is due to the fact that the probability of the formation of an electron-hole pair increases [10].

Thus, the experimental dependences obtained will make it possible to choose a working point for the optimal mode of recording single photons for quantum cryptography systems, based on the permissible level of false positives and quantum efficiency.

3. Conclusions

As a result of experimental studies of the dependences of the SPD parameters on the SPAD temperature and the bias voltage applied to it, it was shown that the parameters of the SPD operating in the gating mode exceed the SPD parameters in the standby mode, since it has a very high quantum efficiency and low noise. In gated mode, the gate pulse reduces the contribution of residual pulses to the overall noise picture. There is a difference in parameters between the waiting detectors due to the precision of the technological fabrication of the SPAD structure. In educational and research installations of quantum cryptography, it is convenient to use SPD in the standby mode, because it is easier to determine the time of arrival of the first pulse in the train and synchronize the rest of the operation of the quantum secure data transmission device.

Acknowledgments
The authors are grateful for financial support of the studies in the framework of the project of the Ministry of Education and Science of the Russian Federation (project 0721-2020-0048), within which a scheme was developed for measuring the SPD characteristics. We would like to express our gratitude to the Russian Quantum Center and CLL «QRate» for consulting on single photon detectors.

References
[1] Izhnin I I *et al.* 2020 *Applied Nanoscience* 1-11 doi: 10.1007/s13204-021-01667-0
[2] Itzler M A *et al.* 2007 *J Mod Opt.* **54** 283–304
[3] Jiang X, Itzler M.A, Ben-Michael R, Slomkowski K 2007 *IEEE J. Sel. Top. Quant. Electron.* **13** 895–905
[4] Itzler M A *et al.* 2011 *J Mod Opt.* **58** 174–200
[5] Cova S *et al.* 1996 *Appl Opt.* **35** 1956–1976
[6] Tosi A, Mora AD, Zappa F, Cova S 2009 *J Mod Opt.* **56** 299–308
[7] Gallivanoni A, Rech I, Ghioni M 2010 *IEEE T Nucl Sci.* **57** 3815-3826
[8] Cova S, Ghioni M, Lotito A, Rech I and Zappa F 2004 *J. Mod.Opt.* **51** 1267
[9] Zavodilenko V, Losev A, Miller A, Kurochkin V, Kurochkin Y 2017 *Journal of Physics* **917** 062034
[10] Trifonov A, Subacius D, Berzanskis A, Zavriyev A 2004 *Mod. Opt.* **251** 1399–1415

Journal of Physics: Conference Series **2140** (2021) 012031 doi:10.1088/1742-6596/2140/1/012031

Empirical dependence of the probability of blocks rotations on the diffusion coefficient in a cellular automaton with a Margolus neighbourhood

N A Shalyapina, M L Gromov, A K Matolygin and S N Torgaev

Tomsk State University, Tomsk, Russia

E-mail: nat.shalyapina@gmail.com

Abstract. The paper considers a cellular automaton approach to modeling physical processes, in particular, the diffusion of pulsed periodic gas discharges in plasma. An attempt to determine the dependence of the probability of the block rotation on the diffusion coefficient for a cellular automaton diffusion model with a Margolus neighborhood is describe. Some features of the tensor approach to the implementation of random block cellular automata are described. The results of computer experiments are presented. The analysis of this results indicates that the data from the literature sources differ from the data obtained during the experiment.

1. Introduction

Modeling is an important part of any natural science discipline. A model of a process allows predicting expected effects and selecting the best parameters (or at least close to the best ones) for future experiments or mechanisms being developed. This reduces the amount of required field experiments, which, ultimately, has a positive effect on the financial and material balance of any research. Another useful aspect of modeling is the ability to predict new, previously unknown or difficult to detect effects. But benefit of modeling arises only if the model is sufficiently adequate and detailed. The classical method of building models is based on solving a large number of differential equations. The more adequately the model is constructed, the more the size of this system of equations grows. When inhomogeneities are introduced into the process under study, the complexity of the system of equations increases especially sharply. As a result, it may turn out that the resulting system of equations is not only unsolvable analytically, but it is difficult to solve the system of the equations numerically. Thus, the researcher faces the problem of finding a balance between the model adequacy (system of differential equations) and the possibility of obtaining a numerical solution to the constructed model (system of differential equations). In some cases, an acceptable balance cannot be found. For example, for processes in the plasma of active optical systems on metal vapors at the moment it has been possible to obtain a model (that is, to find a numerical solution of the system of differential equations) only for a one-dimensional homogeneous case in the radial direction. Such a model cannot in any way give answers to questions about the optimal parameters of the device's operation or allow the discovery of new effects (if only because the environment is heterogeneous).

One of the examples of numerical simulation of physical processes is a model of kinetic processes occurring in the plasma of pulsed periodic gas discharges. These processes are usually modeled by a system of partial differential equations. This system is called the kinetic model of plasma. The model

Content from this work may be used under the terms of the Creative Commons Attribution 3.0 licence. Any further distribution of this work must maintain attribution to the author(s) and the title of the work, journal citation and DOI.
Published under licence by IOP Publishing Ltd

is based on a system of nonlinear differential equations of diffusion type. The physical processes described by these equations include thermodiffusion of atoms of the active substance (copper), buffer gas (neon), ambipolar diffusion of ions, as well as thermal conductivity in electronic and atomic gas. The solution of a system of partial differential equations is found using a difference scheme. At the same time, the kinetic model does not allow an analytical solution, and the numerical solution either takes a long time or is possible only for the one-dimensional case, which significantly limits the scope of its application. Thus, a one-dimensional model of an active medium on copper vapor requires several hours to reach a steady state. It is also not possible to perform parallel calculations using multi-core processors. Such a model can in no way provide answers to questions about the optimal parameters of the device or allow detecting new effects.

Therefore, researchers are actively developing various non-classical approaches and methods for modeling physical processes. One of such non-classical models is the cellular automaton [1, 2].

The cellular automaton (CA) [3, 4] is a discrete dynamic model, the functioning of which occurs in space, which is a set of cells of the same shape and size, and discrete time intervals, called iterations. Each cell of the model space corresponds to a certain area of the simulated physical space and is described by a set of dimensionless characteristics. The main advantage of CAs and mathematical models built on their basis is the parallelism inherent in the structure of automata. This feature allows the use of modern supercomputers, consisting of a large number of cores, to build efficient software implementations.

Cellular automata and models based on them have found application in solving problems of physics of continuous media [1], chemistry [5], physics of semiconductors [6], physics of lasers [7], as well as some social processes [8]. Many researchers have addressed the issue of creating a diffusion cellular automaton model [9–12]. Several models have been proposed based on the degree of knowledge of the properties of various configurations of cellular automata. It should be noted that today a significant contribution to the development of the cellular automata theory, to the study of their spatially distributed processes modeling application, including diffusion, is made by a group of researchers from Novosibirsk: Bandman O.L., Medvedev Yu.G., Sabelfeld K.K., Kireeva A.E. etc. [6, 13, 14]. In their works, the authors, to simulate the diffusion process, use a diffusion cellular automaton model with the vicinity of Margolus [15]. In this model, the diffusion coefficient D is reflected in the probability of turning the automaton blocks p: by changing this probability, you can change the diffusion rate. In works [9, 13] researchers make attempts to establish the dependence of block rotation probability on the diffusion coefficient. In particular, it is indicated that the model (in terms of cells and ticks) diffusion coefficient at probability $p = 0{,}5$ is equal to $D = \frac{3}{2}$, and the dependence of diffusion coefficient probability is linear. Moreover, it is proposed to change the probability only for turns in one direction, and in the other it is indicated equal to $1 - p$. However, the rotation directions are equivalent, so it is not clear why, if one gives preference to one direction (making the probability greater than 0,5), the diffusion rate should increase, and if the other, then it should decrease.

In [10], it is proposed to consider three situations: clockwise rotation of the block, counterclockwise rotation and the block does not rotate. In this work, a model diffusion coefficient is indicated equal to $D = \frac{4}{9}$, and they also speak of a linear dependence of block rotation probability on the diffusion coefficient. However, the authors do not offer theoretical calculations substantiating these conclusions.

In this work, we attempt to determine the dependence of block rotation probability on the diffusion coefficient for a block cellular diffusion automaton with a neighborhood of Margolus.

The structure of the work is as follows: the second section is devoted to a brief introduction to the theory of cellular automata and a description of the diffusion cellular automaton model with a Margolus neighborhood, the features of the block cellular automaton with a Margolus neighborhood software implementation using the tensor approach are described in the third section, the results of computer experiments and the dependence of the model parameters on the diffusion coefficients are presented in the fourth section.

2. Cellular automaton

A *cellular automaton* is a discrete dynamic model whose functioning takes place in a space consisting of a set of cells of the same shape and size, and discrete time intervals called *iterations*. Each cell of the model space, called a *cell*, corresponds to a certain area of the simulated physical space and is described by a set of dimensionless characteristics. The main advantage of CA and mathematical models based on them is the parallelism inherent in the structure of automata. This feature makes it possible to attract supercomputers consisting of a large number of cores to build effective software implementations.

A *block cellular automaton with a Margolus neighborhood* deserves special attention (Figure 1).

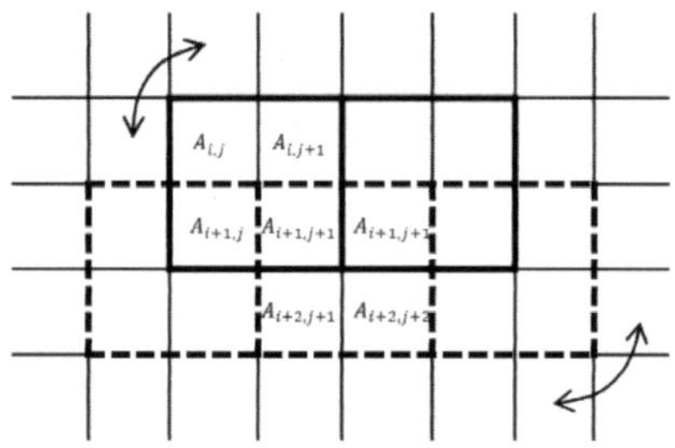

Figure 1. A fragment of the CA with a Margolus neighborhood.

The automaton field is divided into two subsets - blocks in the model: *Even*=$\{A_{i,j}, A_{i,j+1}, A_{i+,j}, A_{i+1,j+1}\}$, *Odd*=$\{ A_{i+1,j+1}, A_{i+1,j+2}, A_{i+2,j+1}, A_{i+2,j+2}\}$. Transition rules: the direction of rotation of the block clockwise or counterclockwise is determined equally likely (p_1=p_2), or with probability p_3=$(1-p_1+p_2)$ rotation does not occur at each clock cycle of the automaton. The automaton operates in synchronous push-pull mode: the transition rules are applied to even blocks on even clock cycles, on odd clock cycles – to odd ones. It is possible to simulate the diffusion process with a diffusion coefficient in a wide range by changing p_1, p_2, p_3 and manipulating the values of steps in time and space [16].

3. Features of the software implementation of probabilistic block cellular automata

We rely on the previously developed tensor approach of describing cellular automata to construct a software implementation of block cellular automata [17, 18]. This approach is focused on the use of multicore graphics adapters and a special working environment (framework) TensorFlow, which allows the user to efficiently build parallel software implementations and there is no need for the user to organize parallel sections of programs and the distribution of data between them. The main data structure of the TensorFlow framework is a tensor (multidimensional matrix). It is necessary to represent a cellular automaton in the form of a tensor, and the transition rules of an automaton from one state to another in the form of some tensor operations to describe the process of functioning of a cellular automaton.

Some difficulties (limitations) in applying the tensor approach were discovered during the construction of the software implementation of the model. The first (not so critical) is that TensorFlow functionality does not allow operations with tensors of arbitrary rank. We are interested in the convolution operation of a tensor with a given kernel of arbitrary rank, but a convolution with a kernel of the same rank as a given tensor in TensorFlow is available (for example, for a tensor of the second rank - the case for a two-dimensional cellular automaton, the kernel must be of the second rank). The second difficulty (more critical) is that the probabilistic nature of the behavior of the model requires the generation of a certain type of convolution kernel at each clock cycle of the automaton (a new clock cycle is a new kernel). This fact increases the complexity of the algorithm and affects the effectiveness of the approach. To overcome these difficulties, the operations of rotating even and odd blocks of the automaton were implemented by us within the framework of the TensorFlow possibility of implementing our own operations on tensors at the level of manipulation of the main data structures (arrays) that represent tensors. TensorFlow developer libraries help implement data operations as programs for the CPU, and as programs for graphics accelerators in NVIDIA's CUDA language. The third difficulty is related to probabilities. Technically, generating a random number is an expensive operation (in a computational sense). $N+M$ random numbers are required to implement one iteration of the automaton in a cellular automaton diffusion model with a Margolus neighborhood, where N is the number of even blocks, M is the number of odd blocks. Consequently, it will take a large amount of time to generate such a large number of random numbers, which significantly affects the effectiveness of the software implementation. In addition, graphics adapter processors are not focused on generating random numbers, without the use of special "complex" algorithms that require large computing resources. We propose to generate random numbers before conducting a computational experiment on diffusion modeling in order to minimize the impact of complexity on the efficiency of implementation. In this case, we may not take into account the time resources needed to generate random numbers.

4. Experimental results

A block cellular automaton with a Margolus neighborhood of 1000 by 1000 cells was taken for experiments. The global state of the cellular automaton is calculated after 30000 cycles of operation. At the same time, the probability of rotations of the blocks changes: $p_3=1-(p_1+p_2)$, $p_1=p_2$. The substance is concentrated in a central rectangle of 1000 by 100 cells in the initial global state. Calculating the evolution of the model for 30000 clock cycles took 61 seconds for our system (Intel Core i7-8700K processor, 24 GB RAM, NVIDIA GeForce RTX 2080 Ti graphics card, 11 GB video card RAM).

To obtain the type of distribution of the concentration of a substance at different values of p_1, p_2, p_3 (Figure 2), the concentration at each point was averaged over a neighborhood with a radius of $r = 5$ cells.

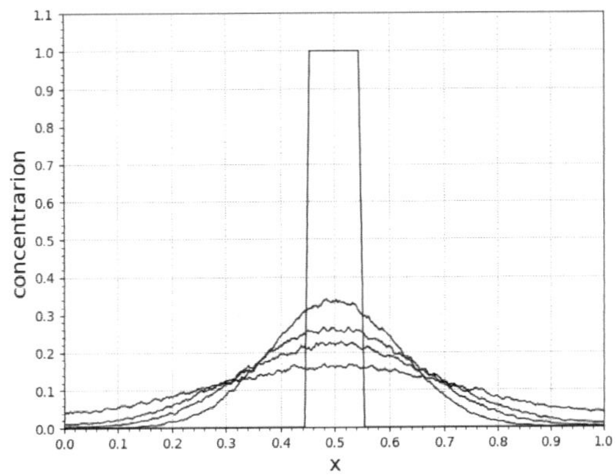

Figure 2. Concentration distribution.

Let us turn to the homogeneous differential diffusion equation for the one-dimensional case in order to obtain an empirical dependence of the probability of rotation of blocks on the diffusion coefficient:

$$\frac{\partial U}{\partial t} = D\left(\frac{\partial^2 U}{\partial x^2}\right) + f(x,t), \tag{1}$$

where $U(x,t)$ – concentration of the diffusing substance, D – diffusion coefficient, $f(x,t)$ – function describing the sources of the substance (there are no sources in our case $f(x,t) = 0$).

The solution of the homogeneous Cauchy problem is given in the form of a Poisson integral in a space with Cartesian coordinates:

$$U(x,t) = \frac{1}{2\sqrt{\pi D t}} \int_{-1}^{1} exp\left(-\frac{|x-y|^2}{4Dt}\right) dy. \tag{2}$$

We obtain the theoretical values of the diffusion coefficient at the concentration values obtained during computer experiments by numerically solving equation (2) (Table 1).

Table 1. Experimental results.

Probability of blocks rotation (practical)	Probability of blocks rotation (regression)	Diffusion coefficient (D)
0	0	0
0.085	0.074	0.09
0.18	0.164	0.21
0.19	0.185	0.24
0.23	0.218	0.29
0.27	0.268	0.26
0.31	0.307	0.24
0.35	0.348	0.22
0.38	0.371	0.21
0.43	0.426	0.19
0.46	0.450	0.18
0.48	0.473	0.17
0.5	0.493	0.16

Figure 3 shows a graph of the dependence of the diffusion coefficient on the probabilities of rotation of the blocks.

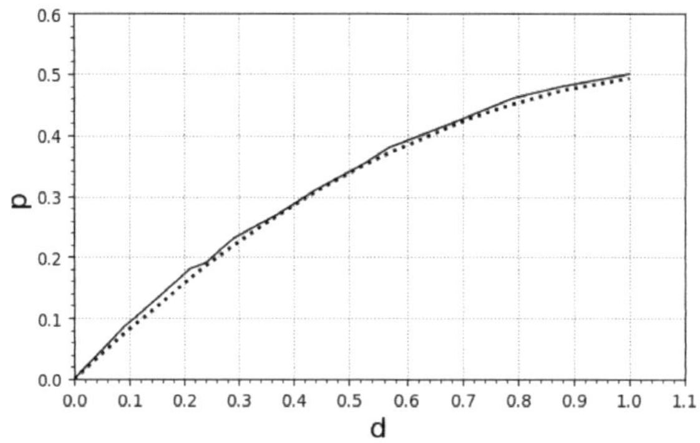

Figure 3. The dependence of the diffusion coefficient on the probability of rotation of the automaton blocks (solid line - experimental dependence, dotted line - regression analysis).

Regression analysis results (dotted line Figure 3) showed that the dependence of the diffusion coefficient on the probability of rotation of the blocks is parabolic and has the form:

$$p = -0.35D^2 + 0.86D. \quad (3)$$

5. Conclusion

According to the results of the experimental study, it was found that the dependence of the diffusion coefficient on the probability of rotation of the blocks is parabolic. We believe that the differences between the obtained results and the known ones [9, 10, 13] are due to the fact that the authors of these works considered the possibility of manipulating the diffusion rate by changing model parameters, and also investigated cellular automaton modeling of a single substance. In this case, a five percent deviation of the cellular automaton model from the model of differential equations, on which the authors relied, is quite acceptable. The purpose of our further research is to create a model of multicomponent diffusion. For this model, the systematic lag or advance of the diffusion of one substance in the model from the actual diffusion can lead to distortion of the overall model, especially for a large number of iterations.

Acknowledgments
The reported study was funded by RFBR according to the research project No 20-31-90070.

References

[1] Medvedev Yu G 2003 *Autometry* **39**(3) 43–50
[2] Bandman O L 2011 *J. Supercomp.* **57**(2) 121–131 doi: 10.1007/sl1227-010-0391-5
[3] Toffolli T 1984 *Physica D: Nonlinear Phenomena* **10**(1-2) 117–127
[4] Wolfram S 1983 *Mod. Phys.* **35** 601–643
[5] Kireeva A E, Sabelfeld K K, Maltseva N V *et al.* 2019 *Vestnik Tomskogo gosudarstvennogo universiteta. Upravlenie vychislitelnaja tehnika i informatika* [*Tomsk State University Journal of Control and Computer Science*] **46** 31–39
[6] Kireeva A E, Sabelfeld K K Cellular Automata 2015 *Proceedings of the 13th International Conference of Parallel Computing Technologies* **9251** 191–200
[7] Jiménez-Morales F, Guisado J L, Guerr J M 2018 *Understanding Complex Systems* **F3** 405–422
[8] Shimura K, Vizzari G, Bandini S *et al.* 2013 *Parallel Computing Technologies* **7979** 369–380
[9] Malineckij G G, Stepancov M E 1999 *Computational Mathematics and Mathematical Physics* **38**(6) 1017–1020
[10] Gubarev S V, Berg D B, Dobryak P V 2013 *Modern problems of science and education* (6) 176
[11] Rubtsov S E, Pavlova A V, Olejnikov A S 2017 *Ecological bulletin of research centers of the black sea economic cooperation* **4**(1) 86–93
[12] Bandman O L 2014 *Bulletin of the Novosibirsk Computing Center. Series: Computer Science* **36** 21–31
[13] Bandman O L 1999 *Parallel Computing Technologies* 395–409 doi:10.1007/3-540-48387-x_41
[14] Medvedev Yu G 2010 *Proc. Intern. Conf. on Methods and Tools of Parallel Programming Multicomputers* **6083** 204–211
[15] Toffoli T, Margolis N 1987 *Cellular automata machines: a new environment for modeling* (Massachusetts: MIT Press)
[16] Bandman O L 2006 *System Informatics* (10) 59–113
[17] Shalyapina N A, Gromov M L 2019 *Proceedings of the Institute for System Programming of the RAS* **31**(3) 217–228 doi: 10.15514/ISPRAS-2019-31(3)-17
[18] Matolygin A K, Shalyapina N A, Gromov M L *et al.* 2020 *Journal of Physics: Conference Series* **1680** 1–6 doi:10.1088/1742-6596/1680/1/012035

Active IR location as the roadway profilometry method

V L Khmelev, A F Fominykh

Tomsk State University, Tomsk, Russia

E-mail: alexandrafedorovna02@gmail.com

Abstract. This article observe a using of active infrared beam location as roadway surface quality control. Changes in the spatial structure of the emitted IR radiation by surfaces within the capture scene allow creating a depth map of this scene. An optical camera makes it possible to use classical computer vision methods for stitching a depth map. For testing the possibility of using this approach, we made statistical studies on a multiple sample of distance measurements. Here we explain two experimental schemes with a programmable mechanical scanning system. The first one, we had determined the distance, which the image is capture accurately. The second, we measure the planar resolution, a minimum size of the defect that recognize by the infrared beam location system.

1. Introduction

The operation of highways eventually leads to their destruction, and, consequently, the appearance of defects in the road surface. Their presence significantly affects to the traffic speed, the safety of drivers and the service life of vehicles. The size of defects directly influences to how big the material damage will be for the car owner and what physical damage the car will get during every day using. The trivial way to reduce the risk is to reduce the speed of the vehicle, which will make it possible to make a maneuver, which will avoiding holes and defects in the road surface. The use the maneuver for small amount of drivers in the traffic will lead to an increase in fuel consumption and a slowdown of all traffic participants.

For minimization of a number of defects and timely renovation and recovering pf the road surface, it is important to collect an information of the road surface quality on all road sections. City infrastructure monitoring services should collect this information and transmit to the repair crews for planned and emergency repairs of a road surface.

The collected information can also be used to prevent accidents and adjust the trajectory and speed of traffic drivers, taking into account the current state of the road surface.

This article discusses one of the aspects of determining the quality of the roadway - profilometry of the roadway Engineers use various types of radiation infrared, ultrasonic and electromagnetic radiation [1–3] for recovering a profilogram of some surface. Because of using of the road surface, the given plane of the profile violates. In addition, the process of internal destruction of the roadway is showing in the change of a road plane. After depressions and swelling of the plane starts to forming. Over time, the destruction process becomes notable via a visual inspection. It indicate about serious violation of the internal structure of the roadway. If a visual inspection was carried out after the asphalt paving, then violations of the profile is an indicator of violations of the technical process of asphalt paving.

Various non-invasive control approaches are applicable to determine the profile of the roadway. The specificity of working with a road surface implies a reflective, not a transmission methods. After the

Content from this work may be used under the terms of the Creative Commons Attribution 3.0 licence. Any further distribution of this work must maintain attribution to the author(s) and the title of the work, journal citation and DOI.
Published under licence by IOP Publishing Ltd

transmitter was emitted the signal, inhomogeneities will reflects the signal back to receiver, and then the received signal will containing information about scattering inhomogeneities. Which have distorted the origin form of the emitted signal. The authors propose to use infrared multi-beam scanning technology to reconstruct a road surface.

2. Active IR location method

The multi-beam infrared optical location use for 3D scanning of the environment at a distance of 0.1 to 12 meters. An optical block enable to obtain the depth map usually have an IR structured light emitter, an IR camera for catching reflected the IR radiation and Grayscale or RGB cameras to catch habitual video sequences for image processing methods.

The optical unit emits infrared spatially structured light, which will reflects from surfaces inside the catching scene and at the end will received with the lens of the infrared camera. The spatially structured light forms an irregular mesh that covers all reflective surfaces in the scene. Depending on the distance between the irradiated surfaces and the optical block, the distance between the grid nodes will differ from the origin one. Data about changes in the inter-node distances in the spatial grid allows you to build a depth map of all reflected surfaces within the captured scene.

Ultimately, after capturing the infrared grid, the optical unit will form an image the captured surface. The image pixels intensity reflects not the relative intensity of the reflected light, but the distance between surfaces in the scene and the lens of the IR camera. Obviously, such an image of the roadway will be synthesized from the set of sub-images that will be obtained while the camera is moving along the roadway. The same board of optical block has Infrared and optical cameras and infrared projector, therefore we will take account them as physical fixed. Images they receive should also be considered physically fixed relative to each other, which allows obtaining pixel correspondence in IR, RGB, Grayscale and Depth images. This correspondence will allow the use of one matrix of coefficients for the perspective or affine transformation of all images, which will performed for stitching, will be the same for all types of images.

For test a possibilities of active IR location as a method for obtaining a road profile image, we had conducted two experiments.

3. Depth resolution of the propose method

Determination of the depth resolution is the purpose of the first experiment. During the experiment, the RGB camera catch the scene image, after it, we detected a QR code into the scene. We take into account for statistical processing an area into Depth image, which are corresponding to the location of the QR code into RGB image. The data obtained after image processing sent to statistical buffer for accumulation the statistical set. The set allows obtaining the average value by of the distance (by this algorithm [4]) from the camera plane to the QR code plane. For accumulation of the statistical set, the experiment had repeated multiple times.

An image of the experiment performed is shown in (Figure 1). The mechanical linear scanner had equipped with an Intel RealSense D435 optical sensor capable to synthesize depth images. During scanning, the camera moved to the object and returned to the outlying position of the mechanical scanner. Each step the optical block capture a RGB image, and synthesize a depth image. The scanning step was one cm. For the object under research had used a flat target with a printed QR code. For QR code recognition in RGB image has use algorithms [5].

When take the shot, the system requires the position of the camera from the mechanical line scanner. After the system has received the scanner position, it sum the position with the distance from the closest scanner position to the object. The same time synthesized depth image moves to the computer via a USB cable. After the image processing has finished both value (position compute with a depth image and position gets from the scanner) will be writing in neighbor fields of statistical set.

We had repeated this experiment several times to accumulate a statistical set and estimate the error at different distances between the optical unit and the object. The error estimate allows us to understand which distance between the object and the optical unit will satisfied GOST requirements [6] in depth resolution.

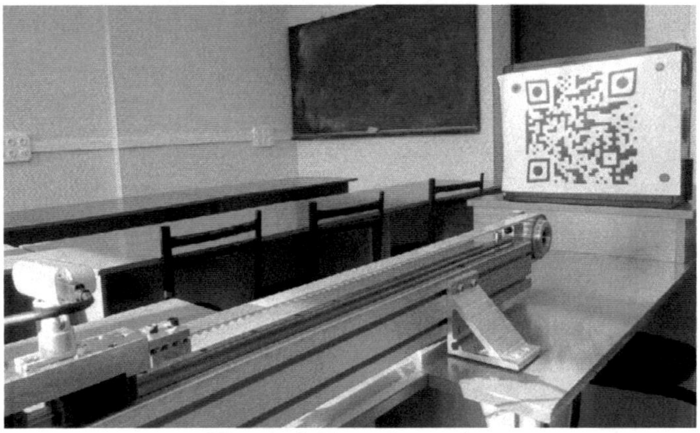

Figure 1. Experimental scene.

4. Spatial resolution and defect size

The second experiment had aimed at identifying the minimal planar size of defects and the planar resolution of the method. For determination of the characteristics, we create special test object with plexiglass. A various defects had simulated on the object. Defects had the form of rectangles and circles with varying distances between centers and sizes of defects. The optical unit had installed parallel to the test object, after images was captured with the subsequence synthetize of depth maps.

Based of experiment is the specially created test object. The object is a rectangular sheet of plexiglass with holes of various shapes, sizes and varying distances, which are modeling defects. Figure 2 shows the test object scheme.

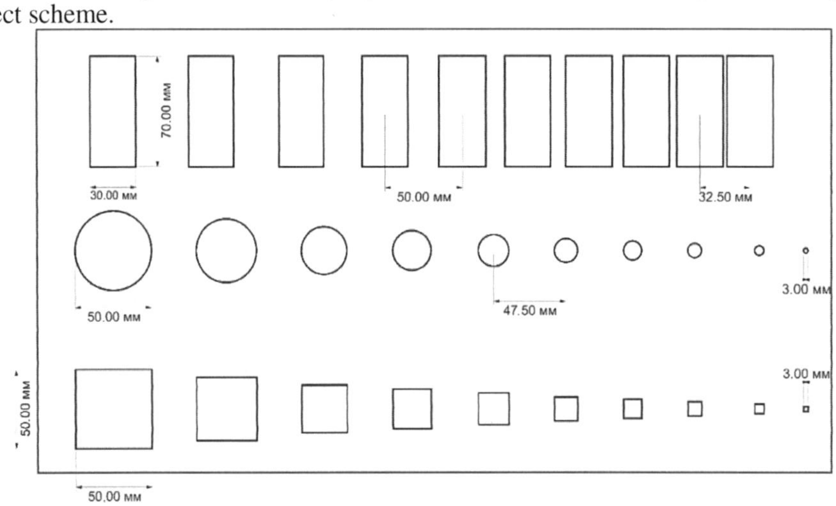

Figure 2. Scheme of the test object.

The test object simulated three sets of defects, each located in its own row. The first row of Figures is rectangular cutouts 30×70 mm and all of them has the same size. In a row, the distance between the centers of the rectangular cutouts changes from 32.5 mm to 65 mm. This set was supposed to show the spatial resolution between objects.

The second row of figure had represented with circular cutouts located at the same distance. The distance between the centers of the circles is 47.5 mm. The cutout diameter changed in the set, which decreases from 50 to 3 mm. This and the next set show the minimum linear size of the defect that the optical unit is able to detect.

The third row is close to the second, but the shape of Figures has been changed to squares. It gives opportunity look in which size of cutoobjects is no longer distinguishable. Instead of the diameter, the side size decreases from 50 mm to 3 mm. The distance between the centers of defects in the third set was also 47.5 mm.

Figure 3 shows the depth map of the test object. The image is the basis for the conclusion about the suitability of the future system for assessing the roadway.

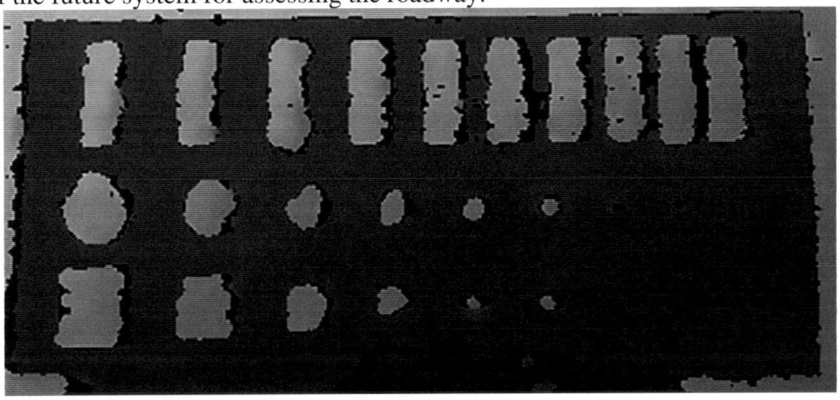

Figure 3. Depth image of the test object.

The distance between the object and the camera lens was 60 cm. The experiment revealed that the camera could recognize a 15 mm defect and distinguish two identical defects at a distance of 2.5 mm from each other. We also note that the distinction between the second and third set of defects is erased and becomes indistinguishable between the 4th and 5th pair of defects in each of the rows.

5. Conclusion

The article proposed the method for surface quality control based on active infrared location. The main element of the proposed system should be the optical module, which synthetize a depth and base on a stereo image in the infrared range. The module is also equipped with an RGB camera, the image from which is pixel by pixel consistent with the depth map. This allows stitch a set of depth maps into a single profilogram using computer vision methods.

The process of statistical experiment had automated with a mechanical linear scanner, what allows us to collect statistical sets.

The first experiment had for a determination of the depth resolution. During the experiment, the RGB camera detected the QR code in a scene and cut out the area, which are corresponding to the location of the QR code in the depth map. The data obtained after processing was statistically processed to obtain the average value of the distance from the camera plane to the QR code for one frame. For accumulation statistical sets of data for each distance with step 1cm, the camera had moved on the mechanical scanner multiple times

The second experiment was aimed to determining the minimum the planar defect size and the spatial resolution. For determination of these characteristics, we create the plexiglass test object, which are simulating various sets of defects. The shape of the defects is quadrangles and circles with variable distances between they centers and linear dimensions.

Preliminary experiments had shown that the technical capability makes it possible to detect defects with the required of GOST. Next, we will work out the possibility of stitching depth maps in motion. The purpose of further experiments will be testing the hypothesis of the pixel matching in the RGB image and the depth map. If it will been confirmed, than we will construct the profilogram with stitching the set of multiple depth maps based an RGB image set.

Acknowledgments
This work was supported by the Grant of the President of the Russian Federation (MK-3578.2021.4)

References

[1] Sukhanov D Y, Kalashnikova M A 2014 *Acoustical Physics* **60**(3) 304–308 doi:10.1134/S1063771014030166

[2] Shipilov S E, Yakubov V P, Satarov R N 2016 *Russian Physics Journal* **58**(9) 1226–1232 doi:10.1007/s11182-016-0636-8

[3] Shipilov S, Satarov R N, Fedyanin I *et al* 2016 *MATEC Web of conference* **79** 01079 doi: 10.1051/matecconf/20167901079

[4] Khmelev V L, Fominykh A F 2021 *J. Phys.: Conf. Ser.* **1843** 012024 doi:10.1088/1742-6596/1843/1/012024

[5] Md Yousuf Ali, Xuan-De Zhang 2019 *Computer Science and Engineering* **9**(1) 12–19 doi: 10.5923/j.computer.20190901.03

[6] State Standard P50597-93 2007 Automobile Roads and Streets. The Requirements to the Level of Maintenance. Satisfied the Traffic Safety *Moscow: Standartinform*

Realization of actions of characters and combustion process in the system of 3D modeling of the evacuation process

M A Tugarinov[1], I D Shulga[1], E A Yurchenko[1] and S N Torgaev[1,2]

[1] National research Tomsk State University, Tomsk, Russia
[2] National research Tomsk Polytechnic University, Tomsk, Russia

E-mail: torgaev@mail.tsu.ru

Abstract. The article presents the results of action modules software implementation in the 3D modeling system of the evacuation process in emergency situations. In particular, the software implementation of such actions of the model characters as organizing a group for evacuation, providing assistance to the victim, taking measures to extinguish the fire is presented. The article also presents the physicochemical properties of materials that are taken into account in model for simulating combustion and smoke formation processes. The software implementation is based on the Unreal Engine program and using automatic descriptions of the physical and psychological states of the characters. The descriptions of the possibility of creating characters in the model with different physical and psychological parameters and collecting statistical data on their condition are given.

1. Introduction

Detailed 3D modeling of various processes, including social, is one of the most effective methods for their study. These processes include the evacuation of people in case of emergencies, in particular in case of fire. Simulation of evacuation processes makes it possible to assess the safety of buildings and structures effectively and to study the influence of various design and technological factors on fire safety. Also, modeling allows us to study in detail people's behavior in different psychological states and temperaments during the evacuation process. Thus, it becomes possible to study the influence of psychophysical characters on the effectiveness of evacuation.

The relevance of such studies is confirmed by many scientific works in this area [1–9]. All models can be divided into several groups depending on the modeling approach: molecular approach [1, 2], route-based approach [3], group-based approach [4, 5], and agent-based approach [6, 7]. It should be noted that the most promising is the agent-based approach. In this case, it is advisable to use 3D modeling tools. It is because 2D models do not consider a large number of characters interactions with each other and with the external environment.

Models that exist today do not consider a wide range of psychological states of characters and interactions. In particular, such processes as organized evacuation as part of a group, aggression of some characters, problems of characters interaction with the environment, and others. Considering such processes and states of characters will allow not only to increase the realism of modeling but also to use the developed system in the fire safety training tasks.

This paper is aimed at the practical implementation in the Unreal Engine program of individual modules of the characters' behavior and their psychological and physical states in the 3D modeling system of the evacuation process. A detailed description of the implemented modules is presented in

Content from this work may be used under the terms of the Creative Commons Attribution 3.0 licence. Any further distribution of this work must maintain attribution to the author(s) and the title of the work, journal citation and DOI.
Published under licence by IOP Publishing Ltd

[10]. In the model, there is a library of various materials which contains their physicochemical characteristics of combustion and smoke formation. This takes into account the physical properties of materials (speed of flame propagation, heat of combustion, smoke formation ability, specific yield of toxic gases, combustion efficiency) in terms of flame propagation speed and toxicity of combustion products.

2. Model description

The 3D modeling system of evacuation processes is being developed in the Unreal Engine software. This software environment allows realizing detailed three-dimensional visualization. The combustion processes and the characters' actions are programmed using the Blueprint visual programming language. The model is developed using advanced automata describing the physical and psychological states of the characters. The extended finite state machine of states is described as

$$M = \{S, X, Y, V, T\},$$

where is a finite set of states; is a set of input symbols; is a set of output symbols; is a set of context variables; is a set of transitions between states [11,12]. An example of an extended automaton describing the psychological state of characters and the transition between them is shown in Figure 1.

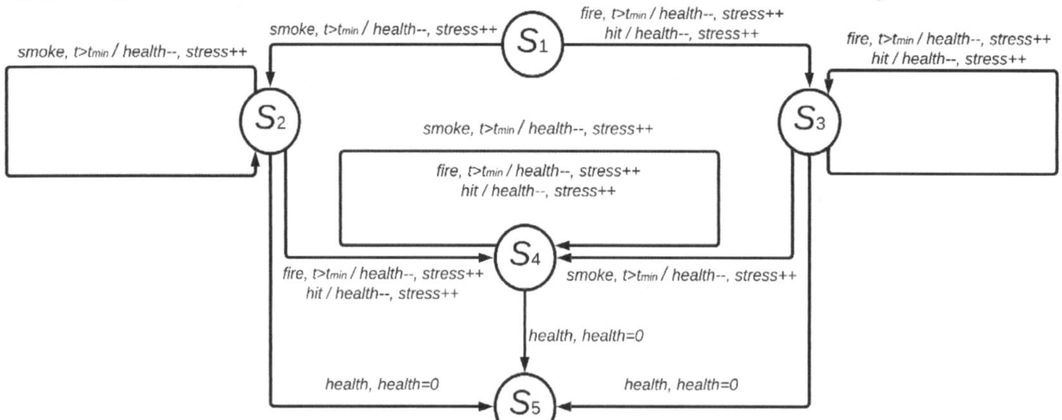

Figure 1. An extended finite state machine describing the psychological state of the model's characters and the transitions between them: S_1 – *Calm;* S_2 – *Panic;* S_3 – *Psychology;* S_4 – *Sympathy;* S_5 – *Group.*

A detailed description of automata models and algorithms for state transitions is presented in [10]. The model provides for the following states of the characters:

- Physical states: Initial state; Intoxication; Injury; Intoxication/Injury; Death.
- Psychological conditions: Calm; Panic; Psychology; Sympathy; Group.

The input parameters of the characters are weight, age, temperament, readiness for emergencies, leadership (the ability of the character to be a leader). Figure 2 shows the interface for determining the parameters of characters. This Figure shows adding a different number of characters with individual parameters to the evacuation model. In particular, Figure 2*a* shows a selection of 10 characters with a Sanguine temperament, a low level of Emergency Preparedness and Leader. In Figure 2*b*, the parameters of two characters with a Choleric temperament and an average level of emergency preparedness are selected. After determining the number of characters and their parameters, they are added to the model (Figure 2*c*). Then is possible to place them inside the space of the simulated building.

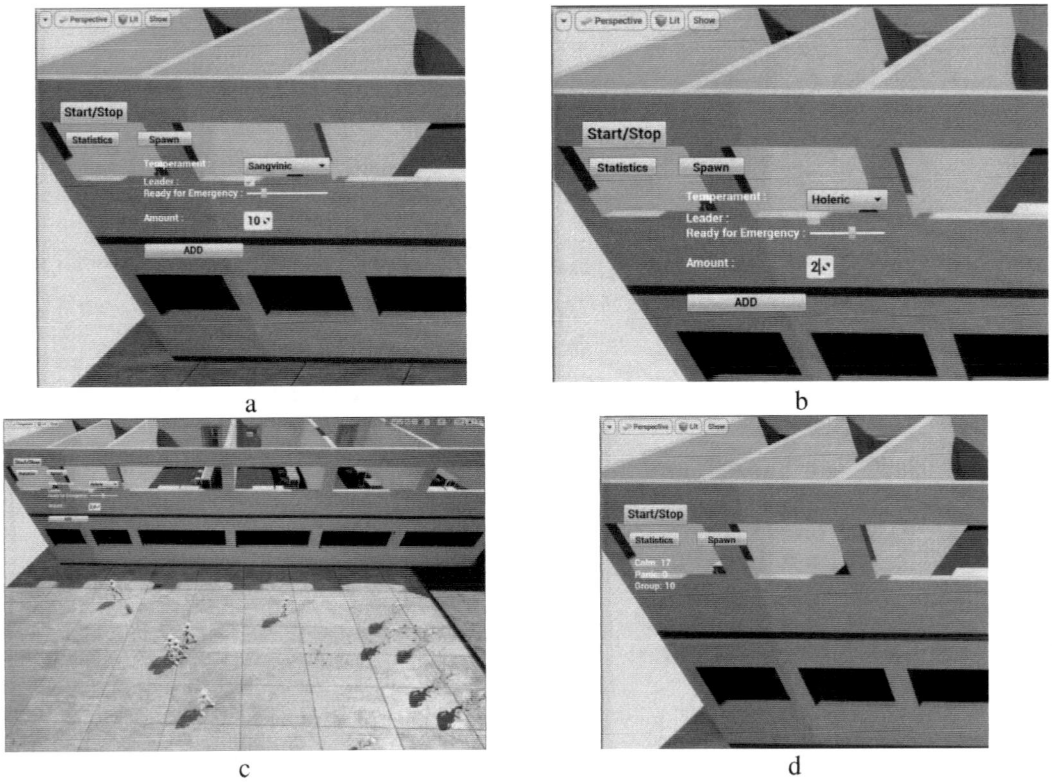

Figure 2. The software interface for determining the parameters of characters.

Depending on the psychological and physical state of the character, he will perform the choice of one or another action. In particular, the following actions are introduced in the modeling system:

- assistance to the victim;
- a collision;
- a running;
- a falling;
- a stupor;
- a panic (throwing);
- a bustle;
- a group;
- an action with a fire extinguisher.

At this stage, the model implements the spread of fire and smoke. The model considers the fire propagation speed and the toxicity of smoke depending on the material. Depending on the toxicity of the smoke, its effect on the character changes; in particular, the more toxic smoke leads to more severe damage to the character per unit of exposure time. We use a foreign classification for the speed of flame propagation. In the normative documentations it is often assumed that the fire in the initial stage develops according to the quadratic law which is similar to the radial flame propagation adopted in the modeling. Depending on the time of reaching the heat power of 1 MW (HRR - Heat Release Rate) the following types of fire are distinguished: superfast; fast; average; slow. The dependence of HRR on the combustion time for the different types of fires are shown in Figure 3.

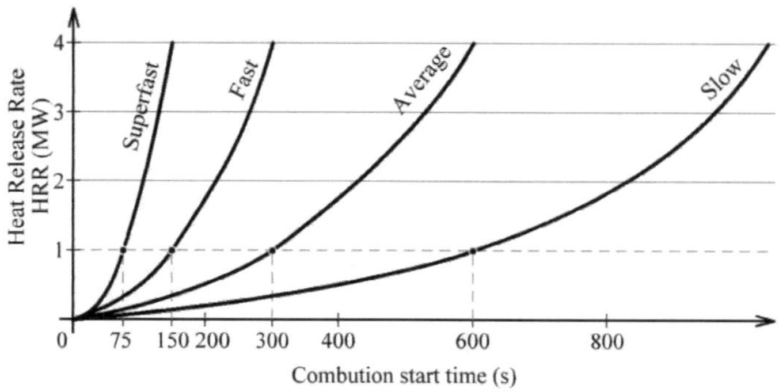

Figure 3. The dependence of HRR on the combustion time.

For modeling the dynamics of hazardous fire factors we use the following material parameters:

- v –the flame propagation speed, m/s;
- Q_n –the lowest heat of combustion, kJ/kg;
- D_m – the smoke-formation ability of the burning material, m^2/kg;
- L – the specific yield of toxic gases during the combustion of 1 kg of the material, kg/kg;
- η – the coefficient of the combustion completeness.

Figure 4 shows examples of flame propagation for materials with different physicochemical parameters. Figure 4*a* demonstrates the initiation of combustion. Figure 4*b* shows the extent of flame propagation over the same period of time. It can be seen that for different physical parameters of the materials the area of flame propagation also differs.

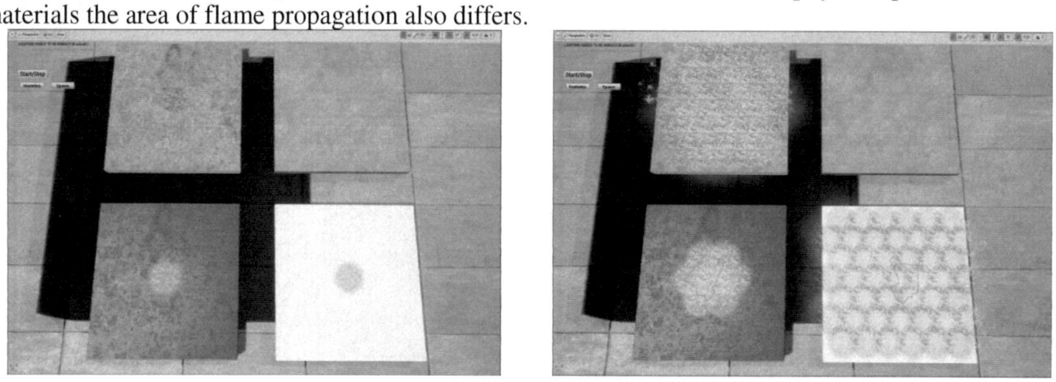

a b

Figure 4. The flame propagation for materials with different physicochemical parameters.

The model also has a module for collecting statistical data. At this stage, statistics determine the number of characters in various psychological states: calm, panic, and group.

3. Model results

This section presents the results of modeling various actions and states of the model characters. Figure 5 shows the results of the implementation of the Group state program module.

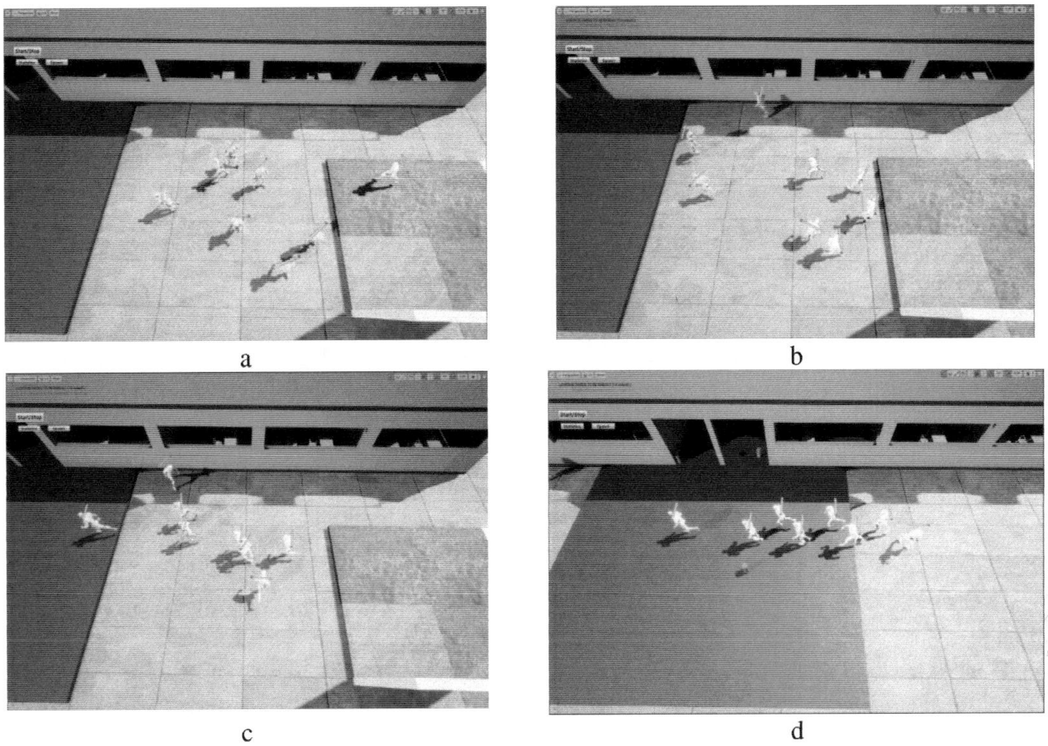

Figure 5. The results of the software implementation of the Group state program module.

The transition to the Group state is carried out if a character with the Leader parameter or a group led by the Leader enters the character's field of view. In Figure 5*a-c*, the characters "see" the Leader and begin to follow him. Moreover, the transition to the Group state occurs regardless of what state the characters are in, i.e., when the Leader appears in the field of view, the Stress parameter decreases. As a result, the model's characters begin to follow the Leader in the group (Figure 5*d*). In this case, the Leader determines the speed of movement of the group and its direction.

Figure 6 shows the results of modeling the process of providing assistance to the victim.

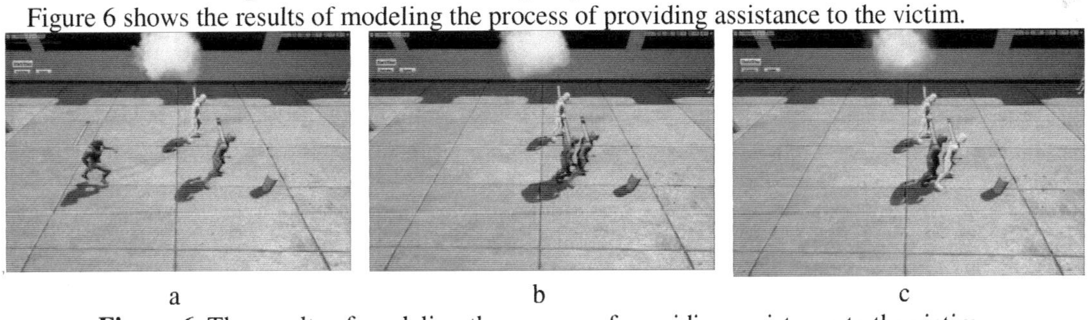

Figure 6. The results of modeling the process of providing assistance to the victim.

A character begins to assist if his "Emergency Preparedness" parameter is not equal to zero, and a character with a "Health" parameter value below the maximum falls into his field of view (Figures 6*a-b*). The character, guided by his knowledge and experience, must know first aid, rules of conduct in case of emergency, and must also know the necessary actions about movement during a fire in order to get out of a burning building (applies to characters who have the parameter "Leader").

A character that took health damage had a reduced speed and increased stress growth before assisting. Also, this character increased the level of stress among others. In the case of helping this character, his stress decreases, and his health increases. Stress is reduced to a value within the minimum value (Figure 6*c*).

Figure 7 shows an example of implementing the fire extinguishing process, i.e., the character taking firefighting actions.

a b c

Figure 7. An example of software implementing the fire extinguishing proces.

The character begins to carry out fire-fighting measures, in particular, extinguishing the fire with a fire extinguisher if his parameter "Emergency preparedness" is not equal to zero and a fire extinguisher enters his field of view (Figure 7*a*). At the same time, the character approaches the fire extinguisher and begins to take measures to extinguish it (Figure 7*b-c*).

4. Conclusion

Thus, the paper presents the results of the practical implementation of some characters' actions of the 3D modeling system of the evacuation process. In particular, such actions as organizing a group for evacuation, taking measures to extinguish, and providing assistance to the victim were implemented. All actions of the model's characters are determined by their physical and psychological states. The model takes into account five physical states and five enlarged groups of psychological states. To describe the states and transitions between them, we used extended automata models. The automatic description will further simplify the process of analysis, testing, and verification of the model through the use of tools of the automata theory [11, 12].

During preliminary experiments, the adequacy of the model was shown in terms of the implementation of the characters' actions, depending on their state and the state of their environment (other characters, fire, smoke, building structures). At the same time, the primary collection of statistical data on the number of characters in one state or another has been introduced into the model. The adequacy of the model was tested based on an expert assessment of psychologists from Tomsk State Pedagogical University.

In the future, it is planned to expand the list of possible actions of the characters and their initial parameters. And also the creation of a library of structures and furniture materials with their real parameters of combustion, smoke generation and toxicity of combustion products. Further improvement of the model will allow carrying out test experiments on evacuating people and comparing them with real incidents, as well as simulating real objects in order to identify possible consequences. Further improvement of the model will allow test experiments on evacuating people, comparing them with real incidents, and simulating real objects to identify possible consequences.

Acknowledgments

This work was supported by a grant from the Innovation Promotion Foundation (project manager – Shulga Irina).

References

[1] Dirk Helbing, Illés Farkas and Tamás Vicsek 2000 *Nature* **407** 487-490 doi: 10.1038/35035023
[2] Yunyun Niu, Yongpeng Zhang, Jieqiong Zhang and Jianhua Xiao 2018 *Simulation International journal of computers communications & control* **13**(5) 865–880 doi: 10.15837/ijccc.2018.5.3223.
[3] Jialiang Kou, Shengwu Xiong, Zhixiang Fang, Xinlu Zong, and Zhong Chen 2013 *Computational*

Intelligence and Neuroscience **2013** 369016 doi: 10.1155/2013/369016

[4] Ping Du, Yan Li, Hong Liu and Xiangwei Zheng 2018 *Proceeding of the 9th International Conference on Information Technology in Medicine and Education* 1018–1026 doi: 10.1109/ITME.2018.00226

[5] Templeton A, Drury J, Philippides A 2015 *Review of General Psychology* **19**(3) 215–229 doi: 10.1037/gpr0000032

[6] Shi J, Ren A, Chen C 2009 *Automation in Construction* **18**(3) 338–347 doi: 10.1016/j.autcon.2008.09.009

[7] Wei Xin-quan and Wang Jian 2015 *Systems Engineering — Theory & Practice* **35**(10) 2473–2483 doi: 10.12011/1000-6788(2015)10-2473

[8] Tugarinov M A, Shulga I D, Yurchenko E A, Ermakov A D 2020 *Journal of Physics: Conference Series* **1680** 012052 doi: 10.1088/1742-6596/1680/1/012052

[9] Tugarinov M A, Shulga I D, Yurchenko E A, Torgaev S N 2020 *Journal of Physics: Conference Series* **1680** 012051 doi: 10.1088/1742-6596/1680/1/012051

[10] Yurchenko E, Shulga I, Tugarinov M, Shelekhov I, Torgaev S 2021 *Proceedings of International Conference on Applied Innovation in IT* **9**(1) 93--100

[11] Gromov M L, Shabaldina N V 2017 *Automatic control and computer sciences* **51**(7) 507–515 doi: 10.3103/S0146411617070094

[12] Gromov M L, Prokopenko S A, Shabaldina N V, Laputenko A V 2019 *Proceedings of the 20th International Conference of Young Specialists on Micro/Nanotechnologies and Electron Devices* 139–142 doi: 10.1109/EDM.2019.8823472

Journal of Physics: Conference Series **2140** (2021) 012034 doi:10.1088/1742-6596/2140/1/012034

Distribution of temperature field in phantom of hand with microwave radiation

A N Nechaev and G E Dunaevsky

National Research Tomsk State University, Tomsk, Russia

E-mail: sasha8227039@gmail.com

Abstract. The present work presents the results of phantom of human hand simulation with the dielectric properties, very similar with human tissues properties. The simulation was conducted with using "CST Studio" software. The phantom was made by a "papier-mâché" method. The thermal field distribution was measured twice: for the simulation of phantom in "CST Studio" software, and for the phantom filled with dispersed meat of pork, the power was 2.45 GHz. The heating profile of certain points on phantom, including fingers, has been studied. It is shown that the heating profile of fingers zone is different from heating profile of wrist and forearm. The obtained experimental data have shown the possibility of using microwave radiation for deep cold injuries treatment.

1. Introduction

From 10 to 15 per cent of all traumatic cases are cold injuries. Deep cold injuries are needed in long-term treatment, which can last as long as two month. Most frequently these cases have resulted in extensive amputation of damaged limbs. The probability of disability reaches 90 per cent of all cases of cold injuries.

At the moment, there is no equipment intended to heat deep cold injuries for treatment. The present work is relevant due to the need to develop such equipment. In 2006, researchers of Tomsk state university have addressed this issue. The tests were performed on animals (rabbits), as a result of which it has demonstrated the high potential of applying microwave heating with frequency of 2.45 GHz, permitted for medical applications [1]. However, application of the present method on patients with cold injuries requires more detailed researches. Thus, one of the main objectives is to examine the inhomogenuity of electromagnetic and thermal fields in the phantom of limb in order to avoid local underheatings and overheatings. The main difficulty is impossibility of conducting human test trials. Therefore it is imperative to create phantom of human limb, with the maximally close dielectric properties to the properties of human tissue.

2. Simulation

Primarily, exposure of high-frequency radiation results in tissues heating due to released heat, produced during absorption of radiation. Conditions of radiation are: incident power density, reflection and refraction coefficients on the boundaries "biological tissue - air", between layers of tissue, and wave decay coefficient [2]. Experimental device based on apparatus "SMVi-200" (as a power source) [3, 4] has been used for examination of temperature distribution in the phantom of limb in this work. Microwave chamber and phantom of human hand with fingers (Figure 1) were simulated in CST-Studio software. Chamber filled with air – grey sector at Figure 1. Dielectric coefficient of phantom is similar

Content from this work may be used under the terms of the Creative Commons Attribution 3.0 licence. Any further distribution of this work must maintain attribution to the author(s) and the title of the work, journal citation and DOI.
Published under licence by IOP Publishing Ltd

to dielectric coefficient of human tissue. The experiment aimed to research the distribution of thermal field in fingers.

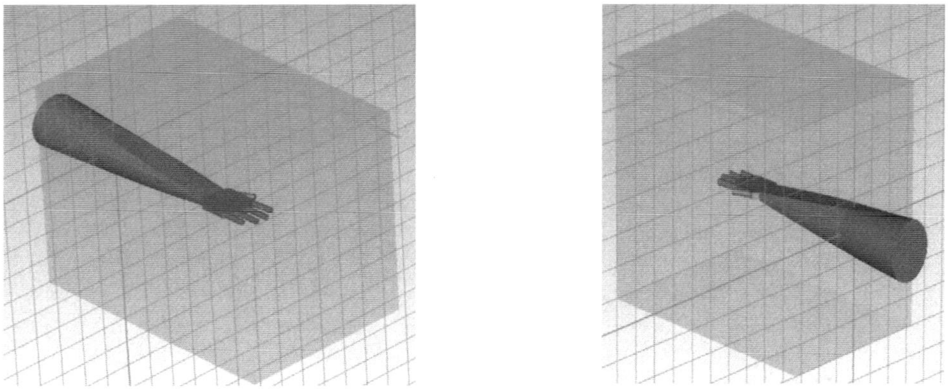

Figure 1. Microwave chamber with phantom of hand and fingers in two planes.

The result simulation of electric energy distribution in microwave chamber and in phantom of limb is provided in Figure 2.

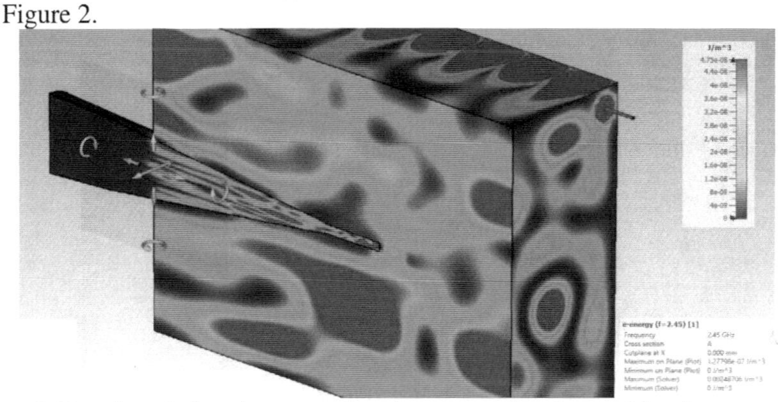

Figure 2. Density of electric energy in microwave chamber and in phantom of limb.

The Figure 3 presents the density of energy in microwave chamber near the whip antenna

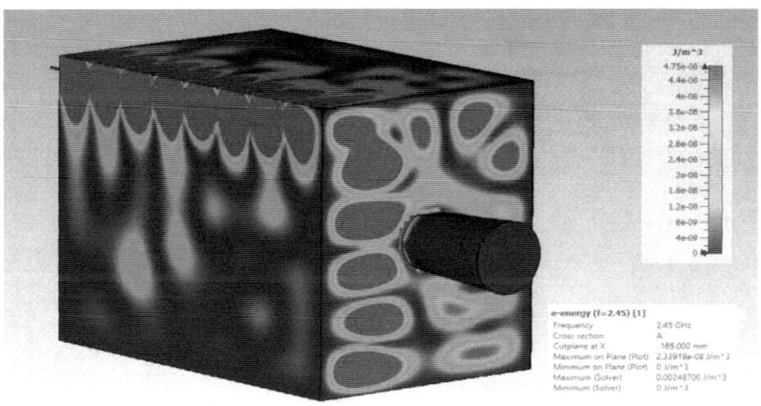

Figure 3. The density of electric energy in microwave chamber near the whip antenna.

There are several types of oscillations in the chamber due to size of chamber (it is much higher than wavelength). The field distribution has complex composition, thus, the density of electric energy has complex composition.

The software solution allows to evaluate the distribution of thermal field in the body of microwave chamber and in the profile (Figure 4).

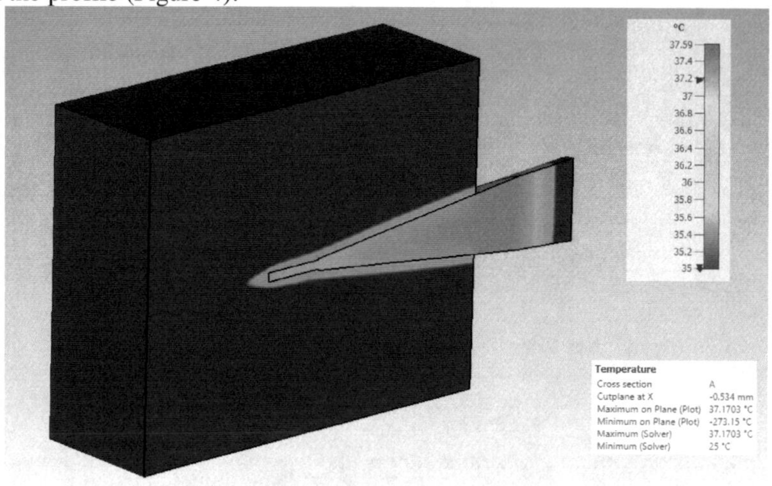

Figure 4. Thermal field in the body of chamber and phantom.

The surface temperature distribution of phantom is presented in Figure 5.

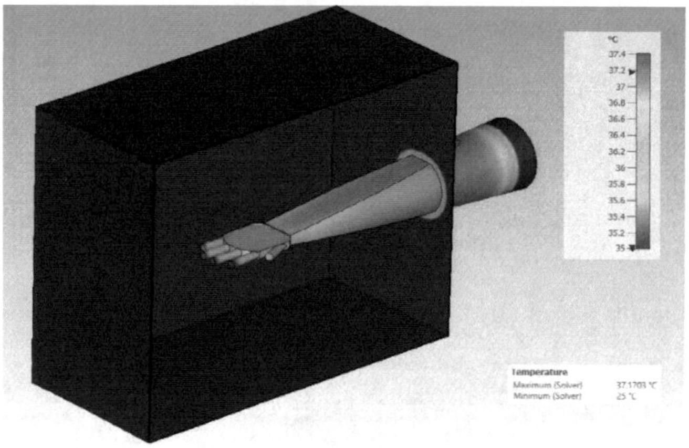

Figure 5. The thermal field of the surface of phantom.

As it is shown in the Figures 2, 4 and 5, despite the high field density in the finger area, due to its high thermal exchange with air in chamber, fingers become less heated. As a result, temperature of fingers is lower than temperature of palm of hand and forearm. The present temperature distribution has a positive impact on cold injury heating.

3. Experimental

The measurements of thermal field distribution in phantom of hand with fingers were conducted in the work. The phantom was made by a "papier-mâché" method, glue and paper were used (Figure 6).

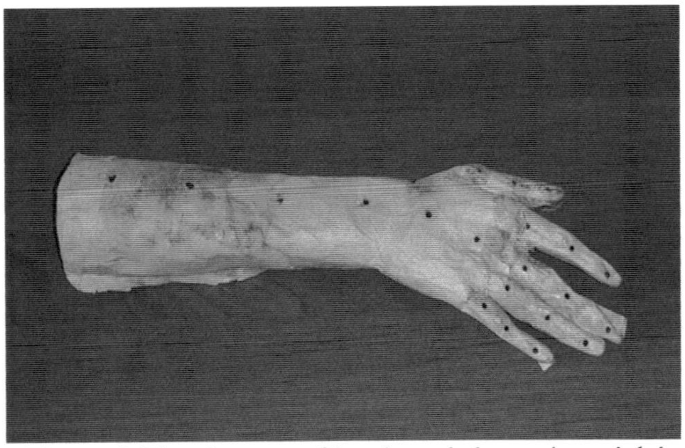

Figure 6. The phantom of hand, made by papier-mâché.

The phantom was filled with dispersed meat of pork and previously was cooled. Temperature distribution was measured in all control points, then 3 sessions of heating were provided, each lasting 6 minutes and 40 W power. The temperature measurements in the body of phantom were conducted with the temperature detector LT-300 [4]. Temperature distribution has been investigated along forearm, wrist, palm of hand and fingers (Figure 7).

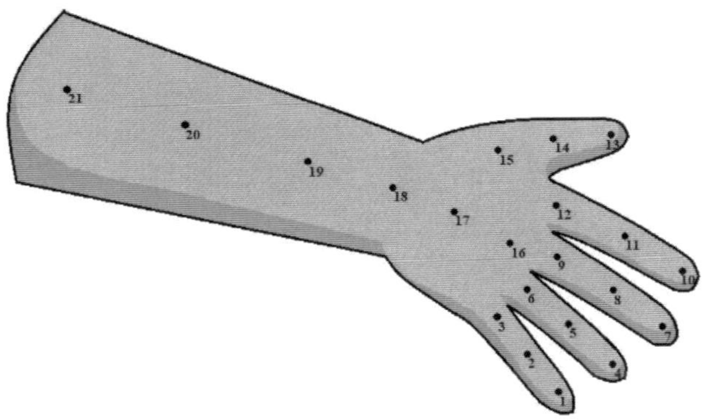

Figure 7. Model of hand phantom with given points of temperature measurements.

4. Results

The Figure 8 presents the results of temperature measurements in the given points of forearm and fingers. In the course of previous experiment [5] cylinder filled with dispersed meat was used as a phantom of limb. In the present work the new phantom, more similar with human hand is used. This phantom has lower volume than cylindrical phantom, thus, more even cooling has been achieved. The temperature rise without heating in fingers area is explained by thermal exchange with air, air temperature is similar to room temperature.

Under the microwave radiation the faster heating of forearm (points 20, 19, 18). Then, temperature increasing is observed in palm of hand area (points 17, 16) and in proximal and middle phalanges of fingers (points 3, 2, 6, 5, 9, 8, 12, 11, 15, 14), and then reducing of temperature in distal phalanges of fingers is observed.

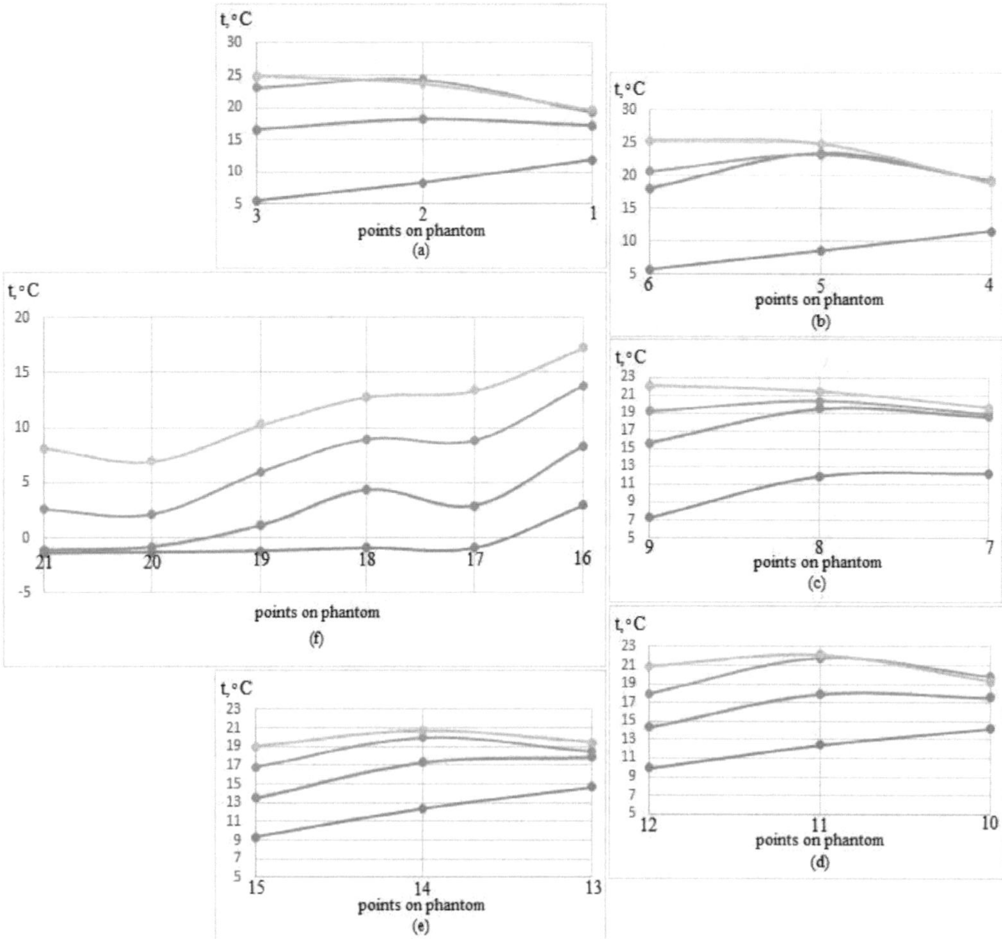

Figure 8. Temperature distribution in the phantom of limb with fingers: little finger (a), ring finger (b), middle finger (c), index finger (d) and thumb (e), forearm (21, 20, 19), wrist (18) and the palm of the hand (17, 16).

Temperature reducing is caused by two reasons: reducing of electromagnetic field strength in this area of microwave chamber due to the position of microwave source along the body frame of chamber, and due to the previously mentioned fast thermal exchange with air.

5. Summary

Thus, the experimental research on field distribution in cylindric phantom presents the possibility of obtaining more homogeneous distribution of thermal field in fingers area of chamber when using distributional whip antenna. Providing certaine level of temperature while heating cold injury will prevent premature heating of digital arteries.

The obtained experimental data have shown the possibility of obtaining more favourable field distribution in microwave chamber, possibility of physical simulation of human hand with the dielectric properties, very similar with human tissues properties, and to ameliorate the modes (power and time of heating) taking into account the particularities the process of fingers heating.

References

[1] Gorelik I E 2010 Prophylaxis of necrosis while limb frostbite in the prereactive and early reactive periods *Abstract of a thesis Tomsk Military Medical Institute*

[2] Antipov V B, Dunaevskii G E, Gavrilin E V 2017 Device for treating limb frostbites *Patent of the Russian Federation*
[3] SMVi-200 "MEDTEKO". Available at: http://medeco-center.ru/smvi-200-medteko.
[4] Electronic laboratory thermometers. Available at: https://termexlab.ru/#!/ru/product/lt-300-n-termometr-laboratornyij-elektronnyij-250300
[5] Nechaev A N, Dunaevsky G E 2021 *Prospects of Fundamental Sciences Development* **1** 250–252

Water body measuring pH automated system based on quadcopter

E V Lanin, K V Dorozhkin, A V Badin, G E Kuleshov, V A Kazanin and M S Yuzhakov

National Research Tomsk State University, Tomsk, Russia

E-mail: zenkalanin@gmail.com

Abstract. The article discusses monitoring water bodies pH level automated system implementation using an unmanned aerial vehicle (UAV). Floating structure prototype for supporting UAV on water surface is described, taking into account creating a rotor thrust theory. Water hydrogen ions monitoring activity concept using a UAV is presented. Testing results water monitoring system in Lake Boyarskoye (Tomsk, Russia) are presented.

1. Introduction

One of important and frequently touched upon problems in modern world is water quality monitoring in reservoirs [1–3]. In a particular case, water acidity (pH) measure is interest due to fact that a low pH water level leads to population and living organisms growth decrease at all trophic levels [4]. pH parameter is hydrogen ions activity, varying in range from 0 to 14, which helps determine acidic (pH <7), alkaline (pH> 7), or neutral (pH = 7).

pH research traditional method using swimming facilities requires significant material costs and time spent large amount on preparing and conducting research, as well as processing the results obtained.

For time resources and reduce financial costs rational use, it is proposed to use an unmanned aerial vehicle (UAV). Recently, UAVs are gaining wide popularity due to their introduction into human activity most areas: monitoring animals population in wild, fire-rescue services, agriculture [5], mapping [6,7], etc. UAVs are classified based on airspace performance such as fixed-wing, rotary and hybrid configurations [8].

This work is devoted system development monitoring acidity in a reservoir using a UAV and a landing platform on water surface for quadcopter.

2. Quadcopter flight principle

To task accomplish, quadcopter was chosen. Quadcopter is an aircraft with four symmetrical rotors driven by motors mounted at two perpendicular axes ends. Electric motors two rotors rotation on one axis is carried out in one direction and in opposite direction for motors pair on other axis to compensate for rotors torque (Figure 1).

Content from this work may be used under the terms of the Creative Commons Attribution 3.0 licence. Any further distribution of this work must maintain attribution to the author(s) and the title of the work, journal citation and DOI.
Published under licence by IOP Publishing Ltd

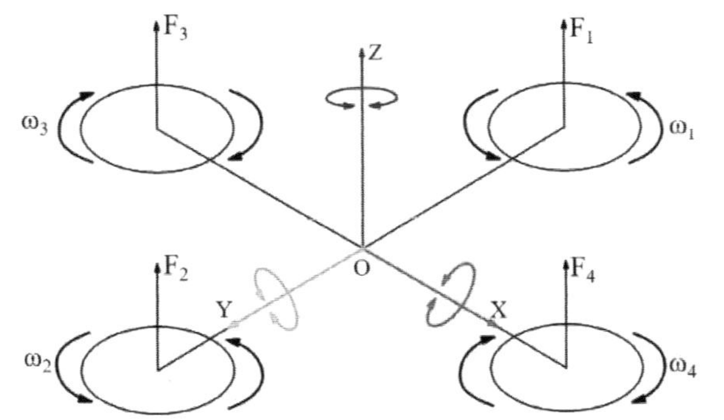

Figure 1. Quadcopter kinematic diagram.
Axes and tilt angles possible directions are illustrated in different colors – green, red and blue.

OXYZ axes form a movable coordinate system. It creates a coherent coordinate system with quadcopter. ω_1, ω_2, ω_3 and ω_4 are angular rotor velocities, which together create thrust forces F_1, F_2, F_3 and F_4. Forces are directed perpendicular to propellers rotation. Quadcopter location is determined by three angles (Euler angles): pitch (θ), roll (φ), and yaw (ψ). Pitch angle sets rotation around *OY* axis, roll angle - rotation around *OX* axis and yaw angle - rotation around the *OZ* axis.

Liftoff and quadcopter control is carried out by sending data via "Oneshot125" protocol to motor's drivers. Drivers send an alternating voltage to motor windings with a frequency that regulates rotor rotational speed. Motors rotation lead to propellers rotation with angular velocities ω_1, ω_2, ω_3 and ω_4. As a result, traction forces F_1, F_2, F_3, F_4 and forces moments are formed. Moment's sum should be equal to zero.

Movement in vertical plane is determined by lift force *F* resulting vector projection onto *OZ* axis. Rotation in angle ψ is carried out by changing rotor angular velocities. So, to turn clockwise, it is necessary to increase rotor angular speed for right-handed electric motors - ω_3 and ω_4 and decrease for left-handed motors - ω_1 and ω_2. This will allow quadcopter not change roll and pitch orientation and to form positive angular momentum.

3. Floating UAV design

When developing a floating structure, creating theory rotor thrust was applied [9]. This theory essence is rotating rotor creates a difference in pressure - a low pressure zone is formed above rotor rotation plane, and a high-pressure zone is formed under plane. This results in a sliding air flow with an inductive velocity V_i. Air flow zone schematic diagram has a symmetrical shape with respect to quadcopter rotor rotation axis (Figure 2*a*). Air pressure behind considered area corresponds to atmospheric pressure.

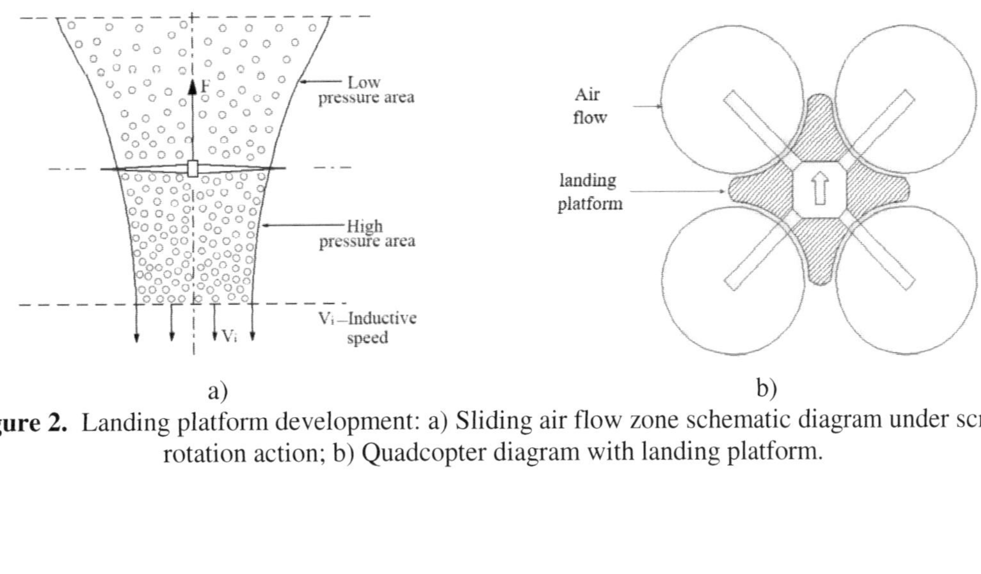

Figure 2. Landing platform development: a) Sliding air flow zone schematic diagram under screw rotation action; b) Quadcopter diagram with landing platform.

Motor's rotors position is symmetrical about quadcopter gravity center. This means that all rotors create same thrust and sliding air flow fields have a uniform shape. Water surface landing platform will take smoothed X-shaped shape form to minimize landing platform impact on movement in airspace (Figure 2*b*).

4. Water pH sensor device

Water pH sensor is combined measuring vessel in which two electrodes are combined: pH electrode and reference electrode [10-12]. Outer casing can be either glass or plastic made (Figure 3).

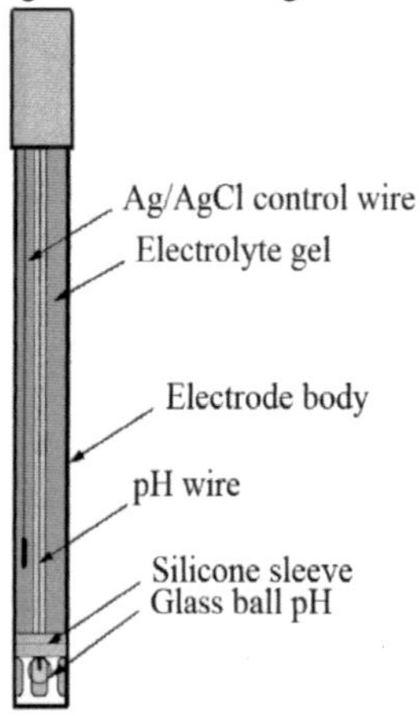

Figure 3. Water pH senso appearance.

Reference electrode is second important component. Electrode electric potential does not change during pH measurement, since it does not allow positive hydrogen ions to pass through.

When pH sensor is immersed in analyzed solution, hydrogen ions in test solution begin to interact with electrode glass surface layers and displace ions contained in the alkali metal glass. This leads to electrochemical potential occurrence. Hydrogen ions are distributed between glass and solution in such a way that these media surfaces acquire opposite charges, and a potential difference forms between glass and analyzed solution. Further, there is an increase in the electrical potential difference between measuring and reference electrodes in proportion to pH units. The difference between these electric potentials determines pH value, since their tendency to pass into glass depends on hydrogen ions concentration in solution. Therefore, emerging potential difference is directed to reduce tendency to pass hydrogen ions into glass.

To obtain reliable results from the pH sensor, debugging is necessary for buffer solutions with reference pH values: 4.00, 6.86 and 10.01. Calibration of the pH sensor consists in sequential immersion of the measuring sensor in buffer solutions. Then, using a potentiometer on the conversion board and based on the measurement data of solutions with a given pH level, a calibration curve (Figure 4) was constructed, an approximation was made by a linear equation ($y = kx + b$) and correction coefficients were found ($k = -8.3254 \pm 0.226$; $b = 7.9951 \pm 0.116$). The obtained coefficients are used to calculate the pH value from the potential difference from the measuring and comparative electrode.

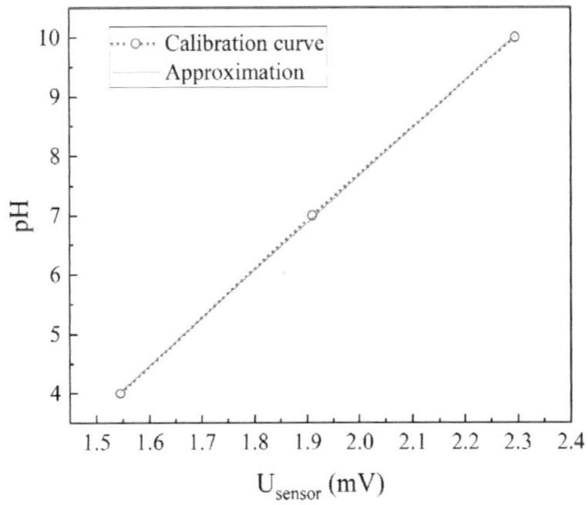

Figure 4. Calibration line.

The permissible error for this type of pH sensor is ± 0.1, stated by the manufacturer. During 10-fold measurements of the pH level of buffer solutions, confidence intervals (equation 1) were revealed for solutions: 6.86 ± 0.025, 4.0 ± 0.029 and 10.01 ± 0.023 (Figure 5).

$$\triangle pH = t_{n,p}\sqrt{\frac{\sum_{i=1}^{n}\triangle pH_i^2}{n(n-1)}} \tag{1}$$

where: ΔpH – confidence interval; n – number of measurements; $t_{n.p}$ – Student's coefficient for 10 measurements with confidence probability 95%; ΔpH_i – deviation from the average value.

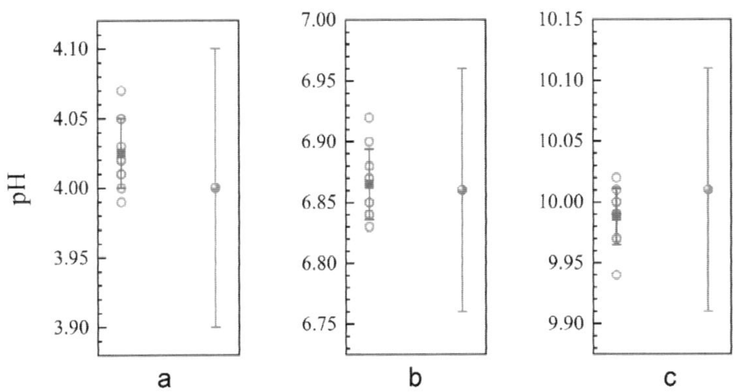

Figure 5. Calibration of the pH sensor in buffer solutions 4.00 (a), 6.86 (b) and 10.01 (c).

The confidence intervals calculated during the calibration of the pH sensor do not go beyond the error stated by the manufacturer.

5. The buffer solution pH dependence on temperature

The pH value directly depends on the temperature, which is explained by the different dissociation of hydrogen ions (H+). When calibrating the pH sensor, the pH temperature of buffer solutions for measurement accuracy should be 25 ° C in order to avoid significant errors in subsequent measurement results.

The activity of hydrogen ions of buffer solution 6.86 was measured in order to accurately measure the pH of the sensor. The change in the pH of the buffer solution according to the documentation attached to them is 0.03 pH in the temperature range of 25 ° C -50 ° C. According to the measurement results, the pH of the sensor in the same temperature range is 0.04 pH (Figure 6).

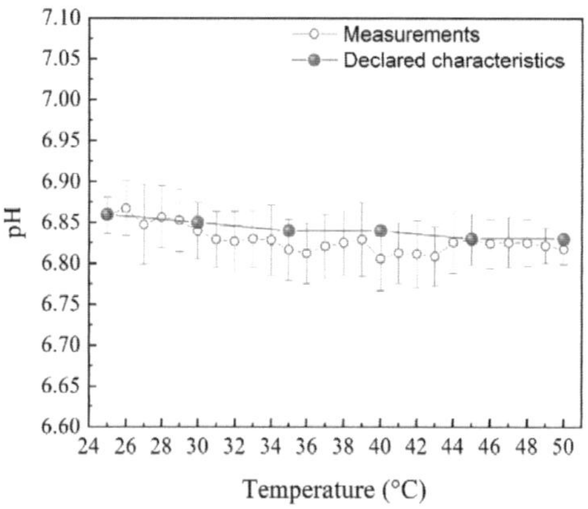

Figure 6. pH dependence of buffer solution 6.86 on temperature.

6. System operation description

Measurement module was developed to control UAV-based pH measurement system. Measuring module is controlled by "Pixhawk 2.4.8" flight controller.

Measuring module contains microcontroller ATmega328P responsible for data processing, a DS3231 real-time clock to create measurment's time stamps and a microSD memory card to store pH measurement results (Figure 7).

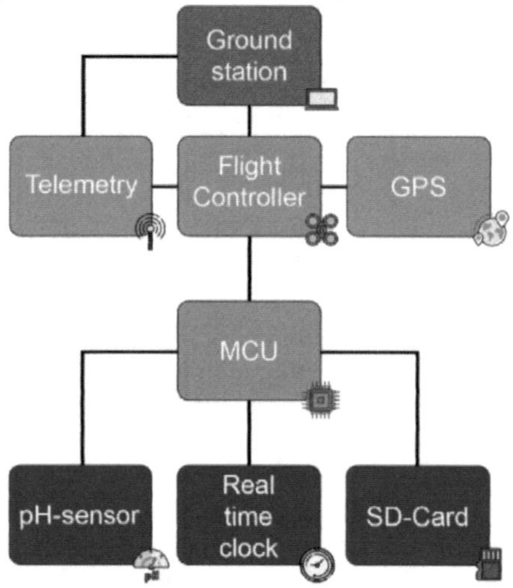

Figure 7. Masurement scheme.

Flight controller "Pixhawk 2.4.8" has an important advantage - automatic mission execution without human intervention. In addition, in force majeure event, such as signal loss from quadcopter telemetry, low battery or GPS signal loss, algorithm for returning to take-off point by GPS coordinates is executed. If a return to base is not possible, an emergency landing will occur.

Ground station can be connected either directly to flight controller or remotely controlled using telemetry. Also, ground station allows planning mission and trajectory for UAV [13]. The mission includes takeoff, flight to measurement point, splashdown and pH sensor activation, waiting for measurement end, takeoff from water surface and moving to next point, repeating above algorithm. Last command will be initial position return by GPS coordinates.

Sensor system microcontroller is connected to Pixhawk flight controller via digital GPIO pins using galvanic isolation with the PC817 optocoupler (Figure 8).

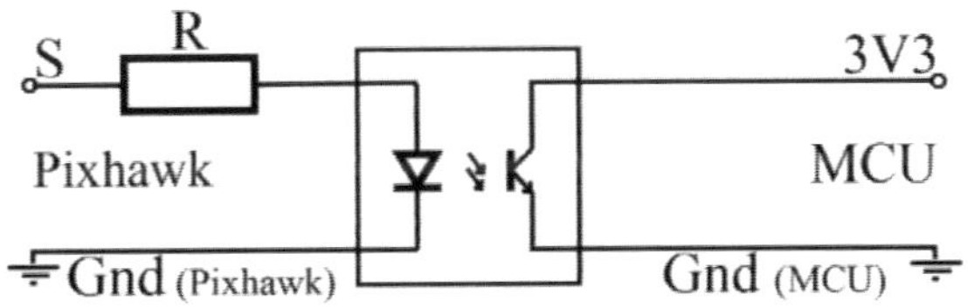

Figure 8. Galvanic isolation between flight controller "Pixhawk " and measurement module.

7. Results

Lake Boyarskoe with coordinates 56°27'07.5" N 84°54'58.9" E was chosen as place for system experimental verification due to its size, access ease and starting point for subsequent changes quality monitoring in water acidity level. Reservoir is undergoing active urbanization due to interuniversity student campus construction of an s (Figure 9).

Figure 9. Measurement site satellite image.

As a result, hydrogen ions in aqueous medium. As a result of three measurements at one point, the average pH value was 7.24. The PH deviations are within the tolerance of the pH sensor field at a temperature of 9.4 °C. Thus, this environment can be characterized as both neutral and suitable for the life of living organisms.

8. Conclusion

Thus, kinematics movement quadcopter its movement in airspace were described. The sing UAVs fundamental possibility to monitoring aquatic environments has been demonstrated. The design for supporting UAV on water surface was developed and tested, taking into account the creating theory of a rotor thrust.

References

[1] Koparan C, Koc A B, Privette C V and Sawyer C B 2018 *Water* **10** (3) 264 doi:10.3390/w10030264
[2] Koparan, C, Koc A B, Privette C V and Sawyer C B 2019 *Water* **11** (3) 604 doi:10.3390/w11030604
[3] Suslyaev V I, Demin V V, Badin A V, Dorozhkin K V, Vorobyov S N, Zhuravlev V A, Polovcev I G, Korovin E U, Olshukov A S, Tretyakov A S, Suslyaevb V V, Pavlova A A, Kolesnichenko Y Y, Uzhakov M S and Pidotova D A 2019 *Proceeding of the 25th International Symposium on Atmospheric and Ocean Optics: Atmospheric Physics* **11208** 112085H doi: 10.1117/12.2540933
[4] Haines T A 1981 *Transactions of the American Fisheries Society* **110** (6) 669-707

doi.org/10.1577/1548-8659(1981)110%3C669:APAICF%3E2.0.CO;2

[5] Adão T, Hruška J, Pádua L, Bessa J, Peres E, Morais R and Sousa J 2017 *Remote Sensing* **9** (11) 1110 doi:10.3390/rs9111110

[6] Reinoso J F, Gonçalves J E, Pereira C and Bleninger T 2018 *Iranian Journal of Science and Technology* **42** (1) 91–96 doi.org/10.1007/s40996-017-0076-x

[7] Erena M, Atenza J F, García-Galiano S, Domínguez J A and Bernabé J M 2019 *Water* **11** (3) 445

[8] Saeed A S, Younes A B, Islam S, Dias J, Seneviratne L and Cai G I 2015 *International Conference on Unmanned Aircraft Systems (ICUAS)* 806-815 doi: 10.1109/ICUAS.2015.7152365

[9] Deshpande M D and Sivapragasam M 2017 *Resonance* **22** (1) 61–77

[10] Wang X, Sun L, Wang T and Shi Y 2020 *Sensors Journal* **20**(24) 9145728 doi:10.1109/JSEN.2020.3010709

[11] Adhipramana M, Mardiati R and Mulyana E 2020 *Proceeding of the 6th International Conference on Wireless and Telematics* doi: 10.1109/ICWT50448.2020.9243614

[12] Nasution T H, Dika S, Sinulingga E P, Tanjung K and Harahap L A 2020 *IOP Conf. Ser.: Mater. Sci* **851** (1) 012053 doi:10.1088/1757-899X/851/1/012053

[13] Kimon P. Valavanis, George J. Vachtsevanos 2015 *Handbook of Unmanned Aerial Vehicles* (luxembourg netherlands)

Radio imaging using Doppler sensors

R N Satarov, I Yu Kyzmenko, S E Shipilov

Tomsk State University, Tomsk, Russia

E-mail: satarov.rail@gmail.com

Abstract. The article shows the multi-angle single-frequency probing of the object with the tactical antenna array, which we install and moves on a two-coordinate scanner. For testing the possibility of using this approach, we made statistical studies. Presented radio images show that using the scanner makes possible to detect metal and dielectric objects. Achieved resolution sufficient to identify forbidden items

1. Introduction

Often, an inspection system use radio waves or X-rays to obtain information about the location of object. The type of radiation determines achievable performance characteristic of a developing inspection system. Radio wave inspection systems comparison with X-ray have a weak penetrating ability, which not enough to detect prohibited objects inside the human body. On the other hand, UWB radiation is less dangerous for human health. The experiments confirm the possibility of reconstructing radio-images of a probed object from data of both multi-frequency [1] and single-frequency radiolocation. The work will use radio emission at a frequency of 24 GHz. With a wide directional pattern of the locating system, the received signal is the result of the interference of many waves reflected by spatially distributed objects. The interference pattern taken with the help of a matrix of microwave sensors is the basis for processing according to the appropriate technology for reconstructing the real distribution of inhomogeneities. In this case, it is possible to adjust the depth of focus and the actual tomography of the image.

2. Antenna

To creating a safe and high-speed inspection device, we propose as a base element the Doppler sensor, which is a miniature radiolocation system. The sensor consists of following constructive electronic components (Figure 1): Transmitting antennas – TX; Receiving antennas – RX; Reference frequency generator – 1; Phase shifter – 2; Mixer – 3, 4; LPF – 5, 6.

An important advantage of the sensor is in-phase (I) and (Q) quadrature components of the received signal.

Content from this work may be used under the terms of the Creative Commons Attribution 3.0 licence. Any further distribution of this work must maintain attribution to the author(s) and the title of the work, journal citation and DOI.
Published under licence by IOP Publishing Ltd

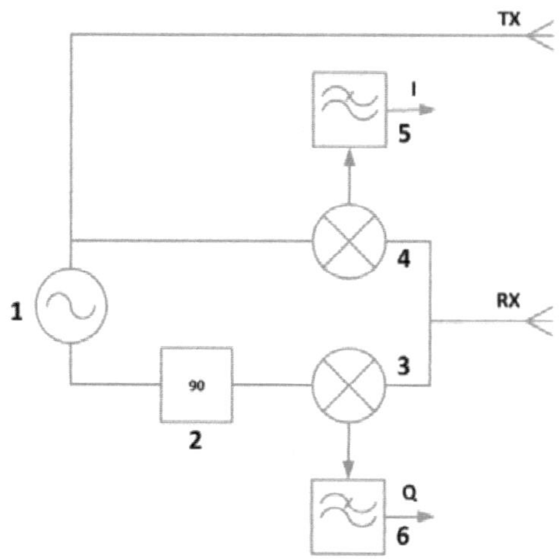

Figure 1. Sensor consists.

Registration of wave projections possible by using one Doppler sensor. If it installs on a two-dimensional scanner system, which moves the sensor along the entire scanning plane, and sequential data acquisition. The need for successive passage of all scan points to get all the necessary wave projections stretch the process for an unacceptably long time. For speed up the process, we propose to use a trade-off between using a single sensor and creating a full-size two-dimensional matrix. The proposed solution is combining Doppler sensors into a one-dimensional array (the linear array). Sensors located in one plane in a line, close to each other. Originally, all sensors in the array switched on simultaneously, i.e. there was no time division of the channels. In testing and debugging processes, we experimentally found that sensors operates in the array have a negative effect on each other. This effect clearly appear when comparing interference patterns. Which had obtained from the same test object with the single sensor and the array of sensors.

The size of the sensor 2.5×2.5 cm does not allow placing sensors in a matrix with the previously calculated step (about 2-5 mm). Therefore, we propose to scan along two axes (Figure 2). In this case, Kotelnikov's theorem will be satisfied and, as a result, there will be no discharged data, which will ensure acceptable image quality.

Figure 2. Two-axis scanner system.

3. Results

The received signal is a result of the interference of many waves, which reflects from spatially distributed objects using. The signal is registered with a wide radiation pattern. The basis for the special processing and recovering of the radio-image the scattering inhomogeneities is the interference picture, which got by using the matrix of microwave sensors. At the same time, it is possible to regulate the focusing depth and actual tomography of the image. Synthetic Aperture Radar (SAR) technology [2] requires the system hardware function (SHF) for processing of radio-image, which is equal to a spatio-temporal response to a point reflector. During SAR image reconstruction we took into account trajectory parameters of the location system. During theoretical SAR image reconstruction, we took into account the trajectory parameters of the location system.

At the end of the scanning process, the wave projections obtained at different trajectories of a matrix move are combining into a coherent interference pattern. An example of such a picture shown in Figure 3*a*, where a probed object was a metal ball, 1 cm in diameter, placed in 15 cm from the scanning plane. Figure 3*b* shown radio-image of the metallic ball.

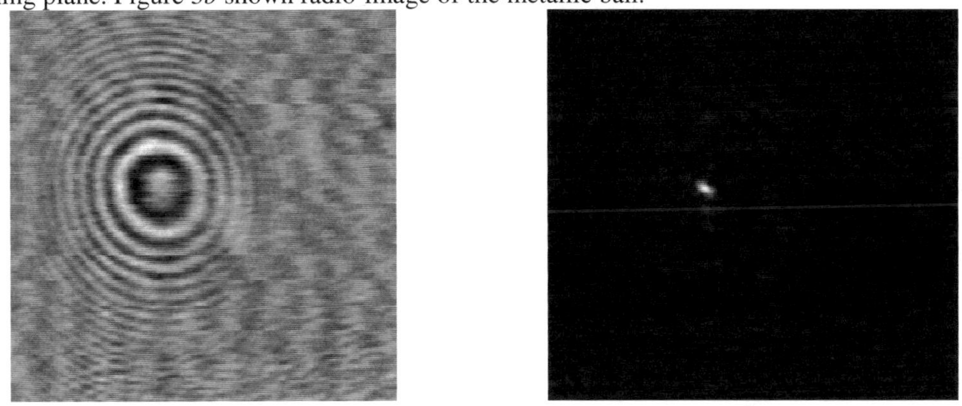

a b

Figure 3. Interference pattern of a metal ball (a) and its radio image (b).

As you can see in these Figures 3, signals reflected from stand are distinguished from the signals reflected from other objects (walls, furniture, etc.). So, it means that, these signals can be processed by using methods of mathematical focusing and synthesis of aperture [3, 4]. Figure 4 shows a reconstructed radio image of a test sample of an explosive simulator (bar of soap), with a size 8×6 cm and placed on a cardboard box. The scan area had 40×40 cm, the scan time about 90 sec.

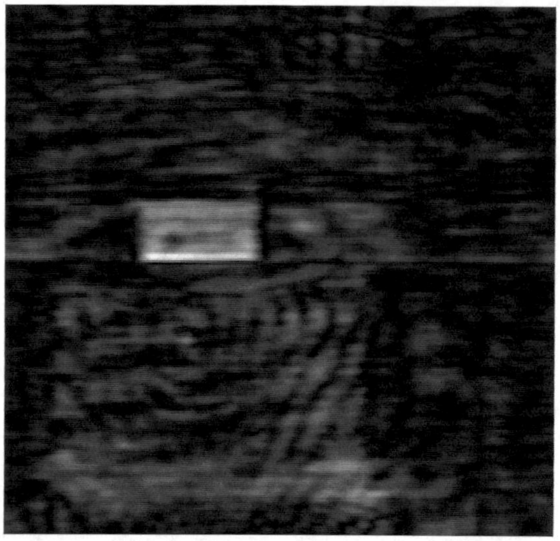

Figure 4. Radio-image of the soap placed on a cardboard.

The test object has clearly seen on the restored radio image. The main task of this technology is restoration of radio images for objects hidden on the human body. Firstly, let's test this technology on a human manikin. Figure 5 shown a reconstructed radio-image of a motionless mannequin with the explosive simulator (a) and (b) without.

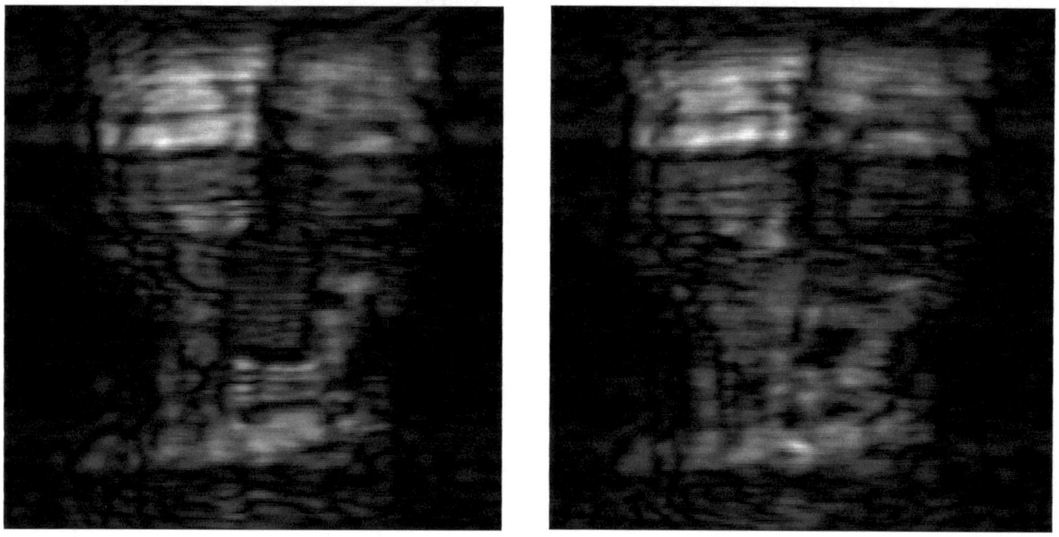

a b

Figure 5. Radio-image of the motionless mannequin with the explosive simulator (a) and without it (b).

A comparison of radio-images shown that the explosive simulator stands out against the background of the mannequin. The situation more complicated when we scan a human because, scanning process require 90 seconds, during this period the subject moves while breathing. Figure 6 shows a radio image of a human's back with fixed the explosive simulator (a) and without it (b). The

Figures show that there is no identical background image of the human body, it associates with his involuntary breathe.

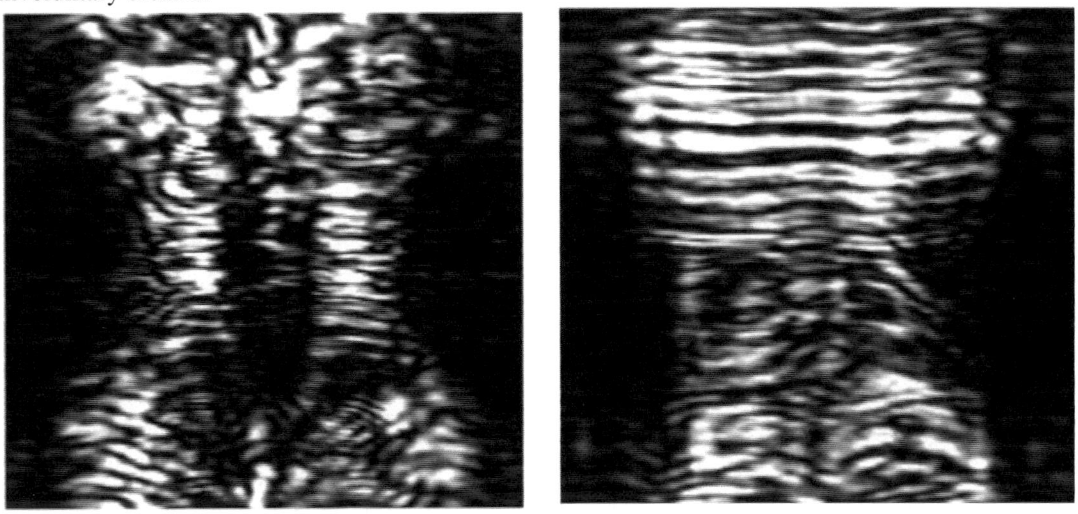

Figure 6. Radio-image with the explosive simulator (a soap bar) (a) and without it (b).

Moreover, the results shown, it allows concluding about the ability of the proposed solutions to restore radio images. These methods work not only in free space but also against the background of other objects, including a human body.

4. Summary

The distribution of the quadrature components of location signals, which is recorded with using the developed 2D scanner, makes it possible to recover radio images of the object by using the aperture synthesis method. The achieved spatial resolution is the order of 1 cm at the sounding frequency of 24 GHz.The distribution of quadrature location signals, recorded using the developed 2D scanner, makes it possible to reconstruct the radio image of an object by the aperture synthesis method, the achieved spatial resolution is of the order of 2 cm.

The presented radio images clearly prove that with the help of the developed scanner it is possible to detect objects made of metals and dielectrics. The achieved resolution is sufficient for identifying hazardous items. At the same time, the scientific groundwork for creating a matrix of Doppler sensors with an offset is shown.

Acknowledgments

This work was supported by the Grant of the President of the Russian Federation (МК-3578.2021.4).

References

[1] Shipilov S, Eremeev A, Yakubov V *et al* 2017 *Medical Physics* **47**(10) 5147–5157 doi:10.1002/mp.14408
[2] Srivastav A, Nguyen P, Mcconnell M *et al* 2020 *System* **69**(10) 7422–7436 doi:10.1109/TIM.2020.2984415
[3] Zyatkov D O *et al* 2016 *Journal of Physics: Conference Series* **671**(1) 012065
[4] Praks J, Kugler F, Papathanassiou K, Hajnsek I, Hallikainen M 2007 *IEEE Transactions on Geoscience and Remote Sensing Letters* **4** 466-470 doi: 10.1109/LGRS.2007.898083

Journal of Physics: Conference Series **2140** (2021) 012037 doi:10.1088/1742-6596/2140/1/012037

Questions of application of fiber-optic sensors for monitoring crack growth during rock deformations

V V Yugay[1], P Sh Madi[1,2], S B Ozhigina[1], D A Gorokhov[2] and A D Alkina[1]

[1]Karaganda Technical University, Karaganda, Republic of Kazakhstan
[2]Tomsk Polytechnic University, Tomsk, Russia
[3]Kazakh Agrotechnical University name of Saken Seifullin, Nur-Sultan, Republic of Kazakhstan

E-mail: perizat1@tpu.ru

Abstract. The paper considers ways to solve the problem of developing a system for monitoring displacement in quarries, which are the main main cause of the collapse of boards and berms in quarries. To ensure safety and constant monitoring during work at the quarry, there are chiseled fiber-optic sensors. The fiber-optic sensor is made on the basis of a single-mode optical fiber, which makes it possible to measure the displacements of the mountain range at distances of about 30 km with high accuracy. Laboratory sample a fiber-optic sensor in its work uses a method for monitoring additional losses that occur during mechanical action on an optical fiber. The fiber-optic sensor was made to show a fairly high linearity and accuracy during measurements and can be used to control the deformation of the array after appropriate refinement of its design. This article is aimed at creating means of controlling the process of deformation and displacement of a mountain massif. Ultimately, the results of the study will help prevent accidents associated with the collapse of the sides. Since the growth of cracks in the rocks of the bort mountain massif leads to its sudden collapse and creates a significant danger for personnel, it also causes the failure of mining equipment.

1. Introduction

Open-pit coal mining is characterized by an increase in the depth of coal sections, since over time the overlying coal seams are worked out and there is a need to deepen the quarry. Accordingly, the modernization of technological processes and ensuring the safety of mining operations, taking into account the peculiarities of the relevant measures, which are one of the main sore issues during operational work at the quarry. It is necessary to develop new methods and means of monitoring the stability of the sides of quarries, since the technical and economic efficiency of development depends on their parameters. In this regard, the control and monitoring of the stability of the sides and the berm at the quarry is an urgent scientific and practical problem. The analysis in [1-3] shows that there is a problem and its solution cannot be approached one-sidedly, it is necessary to use an integrated approach based on high-tech technologies. This requires a technology aimed at developing high-precision digital sensors that use optical fiber (OF). This work is aimed at the development of fiber-optic sensors (FOS), which are able to solve the problem of monitoring the sides and berm of the quarry. The advantages of OF have long been known and the existing FOS are quite successfully used in the aerospace and military industries, as well as in the oil and gas industry. The relevance of the research is to find a new control method compared to fiber-optic systems based on the effect of Bregg

Content from this work may be used under the terms of the Creative Commons Attribution 3.0 licence. Any further distribution of this work must maintain attribution to the author(s) and the title of the work, journal citation and DOI.
Published under licence by IOP Publishing Ltd

gratings, which can significantly reduce the cost of the monitoring system as a whole, which is extremely important for implementation in the mining industry. Reducing the cost of one measurement point will allow building a quasi-distributed system for monitoring the displacement and deformation of rocks on the sides of quarries, as well as tracking the growth of cracks. Since the growth of cracks in the rocks of the bort mountain massif leads to its sudden collapse and creates a significant danger for personnel, it also causes the failure of mining equipment. The safety issues of carrying out installation work are a priority, accordingly, ensuring the safety of employees is a very urgent issue that needs to be addressed immediately by searching for new methods and means.

2. Question theory and literature analysis

Taking into account the world experience and using the results of the analysis, we can say that this direction is very promising for the mining industry. The paper takes the basic principle of setting up data processing experiments and building a mathematical model and apparatus. And the authors in [4,5] conducted a number of experiments on modeling fiber-optic sensors and using them to diagnose the stress-strain state of lifting machines. Currently, various methods are used, which in turn are also effective, but the question arises of developing a new method for controlling displacements and deformations based on the use of fiber-optic technology, which has the advantages previously described in [6,7], where experiments are conducted to study additional losses during mechanical action on an optical fiber. The paper [8] describes the research of the development of the physical foundations for the creation of sensors and a physical and mathematical model of the optical signal parameter control system. Also, the authors [9] considered issues related to the development of a system for monitoring the deformation and displacement of the mountain range, leading to the collapse of the sides of the quarry. Where the monitoring system uses point fiber-optic sensors. There are quite successful studies of a laboratory sample of point-type FOS, which is made on the basis of a two-beam Mach-Zehnder interferometer using a single-mode optical fiber to control deformation (displacement) with a change in sensitivity and a reduced influence of temperature interference leading to zero drift. In addition, a scientific analysis of similar works by foreign authors who work with optical fiber and develop fiber-optic sensors is carried out. Scientists Liu X, Wang C, Liu T, Wei Y, Lv J manufactured a new water pressure sensor based on a fiber Bragg lattice, where the applicability of the sensor for monitoring water pressure was confirmed [10]. In 2011, scientists Kumar Atul, Kumar Dheeraj, Singh U., Gupta P. S., Shankar Gauri illustrated the development of a system for continuous monitoring of roof pressure and closure of the support with progressive advancement of the face in lava workings. An example of the operation of a long face is illustrated to assess the effectiveness of the monitoring system when evaluating the operational characteristics of the support to ensure safer and uninterrupted operation of the roof supports in these workings in various geological and mining conditions [11]. The work of the authors Madjdabadi, B, Valley, B, Dusseault, MB & Kaiser, PK attracts attention in the field of measuring deformations in underground infrastructure, including the mining industry [12], where the measurement of deformation over an almost unlimited length makes it possible to fix the deformation field caused by underground workings outside the damage zone of the excavation. The potential benefit of such monitoring of the deformation field is the deformation of the continuum under tension, compression or shear, or as a fracture deformation due to local shear or expansion of inhomogeneities. In 2016, scientists Yiming Zhao, Nong Zhang and Guangyao Si in their work [13] showed a laboratory sample of FOS for a safety monitoring system at a quarry based on the Bragg fiber lattice effect. Compared to traditional monitoring equipment, the developed new monitoring system has the advantages of providing accurate, reliable and continuous online monitoring of quarry activities. Unfortunately, this product has not yet been put into production, since there are a number of technical problems with its adoption and the issue of reducing its cost has not yet been resolved. Scientists Tao Hu, Gongyu Hou and Zixiang Li in their work shared with the results of tests where a high-strength optical fiber with an alloy coating made of stainless steel has high deformation transmission characteristics, which allows it to be connected to a concrete anchor with uniform deformation. This demonstrates the possibility of using fiber for theoretical and experimental monitoring of reservoir movement [14]. Later, other scientists Chiara Of Lanciano, Riccardo Salvini in their work [15] dealt with the issue of work safety in the conditions of a mining enterprise in China. They performed checks of the stability conditions of the mountain range during various mining operations. An innovative analysis method based on a combination of distributed fiber-optic sensors, digital photogrammetry using unmanned aerial vehicles, topographic and geotechnical monitoring systems is proposed. The previously considered automated fiber-optic system for monitoring the stability of the instrument array of the quarry and dumps [9] has a number of disadvantages inherent in the optical two-beam Mach-Zehnder interferometer. The desire to improve the design of FOS to control the displacement and deformation of rocks, led to the conclusion that it is necessary to abandon the use of the Mach-Zehnder interferometer. The length of the measuring channels in the conditions of open development can be from one to several tens of kilometers. Therefore, the influence of temperature on the measurement accuracy at a large length of the OF is significant. The first thing that field tests have shown is that it is necessary to protect the OF from the effects of external temperature fluctuations, since these fluctuations lead to a change in the size of the OF and, accordingly, change the refractive index [9]. This circumstance is very important in terms of the accuracy of the FOS operation and the noise immunity of the measuring channel as a whole. To solve this problem, a method of protecting the OF using a multilayer shell with external armor was proposed, which allows significantly reducing the linear expansion of the OF during heating. Since the temperature increases by 1 ^{0}C already causes a change in the phase of the propagation of the light wave through the core of the OF. All types of fiber-optic sensors have certain disadvantages, for example, sensors based on interferometers may change sensitivity, and also with extended measuring channels, the influence of temperature interference and a change in the refractive index are observed, which leads to signal fading due to the drift of the initial value of the phase of light wave propagation through the core of the optical fiber. Such problems are more typical for single-mode optical fibers, since they are used to control extended objects in 30-50 km, multimode fibers have less influence of temperature interference, but are able to work at short distances in the range of 500-1000 m. A solution that allows almost completely eliminating the disadvantages of two-beam fiber-optic interferometers. Taking into account the above, when creating FOS, the method of controlling the additional losses introduced was used, since this method is more independent of the influence of external temperature.

3. Materials used, methods and measurement scheme

Changes have been made to the existing design of the well-known rock displacement and deformation meter based on the Mach-Zehnder interferometer [9] in order to simplify the design and increase the noise immunity of the measuring channels. In the considered FOS, spatial localization is achieved, which reduces the drift of the operating point of the interferometer while maintaining the length of the measuring base, but a number of problems remain. The main identified problem in the operation of the Mach-Zehnder interferometer is a strong dependence on changes in the ambient temperature, which ultimately affects the accuracy of measuring crack growth [9]. Since this device is supposed to be used to measure the deformation of rocks in which cracks are formed, this is an important point in the operation of the measuring system. Since the outdoor temperature is subject to fluctuations, it is necessary to take into account this factor, which will negatively affect the operation of the displacement sensor, in turn, the sensor will transmit incorrect data to the information and measurement system. With an increase in the length of the FO in the measuring channel, the influence of temperature on the measurement accuracy increases. Some changes were made to the design, which mainly affected the measuring part. The main part of the sensor remained the same. As a sensitive element, a single-mode FO is used, which is wound on two half-disks in several turns. The sensor body and the half-disk are made of plastic. In contrast to the well-known FOS [9], a method for controlling the introduced additional losses is used, which ensures a smaller influence of temperature on the OF of the measuring channel. Figure 1 shows a FOS scheme for measuring the deformation of rocks and monitoring the growth of cracks. Where 1 is the radiation system with a wavelength of

1310-1625 nm, 2 – optical connector, 3 – optical fiber, 4 – movable semi-tri-triys, 5 – fixed semi-finished, 6 – damper, 7 – Mobile Rigel, 8 – Rigel fixation point 9 FOS corps

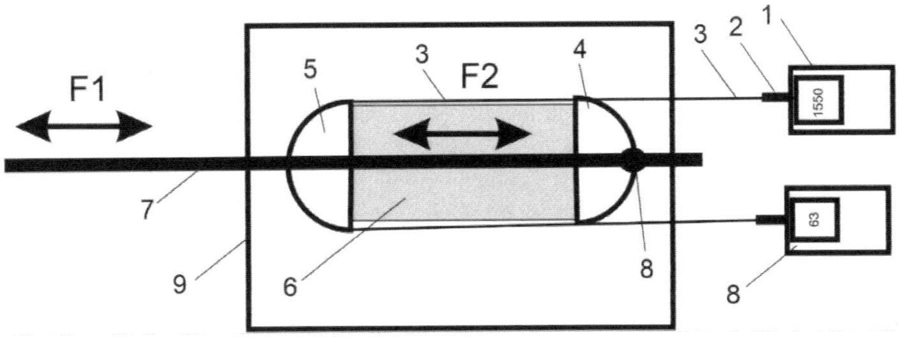

Figure 1. FOS flow diagram for measuring the deformation of rocks and monitoring the growth of cracks.

A number of experiments were carried out to determine the power losses of the optical study passing along the shoulders of the FOS at different values of the reference point offset indicated in the direction F1. As a result of mixing the crossbar, the damper was compressed when moving to the left and unclenched when moving the crossbar to the right, while an elastic force F2 was generated. The values of the crossbar movements were measured using a ruler. The measurements were carried out repeatedly with subsequent processing of the experimental data and averaging of the obtained values. To measure the power loss of optical radiation, the VIAVI optical power meter (JDSU) SmartPocket OLP-38 was used, operating in the dynamic range from -60 to +26 dB, with a wavelength range of 780-1650 nm. The SmartPocket OLS-34/35/36 with built-in Auto-λ and Multi-λ options was used as an optical radiation source, the SmartPocket OLP-38 can automatically measure the power level and insertion losses in a single-mode and multimode optical cable. The connection to the optical fiber was made through a universal UPP 2.5 mm adapter and optical connectors of the SC type. Figure 2 shows a photo of a laboratory sample of FOS and the equipment used for conducting experiments.

Figure 2. A laboratory sample of FOS.

All experiments were carried out in laboratory conditions at an air temperature from 22 to 23 ^{0}C at a relative humidity within 60%. A quartz single-mode optical fiber 9/125 microns (OS2) Corning SMF-28e+® with a low "water peak" (ITU-T G. 652.D standard) was used as a water sensor. The optical fiber has a primary coating of 245 microns (with an outer shell). A coil with an optical fiber with a length of 2.03 km was used, its connection was made using a standard patch cord 20 meters long, edged on both sides with SC connectors. The use of single-mode fibers, as opposed to multimode ones, will increase the length of the sensors to 30 km, which allows you to place the signal processing device at a considerable distance from the FOS.

4. Results of studies of a laboratory sample of FOS

A laboratory sample of FOS based on the method of monitoring additional losses showed a fairly high linearity and accuracy during measurements and can be used to control the deformation of the array after appropriate refinement of its design, by conducting field tests. The experimental results are represented by a graph of the dependence of optical losses on the value of the reference point displacements, shown in Figure 3. The losses increased with an increase in the load on the optical fiber, since we had a photoelastic effect and the refractive index changed, respectively, additional losses increased, which are indicated by the letter A. at the same time, the displacement of the reference point ΔL was controlled.

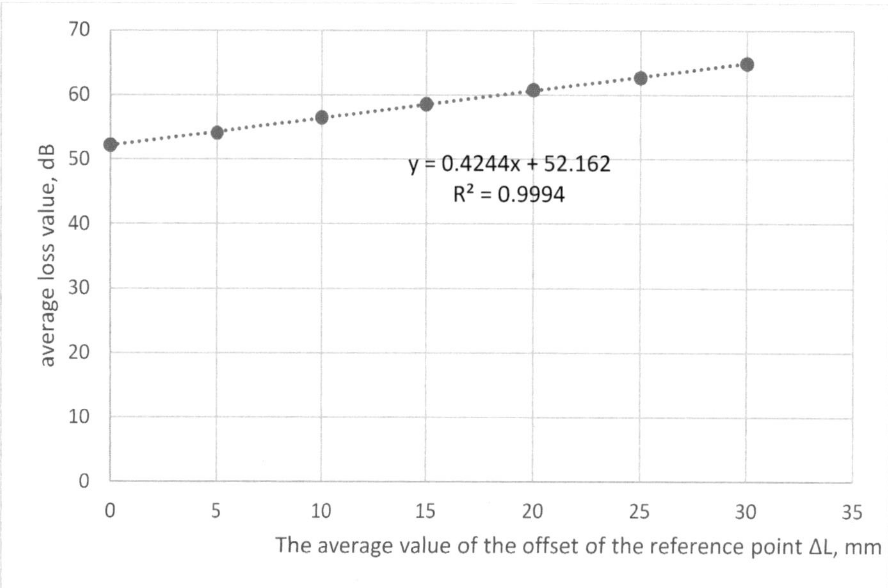

Figure 3. A laboratory sample of FOS.

Based on the measurement results, the absolute error and relative error were calculated, which amounted to 2.06 mm and 4.634%, respectively. The Student's coefficient was 2.602 with a confidence probability of 0.95.

5. Scheme of inclusion of a fiber-optic sensor for monitoring the growth of cracks during deformations of rocks

As mentioned earlier, the basis of the FOS (Figures 1 and 2) is a well-known method for controlling additional losses that are formed during the bending of the OF. The scheme works as follows: a source of coherent radiation of the optical range with a wavelength of 1310 or 1550 nm generates a light wave that propagates through the OF, after which it falls on the surface of the photodetector, which records a change in its intensity. Accordingly, when the crossbar moves in the direction indicated by the arrow, when the crack opens, the pressure on the OF increases and the additional losses introduced increase. A semiconductor laser with a power of 50 mW is used as a light source 1. This light source is used in telecommunications information transmission systems. The wavelength is selected according to the distance from the FOS to the data processing unit, in this sample the wavelength of 1310 and 1550 nm is used, which corresponds to the second and third transparency windows of the OF. This allows the FOS to work at a distance of up to 50 km or more from the data processing unit. Figure 4 shows the inclusion scheme for the fiber-optic crack control sensor [9] during rock deformities

The radiation source 1 is connected to the optical conductor 3 using SC type connectors 2. To do this, you can use standard telecommunications patch-cords. The OF enters the FOS housing through inlet bushings 4 having seals that prevent the fiber from exiting back from the sensor housing. The OF is wound on two half-disks 5 and 6 with a diameter of 2 cm, the use of a smaller diameter is unacceptable due to the critical radius, which leads to damage to the OF when it is bent. The damper 7 is necessary to return the disks to the original position of the half-disk 6, which moves on the axes 8. The half-disk 5 is stationary, and the movable 6 moves along the guide, while the axes 8 are necessary to limit its movement. The sensor is able to track movements in the range from 0 to 50 mm, if necessary, the measurement range can be expanded. It is also possible to connect two FOS in series, and their readings can be summed up. Also, the half-disk 6 is rigidly connected to the bolt 9 and when it moves in the direction shown by the arrow, it moves with it.

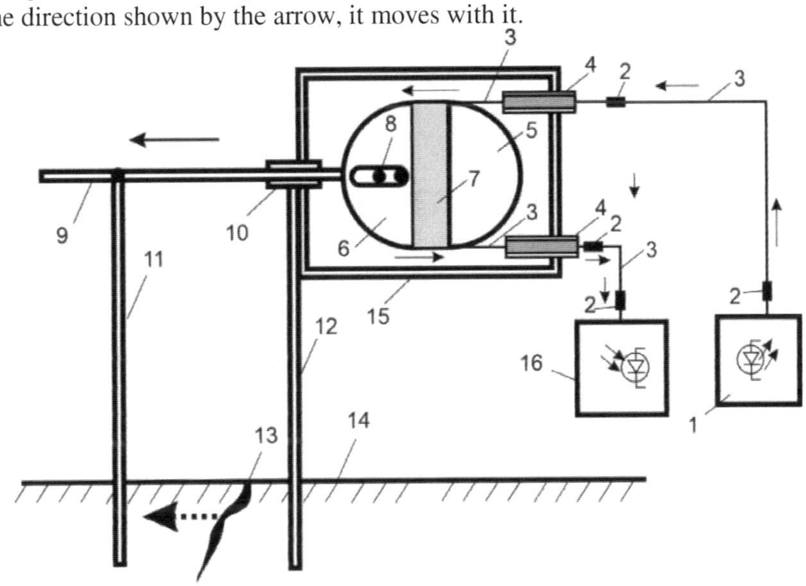

Figure 4. Scheme of inclusion of a fiber-optic sensor for monitoring the growth of cracks during rock deformations.

The crossbar 9 moves freely in the guide sleeve 10. Two reference points 11 and 12 are immersed in the controlled rocks 14, when the crack 13 is opened. The direction of its opening is shown by a dotted arrow, the distance between 11 and 12 increases, respectively, the bolt 9 moves and pulls the disk 6, which in turn affects the OF. The OF fixed between two points is deformed and lengthened by the value ΔL, its length increases accordingly, and the intensity of the incident light wave on the surface of the photodetector changes accordingly. A photoelastic effect occurs and the refractive indices of the OF change, which causes an increase in the additional losses introduced. Since the initial losses are measured when setting up the sensor and are the characteristics of the OF, their change can be relative units of elongation. In the case of closing the crack 13 and returning the reference point to its original position, the damper 7 allows you to set the position of the movable half-disk to the original position "0". At the exit from the FOS of the interferometer, the light waves are combined using a splitter 4, after which the light wave enters the photodetector of the data processing unit with the photodetector 16. After data processing, the numerical value of the reference point offset in millimeters appears on the screen of the data processing unit. If the FOS is more than 100 meters away from the data processing unit, an armored fiber-optic cable should be used, which avoids the negative influence of temperature.

6. Discussion of the results

As mentioned earlier, the fiber-optic system for ensuring the safety of mining operations at open-pit mines consists of three main parts: a radiation source, a fiber-optic sensor, a device for data processing and information output. The first two elements are described in sufficient detail and studied in the scientific literature. The most important element is the latter, since the effectiveness of measurements and the reliability of the results obtained depend on its operation. To develop a device for data processing and information output that meets all modern requirements, it is necessary to use high-tech technologies related to artificial intelligence algorithms. An analysis of the literature showed that this particular device is least covered in open sources, and the developed software is not freely available. The more complex the devices for data processing and information output, as well as its algorithms related to decision-making, the more accurately it provides measurements and eliminates all kinds of errors associated with various interference. For example, in a distributed fiber-optic system, there may be a significant influence of temperature interference, leading to fading or fading of the signal. Complicating the algorithm of data processing and decision-making is one of the ways to create devices that are effective in all respects for data processing and information output, but this requires certain hardware and software, which seriously increases the cost of the monitoring system. A condition is created when the buyer must pay for expensive and long-term research work in addition to the costs of manufacturing the monitoring system. Given the lack of widespread demand and limited application, this ultimately creates a fairly high cost of a measuring system based on fiber-optic technologies, which limits their use. Also, a number of technical problems have not been fully solved, which do not allow for their widespread implementation at mining enterprises. Taking into account the above, work has begun on creating its own hardware and software complex for four measuring channels. So far, the physical number of channels is limited to four due to the limited computing capacity of the computer and the performance of the video card, as well as the created algorithms of the program, but in the future their number will be increased to the real needs of the enterprise. An increase in the number of channels will definitely lead to an increase in cost, primarily due to an increase in the length of fiber-optic control cables, but security issues are a higher priority. If there is a production need, the monitoring system can simultaneously monitor several thousand reference stations. The FOS presented in this article is a point sensor, for the organization of several measurement points, a serious revision of the presented system is necessary, since it requires expanding the measurement capabilities and increasing the control points. To do this, in the future it is necessary to develop a multi-channel data processing system received from several dozen sensors, which will allow creating a quasi-distributed measuring network capable of monitoring the stability of the sides of the entire quarry as a whole. It is also necessary to create a hardware and software complex for the unified management of the entire measurement system and data storage in the computer memory.

7. Conclusion and conclusions

The work is aimed at solving production problems related to improving the safety of mining operations. The use of FOMS (fiber-optic monitoring systems) of distributed and quasi-distributed types will allow remote monitoring of the stability of the sides of the quarries in real time. The obtained results of laboratory studies allow us to assert that the developed fiber-optic sensor, made on the basis of the method of monitoring the introduced additional losses, has a fairly good linearity of characteristics and low power consumption. The method used allows to reduce the negative impact of external factors on the accuracy of the measuring system. It is also possible to carry out measurements at a distance of about 30 km with a laser energy consumption of less than 50 mW, compared with electric measuring systems, this is a very high indicator of energy consumption efficiency. The use of the additional loss control method, the development of our own circuit solutions and a hardware and software complex allows us to significantly reduce the cost of FOS as a whole, compared to fiber-optic systems based on the effect of Bregg gratings, which is extremely important for implementation in the mining industry. Reducing the cost of one measurement point will allow building a quasi-distributed system for monitoring the displacement and deformation of rocks on the sides of quarries, as well as tracking the growth of cracks.

Acknowledgments
The publications are based on the results of scientific research carried out at the expense of the IPN project No. AP09562222 "Information and measurement system of a new generation based on fiber-optic sensors" (contract 254/12-2 of 15.07.21).

References

[1] Ozhigin S, Ozhigina S, Ozhigin D 2018 *Journal of the Polish Mineral Engineering Society Inzynieria Mineralna* **41**(1) 203–207 doi:10.29227/IM-2018-01-32
[2] Dorokhov D V, Nizametdinov F K, Ozhigin S G, Ozhigina S B 2018 *Journal of Mining Science* **54**(5) 874–882 doi: 10.1134/S1062739118055011
[3] Sannikova A P, Bazykina L R, Ozhigin D S 2017 *Journal of Industrial Pollution Control* **33**(1) 852 – 855
[4] Kalytka V, Korovkin M, Mekhtiyev A, Yurchenko A 2018 *Russian Physics Journal* **61**(4) 757 – 769 doi:10.1007/s11182-018-1457-8
[5] Yurchenko A, Mekhtiyev A, Bulatbayev F, Neshina Y, Alkina A 2018 *Russian Journal of Nondestructive Testing* **54**(7) 528 – 533 doi: 10.1134/S1061830918070094
[6] Madi P, Kalytka V, Alkina A and Nurmaganbetova M 2019 *Journal of Physics: Conf. Series* **1327** 012036 doi:10.1088/1742-6596/1327/1/012036
[7] Madi P, Gorokhov D, Mekhtiyev R and Nurmaganbetova M 2021 *Journal of Physics: Conf. Cep.* **1843** 012016 doi:10.1088/1742-6596/1843/1/012016
[8] Mekhtiyev A, Yurchenko A, Neshina Y, Alkina A, Madi P 2020 *News of higher educational institutions: Physics* (2) doi: 10.17223/00213411/63/2/129
[9] Mekhtiyev A, Neshina Y, Madi P, Gorokhov D 2021 *Occupational safety in industry* **4** 19–26 doi:10.24000/0409-2961-2021-4-19-26
[10] Liu X, Wang C, Liu T, Wei Y, Lv J 2009 *ACTA Photonica Sinica* **38** 112–114
[11] Kumar Atul, Kumar Dheeraj, Singh U, Gupta P S, Shankar Gauri 2011 *Journal of Control and Automation* **3** 63–70
[12] Madjdabadi B, Valley B, Dusseault M B, Kaiser P K 2014 *Proceedings of the Seventh International Conference on Deep and High Stress Mining, Australian Centre for Geomechanics* 457–468 doi:10.36487/ACG
[13] Yiming Zhao, Nong Zhang and Guangyao Si 2016 *Journal List Sensors (Basel)* **16**(10) 1759 doi:10.3390/s16101759
[14] Tao Hu, Gongyu Hou and Zixiang Li 2020 *Sensors* **20**(5) 1318 doi: 10.3390/s20051318
[15] Chiara Lanciano, Riccardo Salvini 2020 Sensors **20**(7) 1924 doi:10.3390/s20071924

A feature of organization the regional digital network of agrometeorological observations in the Tomsk region

A S Myagkov[1,2], A V Solovyov[2] and D A Fomin[3]

[1]Institute of Monitoring of Climatic and Ecological Systems Siberian Branch of the Russian Academy of Sciences, Tomsk, Russia
[2]Tomsk Scientific Center of the Siberian Branch of the Russian Academy of Sciences, Tomsk, Russia
[3]UMIUM LLC, Tomsk, Russia

E-mail: asms@imces.ru

Abstract. The article describes the organization of a digital network for collection and analysis of agrometeorological data. This digital network is based on the concept of an integrated approach to the collection, systematization and analysis of agrometeorological data for forecasting and developing recommendations for agricultural producers. The digital network is three-tiered and consists of meteorological stations, agrometeorological posts and agrometeorological probes installed directly on the fields. The presented modern network for monitoring meteorological parameters is an important tool of modern agricultural production, which is especially relevant in a dynamically changing climate. It is reasonable to build such systems in the ideology of the Internet of Things (IoT).

1. Introduction

Agrometeorology is the applied branch of meteorology aimed at collection, systematization and processing of meteorological information in relation to agricultural production. First of all, this applies to crop production, since the growth and development of plants is mainly due to weather influences [1]. Various weather risks in a dynamically changing climate are increasing significantly. Therefore, timely and accurate management decisions are needed to adjust current agricultural production. This, in turn, the modern meteorological measuring systems with sufficient spatial and temporal resolution are requires. The task of providing agrometeorological observations is attributed to Roshydromet activities. However, the measuring meteorological network of this structure is significantly fragmented, which does not allow solving the problem of actual agrometeorological support of the agricultural industry in full. Thus, the grouping of measuring meteorological stations of Roshydromet in the Tomsk region consists of 24 units, and only 5 of them partially conduct agrometeorological observations. Such a number of posts is not enough to solve agrometeorological tasks. Therefore, it is necessary to develop departmental agrometeorological observation networks. In Tomsk region the concept of agrometeorological monitoring is implemented on the basis of the regional digital network «Agropogoda.rf» (https://agropogoda.com) [2]. This concept is an integrated approach to the collection, systematization, analysis, forecasting and development of recommendations for agricultural producers.

Content from this work may be used under the terms of the Creative Commons Attribution 3.0 licence. Any further distribution of this work must maintain attribution to the author(s) and the title of the work, journal citation and DOI.
Published under licence by IOP Publishing Ltd

2. Technical configuration

The «Agropogoda.rf» network is three-tiered and consists of meteorological stations, agrometeorological posts and agrometeorological probes installed directly on the fields (Figure 1). As of the end of 2021, the network is a complex of more than 120 measuring systems (including 13 weather stations) located in seven districts of the Tomsk region (Asinovsky, Pervomaysky, Krivosheinsky, Zyryansky, Tomsky, Kozhevnikovsky and Bakchar districts).

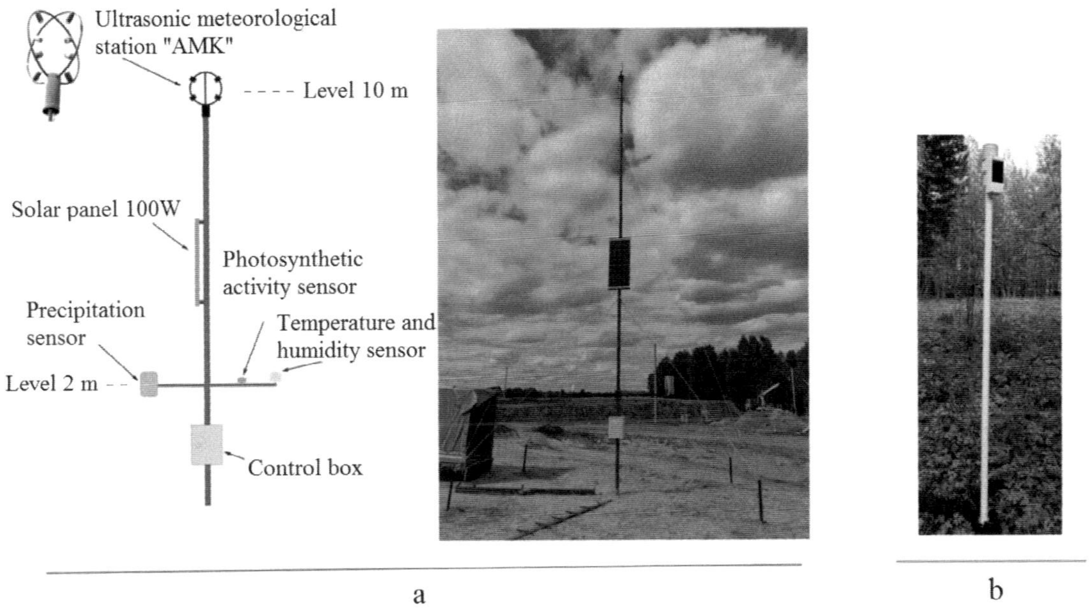

Figure 1. Measuring systems «agrometeorological post» (a), «agrometeorological probe » (b).

The measuring equipment continuously monitors basic meteorological characteristics, such as air temperature and humidity, wind speed and direction, soil temperature and moisture to a depth of 50 cm, calculates the amount of liquid precipitation and monitors the level of photosynthetic activity. All measuring equipment's are certified. The measured meteorological characteristics are transmitted via the networks of mobile operators to the server for storage and further processing every hour. All measuring equipment's are autonomous. Their sensors are powered by built-in batteries, which are charged by solar panels. Detailed technical description of the equipment, its metrological and operational characteristics are available for review [2].

3. A feature of organization the regional digital network

Nowadays, the agrometeorological observation network is actively developing in Tomsk region. But it is necessary to take into account the specifics and the limitations associated with them. The main limitation is related to insufficient coverage of the territory by mobile operators' networks. Often there is no mobile communication on remote agricultural land plots. Agrometeorological network requires a stable coverage of the territory by the GSM network. The territory of Tomsk region is characterized by uneven landscape and forest areas, which does not provide sufficient mobile network coverage. The authors and technical experts are considering the option of completely abandoning the use of mobile operator infrastructure. This requires a special configuration of the data network.

For example, such a measurement network can be built using the LoraWan or NbIot protocols [3]. Each meteorological post is both a point of measurement of meteorological information and a point of receiving-transmitting information from neighboring meteorological posts in this configuration (Figure 2). In the Tomsk region the areas of sown land are on average 5–15 thousand hectares, and farms are located within a radius of 5–20 km. This makes it possible to build an IoT network with a central hub located within the boundaries of administrative areas (villages, local administrations). Agrometeorological information can be collected using LoraWan protocols at a distance of 5–15 km from the central node.

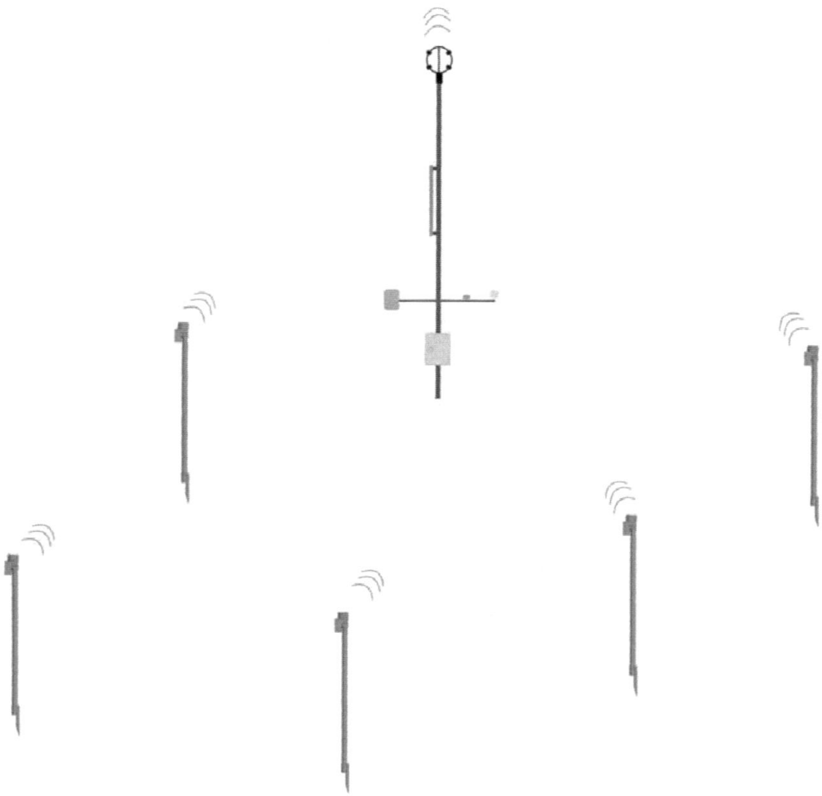

Figure 2. Measuring systems in the ideology of the Internet of Things (IoT).

An important feature in constructing a quality agrometeorological observation is the network density. According to our estimates, the agrometeorological observation network for Tomsk region should contain up to 100 measuring posts at the boundaries of administrative areas and at least 500–1000 weather probes on agricultural lands. The proposed network density will allow to solve such tasks as: qualitative short-term weather forecasting, development of digital assistants for agronomists, monitoring the expansion of dangerous agricultural pests, plant disease monitoring, risk assessment of dangerous meteorological phenomena (such as hail, heavy rainfall, return frost, etc.). Our estimates indicate that for quality provision of agrometeorological observations it is necessary to install at least 1 agrometeorological post and 5-10 agrometeorological probe s for every 500 km^2 of territory. It is necessary to consider the landscape to build the network. It is advisable to follow the "three-point rule" for equipment placement. According to this rule, the first probe should be placed at the lowest point of the landscape, the second at the highest point, and the third at the median. This will make it possible to track meteorological trends in landscape features.

The specially developed information platform «Agropogoda.rf» (see https://agropogoda.com) is used for storage, processing and visualization of accumulated agrometeorological information. Each user of this platform has the corresponding access rights to different services (Figure 3). All measuring systems are mapped on the digital map. Information about the current meteorological situation is available to the user. The user can request a variety of meteorological information, analyze it himself or use ready-made services which are called "digital assistants". Digital assistant is a calculation algorithm that builds a

Journal of Physics: Conference Series **2140** (2021) 012038 doi:10.1088/1742-6596/2140/1/012038

certain scenario of events from the available agrometeorological data and then informs the user about it. For example, it can be a morning frost forecast or fog probability. It can be a calculation of the pest appearance time. For example, calculating the time of rapeseed moth emergence allows the agronomist to decide on timely insecticidal treatment. You can learn more about the features of the information platform in demo mode (login and password "demo" on the user authorization page) [2].

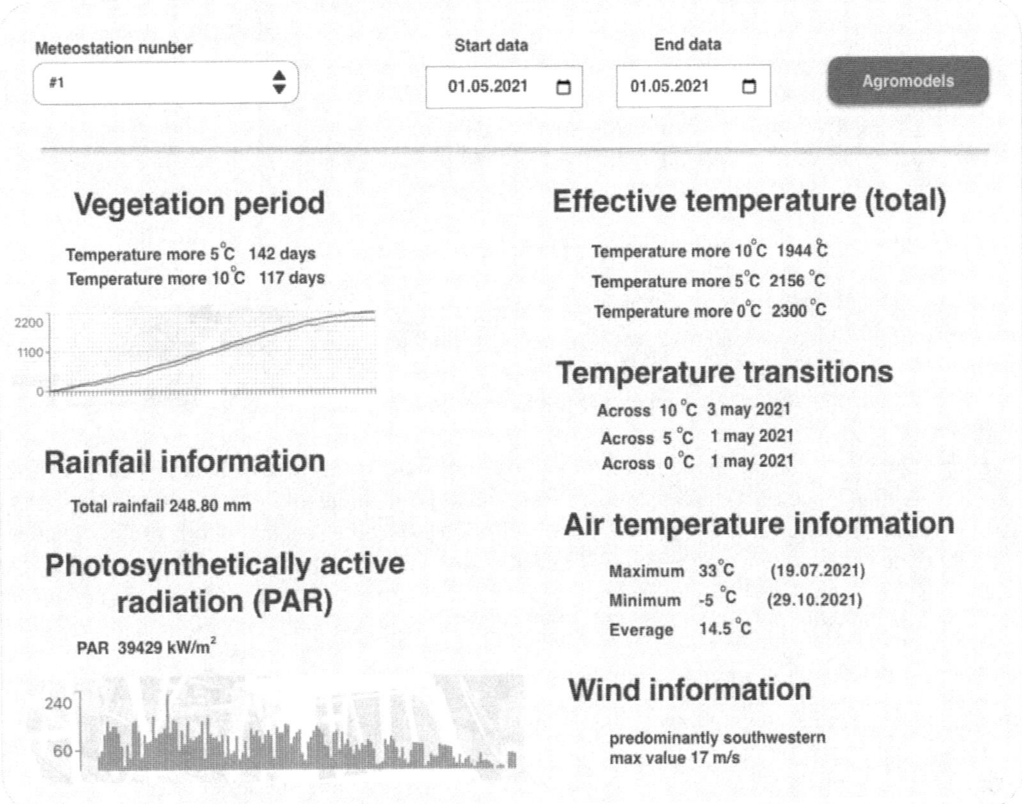

Figure 3. An example of information presentation in the agrometeorological information platform.

4. Results and discussion

The most significant and valuable is qualitative agrometeorological monitoring in the pre-sowing and harvesting periods. Figure 4 shows the results of agrometeorological observations of soil temperature at depths of 10, 30 and 60 cm in Kozhevnikovsky district of Tomsk region during the sowing season of 2020. For the analysis, major agricultural activities such as moisture addition, beginning and ending of grain seeding are superimposed on the graph. The graph of soil temperature at a depth of 10 cm clearly shows a return frost during the main sowing period, which significantly slowed down the starting development of plants. Obviously, the increase in moisture and the start of the sowing campaign is late by about 10-12 days. In fact, this means additional yield losses.

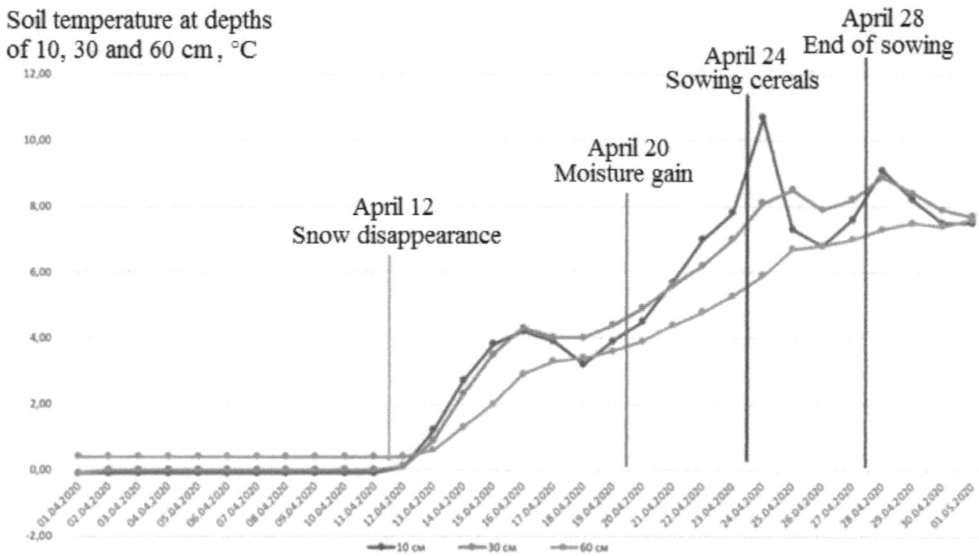

Figure 4. Agrometeorological situation in the spring of 2020.

The network operates in the mode of collecting series of meteorological data, which in itself is extremely valuable, since contact measurements are the most accurate and important. We live in the age of space monitoring of weather, but no matter how efficiently remote space methods of its control would develop, networks of ground-based meteorological observations will always be and only they can be used to correct any calculated climatic models. The data from the network is applicable in other sectors of the economy, for example, for road services, scientific research and ecology or wind research for alternative energy systems etc.

Of course, there are various measuring systems for agriculture. A variety of meteorological measuring equipment can be purchased, but individual systems, no matter how good they are, do not solve the global problems of agrometeorology. Only the construction of a distributed and well-developed network will make it possible to use the potential of agricultural meteorological technologies in full, and the creation of such networks is of an infrastructural nature. It is the regional provision of the agricultural industry with high-quality agrometeorological information that will seriously strengthen the industry and obtain a significant economic effect.

5. Conclusion

As a conclusion, we can point out that the development of modern networks for monitoring meteorological parameters is an important tool of modern agricultural production, which is especially relevant in a dynamically changing climate. At the same time, it is advisable to build such systems in the ideology of the Internet of Things (IoT), and measurement systems should be made as cheap as possible by using modern digital approaches for the collection, transmission and further processing of meteorological information.

References

[1] Losev A P 1994 *Practicum on agrometeorological support of crop production* (St. Petersburg: Gidrometeoizdat)

[2] *Information platform «Agropogoda.rf»* Available at: https://agropogoda.com

[3] Anisimov A *LoRaWAN specification* Available at: https://habr.com/ru/post/316954/

Journal of Physics: Conference Series **2140** (2021) 012039 doi:10.1088/1742-6596/2140/1/012039

The optical and electrical non-invasive methods of measurement for glucose concentration in biological liquids

Y K Zhexenbayev[1], T D Bulembayev[1], A V Gorst[2], K V Zavyalova[2], A S Mironchev[2], and E V Yurchenko[3]

[1]National Research Tomsk Polytechnic University, Tomsk, Russia
[2]National Research Tomsk State University, Tomsk, Russia
[3]Siberian State Medical University, Tomsk, Russia

E-mail: 2774547@mail.ru

Abstract. The article describes the radio-wave and optical methods of determining glucose concentration. The radio-wave method is based on the use of a sensor with a resonant frequency that is displayed when in contact with highly lossy materials and with an extended near-field zone in the resonance area. The optical method is based on studying the influence of glucose concentration (0–20 mmol/l) in bidistilled water on absorption spectra in the range of 190-1000 nm. The article presents the results of the experimental test of the near-field sensor with the pre-produced biological media imitating the human body tissues, and the results of the optical method demonstrate the possibility to measure the concentration with the use of an optical emitter with a wave length of 830 nm.

1. Introduction

Diabetes mellitus is one of the three diseases most frequectly leading to incapacitation of the population and death. According to the World Health Organization (WHO), in 2010 the number of diabetic people equaled 284.6 million people , which is 6% of the population aged between 20 and 79. In Russia, in 2010 there were registered over 9.6 million diabetic patients. According to the WHO prognosis, for the period between 2010 and 2030, the number of deaths from diabetes will double [1].

The way to fight the disease is to follow a strict diet with regular self-control of the blood glucose concentration. However, the modern invasive glucometers have a number of disadvantages, such as: painful procedure of finger stick, threat of infection penetration when doing the procedure in non-sterile conditions, as well as the high cost of consumables. All these disadvantages have led to the necessity to develop a new type of glucometers that would not require a blood sample to determine the glucose level. This type of devices was called a non-invasive glucometer. The most well-known non-invasive methods of blood glucose determination are: Raman spectroscopy, impedance spectroscopy, near infrared spectroscopy, photoacoustic spectroscopy, etc. The Raman spectroscopy [2] is based on measuring scattered light. The impedance spectroscopy is based on measuring resistance when the radiation frequency is changed. Measurement of glucose level requires several sensors placed in the vein area on human arms [3]. The near infrared spectroscopy is based on transmission of near infrared radiation through the vascular area of the body (a finger, an ear lobe, etc.). In this case, the glucose concentration is calculated on the basis of the obtained spectral information [4]. All measurements in the near infrared spectroscopy are based on passing of light radiation through or depthward the sample and measuring the intensity of (passed or reflected) beam. Spectrometers for measuring in the near

Content from this work may be used under the terms of the Creative Commons Attribution 3.0 licence. Any further distribution of this work must maintain attribution to the author(s) and the title of the work, journal citation and DOI.
Published under licence by IOP Publishing Ltd

Journal of Physics: Conference Series **2140** (2021) 012039 doi:10.1088/1742-6596/2140/1/012039

infrared spectroscopy have an adequate light source (such as a high stability quartz tungsten lamp), a monochromator or an interferometer, and a detector. Regular monochromators are optoacoustic tunable filters, diffraction grids, or prisms. Spectroscopy in the middle infrared region is based on light absorption by glucose molecules [5]. This method uses a light beam passing through a crystal contacting the skin, thus guaranteeing that the electromagnetic field created by reflected light reaches the dermis (the skin layer that contains the largest amount of glucose). There is a technology based on ultrasonography – photoacoustic spectroscopy [6]. This method is based on an acoustic response of an activated liquid with the use of laser light. This method is similar to spectroscopy in the middle infrared region. There are also other, less known methods for determination of blood glucose level [7].

2. Experiment

Determination of glucose solution absorption in bidistilled water 20 mmol/l, 10 mmol/l, 5mmol/l, and 0 mmol/l. The solution concentration is chosen so that the density on the maximum band was within the optimal range of photometric measurements (0.3–1.5 of optic density). The solutions are prepared in measuring flasks according to the exact sample weight. For standard measuring cells (l = 1 cm) and values of $\varepsilon=10^3$–10^4, regular solution concentrations are about 10^{-2} mol/l. The process of solution preparation is shown on Figure 1.

Figure 1. Solution preparation.

Both cells with the solution are placed inside, and parameters are set (Figure 2) with the help of the software installed on a personal computer.

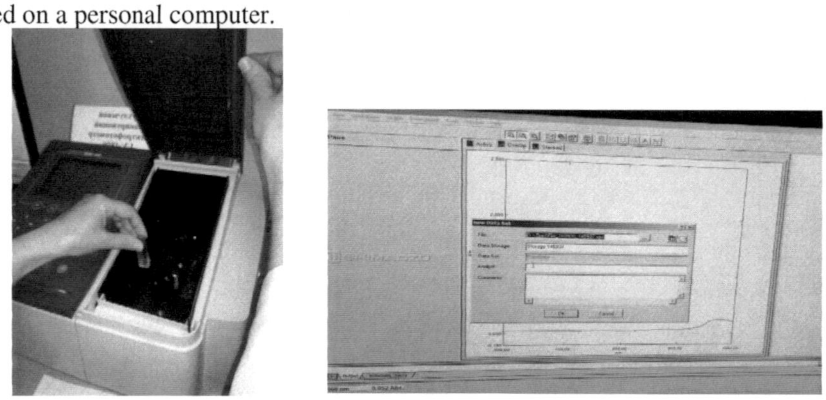

Figure 2. Cell installation and parameter setting.

The results of measurements are glucose solution absorption values within the range of 190 to 1.000 nm, 190 to 450 nm, and 600 to 1.000 nm. The series of absorption spectra measurements conducted on glucose solutions showed dependence of absorption level with the wave length of

825.5nm (peak 3); on all other peaks no changes were detected. Figure 3. shows absorption spectra changes that are detected when glucose concentration is changed. It is evident that the absorption level increases with the increase of glucose concentration.

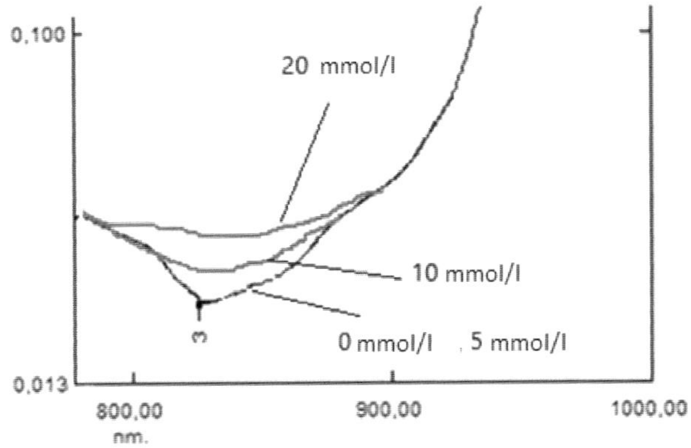

Figure 3. Absorption spectra for solutions with glucose concentration of 0 mmol/l, 5 mmol/l, 10 mmol/l, and 20 mmol/l.

Table 1 shows the results of optical measurements of absorption levels on glucose (3) and water (1) peaks. The detected glucose absorption peak allowed conducting tests of absorption levels in glucose solutions for the same concentrations with monochromatic light-emitting diode irradiation with the wave length of 830 nm (which is necessary for production of a portable glucose level meter).

Table 1. Glucose solution absorption.

Device	Concentration / Wave length, nm	0 mmol/l	5 mmol/l	10 mmol/l	20 mmol/l
Light-emitting diode I/Io	825.5	0.036	0.037	0.053	0.067
Spectrophotometer	976	0.247	0.255	0.255	0.255
	825.5	0.033	0.035	0.039	0.052

Production of a sensor was based on a calculated model [8]. The sensor is a combined split antenna based on a flexible substrate RO3003; it was produced 25mm in diameter with a 50-Ohm port brazed to it. This sensor has high penetration of electromagnetic waves into highly lossy media due to the near field length. Production of a biological phantom was based on graphite, polyurethane, and acetone [9,10]. This structure is sufficiently strong, which allowed creating thin materials, such as horny layer, epidermis, and capillary layer. The mentioned layers are the thinnest. The produced material was measured with the use of the previously developed method for measuring plane-layered materials [11,12]; charts of the real part of frequency-dependent dielectric permittivity were plotted for each material (Figure 4). As the charts show, the values of the real part of dielectric permittivity are similar to the materials demonstrated in modelling.

Journal of Physics: Conference Series **2140** (2021) 012039 doi:10.1088/1742-6596/2140/1/012039

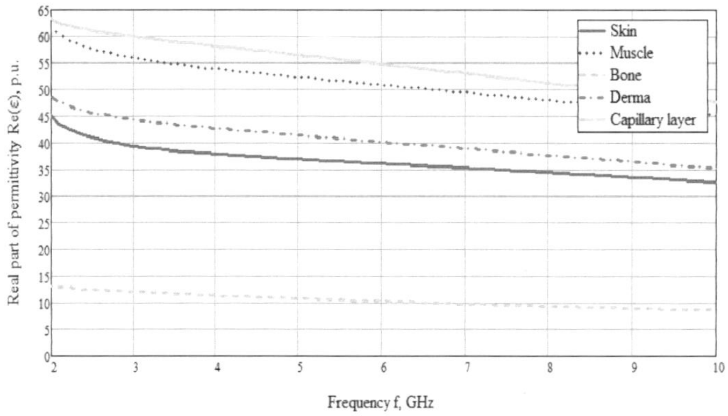

Figure 4. Real part of dielectric permittivity of the model materials and phantoms.

The differences in the values of real part dielectric permittivity were in average 1-2 relative units. The used data allowed creating a phantom of human biological tissues. The created region is forearm, since there is a large diameter vein there. Each layer is produced separately in molds of different thicknesses. Subsequently, each layer was cut to similar parts sized 150×150 mm (Figure 5). The size was chosen according to the size of antennas for blood glucose concentration measurement [13].

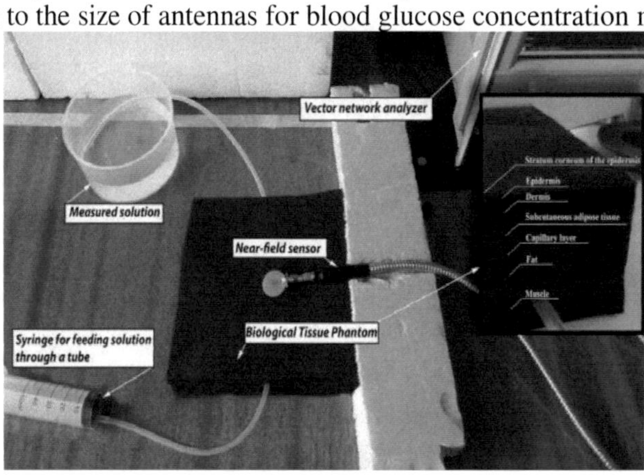

Figure 5. The graphite-based phantom and the measuring installation for determination of concentration in normal saline.

For determination, an installation based on a vector network analyzer was assembled. (Figure 5). A silicone tube with the internal diameter of 5mm was used as a venous vessel. A syringe with normal saline and the necessary glucose concentration was placed inside the tube, and when pressed, the liquid passed through the tube thus modelling blood flow inside the human veins. This choice allowed changing glucose concentration without physical effect on the antenna and thus conducting more accurate measurements.

The experimental research showed that the maximum amplitude difference is within the frequency range of 1.45–1.55 GHz. Based on the obtained results, the curve of reflected signal S11 to different glucose concentrations in normal saline was plotted (Figure 6). The dots show the data obtained experimentally on frequency 1.53 GHz. It is evident that in minimal glucose concentration the obtained values line up in a regressive curve. Also, the concentration of 10 mmol/l coincides with the regression curve. For concentrations of 7 and 9 mmol/l, the values differ from the supposed ones. This deviation is associated with the nonlinear behavior of dielectric permittivity values in low frequencies [14].

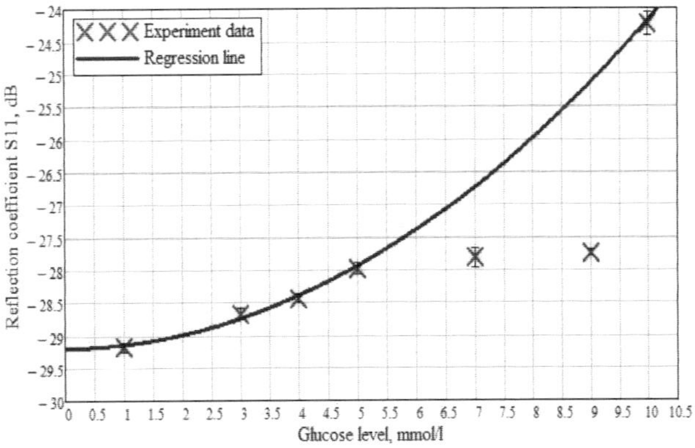

Figure 6. Dependence of the reflected signal on glucose concentration in normal saline in frequency of 1.53 GHz.

3. Conclusion

This thesis described the sensor based on a combined split antenna with an extended near field with high penetration depth of electromagnetic field. The reflected signal from an arm model with superficial veins was measured. The sensor demonstrated high reflected signal response in small changes of glucose in water solution. The obtained data show the possibility to determine blood glucose with high accuracy. The average reflected signal data variation with a 1 mmol/l change equaled 0.1–0.15 dB, which is a high value for a small glucose level change.

Also, there were conducted experimental measurements of glucose level in biological liquids by NIR-methods in the range of 190 to 1100 nm in solutions with different glucose concentrations — 20 mmol/l, 10 mmol/l, 5 mmol/l, and 0 mmol/l, and spectra of those solutions were obtained.

However, due to the fact that glucose concentrations in these solutions are low, similar curves of solution absorption spectra were obtained. A possibility to measure glucose concentration at λ=830nm was demonstrated.

References

[1] WHO, 2016, *The World Health Organization media centre*, November factsheet 312
[2] Forst T 2006 *Diabetes Technology & Therapeutics* (1) 94–101
[3] Caduff A 2003 *Biosensors and Bioelectronics* (3) 209–217
[4] Khalil O S 1999 *Clinical chemistry* (2)165–177
[5] Roychoudhury P, Harvey L M, McNeil B 2006 *Analytica chimica acta* (1-2) 218–224
[6] Waynant R W, Chenault V M 1998 *IEEE LEOS Newsletter* (2) 3–6
[7] Guo X, Mandelis A, Zinman B 2012 *Biomedical optics express* (11) 3012–3021
[8] Gorst A 2020 *Bioengineering* (7) 0149 doi:10.3390/bioengineering7040149
[9] Oliveira B L 2018 *Biomedical Physics & Engineering Express* (2) 025036
[10] Shipilov S 2020 *Medical Physics* (10) 5147–5157
[11] Gorst A 2020 *Applied Sciences* (24) 8936
[12] Mekhtiyev A D, Yurchenko A V, Bulatbayev F N, Neshina Y G, Alkina A D 2018 *News of the National Academy of Sciences of the Republic of Kazakhstan, Series of Geology and Technical Sciences* **5**(431) 66–75
[13] Serikov T G, Yakubova M Z, Mekhtiev A D, Yurchenko A V, Alkina A D 2017 *Proceedings of the 2016 11th International Forum on Strategic Technology* 510–515
[14] Shipilov S E, Satarov R N, Yakubov V P 2017 *Optical and Quantum Electronics* **49**(10) 339 doi: 10.1007/s11082-017-1172-7

AUTHOR INDEX